LONDON MATHEMATICAL SOCIETY LECTURE NOTE SERIES

Managing Editor: Professor J.W.S. Cassels, Department of Pure Mathematics and Mathematical Statistics, University of Cambridge, 16 Mill Lane, Cambridge CB2 1SB, England

The books in the series listed below are available from booksellers, or, in case of difficulty, from Cambridge University Press.

34 Representation theory of Lie groups, M.F. ATIYAH *et al*
36 Homological group theory, C.T.C. WALL (ed)
39 Affine sets and affine groups, D.G. NORTHCOTT
40 Introduction to H_p spaces, P.J. KOOSIS
43 Graphs, codes and designs, P.J. CAMERON & J.H. VAN LINT
45 Recursion theory: its generalisations and applications, F.R. DRAKE & S.S. WAINER (eds)
46 p-adic analysis: a short course on recent work, N. KOBLITZ
49 Finite geometries and designs, P. CAMERON, J.W.P. HIRSCHFELD & D.R. HUGHES (eds)
50 Commutator calculus and groups of homotopy classes, H.J. BAUES
51 Synthetic differential geometry, A. KOCK
57 Techniques of geometric topology, R.A. FENN
59 Applicable differential geometry, M. CRAMPIN & F.A.E. PIRANI
62 Economics for mathematicians, J.W.S. CASSELS
66 Several complex variables and complex manifolds II, M.J. FIELD
69 Representation theory, I.M. GELFAND *et al*
74 Symmetric designs: an algebraic approach, E.S. LANDER
76 Spectral theory of linear differential operators and comparison algebras, H.O. CORDES
77 Isolated singular points on complete intersections, E.J.N. LOOIJENGA
78 A primer on Riemann surfaces, A.F. BEARDON
79 Probability, statistics and analysis, J.F.C. KINGMAN & G.E.H. REUTER (eds)
80 Introduction to the representation theory of compact and locally compact groups, A. ROBERT
81 Skew fields, P.K. DRAXL
82 Surveys in combinatorics, E.K. LLOYD (ed)
83 Homogeneous structures on Riemannian manifolds, F. TRICERRI & L. VANHECKE
86 Topological topics, I.M. JAMES (ed)
87 Surveys in set theory, A.R.D. MATHIAS (ed)
88 FPF ring theory, C. FAITH & S. PAGE
89 An F-space sampler, N.J. KALTON, N.T. PECK & J.W. ROBERTS
90 Polytopes and symmetry, S.A. ROBERTSON
91 Classgroups of group rings, M.J. TAYLOR
92 Representation of rings over skew fields, A.H. SCHOFIELD
93 Aspects of topology, I.M. JAMES & E.H. KRONHEIMER (eds)
94 Representations of general linear groups, G.D. JAMES
95 Low-dimensional topology 1982, R.A. FENN (ed)
96 Diophantine equations over function fields, R.C. MASON
97 Varieties of constructive mathematics, D.S. BRIDGES & F. RICHMAN
98 Localization in Noetherian rings, A.V. JATEGAONKAR
99 Methods of differential geometry in algebraic topology, M. KAROUBI & C. LERUSTE
100 Stopping time techniques for analysts and probabilists, L. EGGHE

101 Groups and geometry, ROGER C. LYNDON
103 Surveys in combinatorics 1985, I. ANDERSON (ed)
104 Elliptic structures on 3-manifolds, C.B. THOMAS
105 A local spectral theory for closed operators, I. ERDELYI & WANG SHENGWANG
106 Syzygies, E.G. EVANS & P. GRIFFITH
107 Compactification of Siegel moduli schemes, C-L. CHAI
108 Some topics in graph theory, H.P. YAP
109 Diophantine Analysis, J. LOXTON & A. VAN DER POORTEN (eds)
110 An introduction to surreal numbers, H. GONSHOR
111 Analytical and geometric aspects of hyperbolic space, D.B.A.EPSTEIN (ed)
112 Low-dimensional topology and Kleinian groups, D.B.A. EPSTEIN (ed)
113 Lectures on the asymptotic theory of ideals, D. REES
114 Lectures on Bochner-Riesz means, K.M. DAVIS & Y-C. CHANG
115 An introduction to independence for analysts, H.G. DALES & W.H. WOODIN
116 Representations of algebras, P.J. WEBB (ed)
117 Homotopy theory, E. REES & J.D.S. JONES (eds)
118 Skew linear groups, M. SHIRVANI & B. WEHRFRITZ
119 Triangulated categories in the representation theory of finite-dimensional algebras, D. HAPPEL
121 Proceedings of *Groups - St Andrews 1985*, E. ROBERTSON & C. CAMPBELL (eds)
122 Non-classical continuum mechanics, R.J. KNOPS & A.A. LACEY (eds)
124 Lie groupoids and Lie algebroids in differential geometry, K. MACKENZIE
125 Commutator theory for congruence modular varieties, R. FREESE & R. MCKENZIE
126 Van der Corput's method for exponential sums, S.W. GRAHAM & G. KOLESNIK
127 New directions in dynamical systems, T.J. BEDFORD & J.W. SWIFT (eds)
128 Descriptive set theory and the structure of sets of uniqueness, A.S. KECHRIS & A. LOUVEAU
129 The subgroup structure of the finite classical groups, P.B. KLEIDMAN & M.W.LIEBECK
130 Model theory and modules, M. PREST
131 Algebraic, extremal & metric combinatorics, M-M. DEZA, P. FRANKL & I.G. ROSENBERG (eds)
132 Whitehead groups of finite groups, ROBERT OLIVER
133 Linear algebraic monoids, MOHAN S. PUTCHA
134 Number theory and dynamical systems, M. DODSON & J. VICKERS (eds)
135 Operator algebras and applications, 1, D. EVANS & M. TAKESAKI (eds)
136 Operator algebras and applications, 2, D. EVANS & M. TAKESAKI (eds)
137 Analysis at Urbana, I, E. BERKSON, T. PECK, & J. UHL (eds)
138 Analysis at Urbana, II, E. BERKSON, T. PECK, & J. UHL (eds)
139 Advances in homotopy theory, S. SALAMON, B. STEER & W. SUTHERLAND (eds)
140 Geometric aspects of Banach spaces, E.M. PEINADOR and A. RODES (eds)
141 Surveys in combinatorics 1989, J. SIEMONS (ed)
142 The geometry of jet bundles, D.J. SAUNDERS
143 The ergodic theory of discrete groups, PETER J. NICHOLLS
144 Introduction to uniform spaces, I.M. JAMES
145 Homological questions in local algebra, JAN R. STROOKER
146 Cohen-Macaulay modules over Cohen-Macaulay rings, Y. YOSHINO
147 Continuous and discrete modules, S.H. MOHAMED & B.J. MÜLLER
148 Helices and vector bundles, A.N. RUDAKOV et al
149 Solitons, nonlinear evolution equations and inverse scattering, M.A. ABLOWITZ & P.A. CLARKSON
150 Geometry of low-dimensional manifolds 1, S.K. DONALDSON & C.B. THOMAS (eds)
151 Geometry of low-dimensional manifolds 2, S.K. DONALDSON & C.B. THOMAS (eds)
152 Oligomorphic permutation groups, P.J. CAMERON
153 L-functions in Arithmetic, J. COATES & M.J. TAYLOR
154 Number theory and cryptography, J. LOXTON (ed)
155 Classification theories of polarized varieties, TAKAO FUJITA
156 Twistors in mathematics and physics, T.N. BAILEY & R.J. BASTON (eds)

London Mathematical Society Lecture Note Series. 150

Geometry of Low-dimensional Manifolds

1: Gauge Theory and Algebraic Surfaces

Proceedings of the Durham Symposium, July 1989

Edited by
S. K. Donaldson
Mathematical Institute, University of Oxford
C.B. Thomas
Department of Pure Mathematics and Mathematical Statistics, University of Cambridge

CAMBRIDGE
UNIVERSITY PRESS

CAMBRIDGE UNIVERSITY PRESS
Cambridge, New York, Melbourne, Madrid, Cape Town, Singapore, São Paulo

Cambridge University Press
The Edinburgh Building, Cambridge CB2 2RU, UK

Published in the United States of America by Cambridge University Press, New York

www.cambridge.org
Information on this title: www.cambridge.org/9780521399784

First published 1990
Reprinted 1992

A catalogue record for this publication is available from the British Library

ISBN-13 978-0-521-39978-4 paperback
ISBN-10 0-521-39978-5 paperback

Transferred to digital printing 2006

CONTENTS

CONTENTS OF VOLUME 2

CONTRIBUTORS

I. R. Aitchison, Department of Mathematics, University of Melbourne, Melbourne, Australia

M. F. Atiyah, Mathematical Institute, 24-29 St. Giles, Oxford OX1 3LB, UK

F. E. Burstall, School of Mathematical Sciences, University of Bath, Claverton Down, Bath, UK

Ralph E. Cohen, Department of Mathematics, Stanford University, Stanford CA 94305, USA

S. K. Donaldson, Mathematical Institute, 24-29 St. Giles, Oxford OX1 3LB, UK

Yakov Eliashberg, Department of Mathematics, Stanford University, Stanford CA 94305, USA

Ronald Fintushel, Department of Mathematics, Michigan State University, East Lansing, MI 48824, USA

A. Floer, Department of Mathematics, University of California, Berkeley CA 94720, USA

Mikio Furuta, Department of Mathematics, University of Tokyo, Hongo, Tokyo 113, Japan, *and*, Mathematical Institute, 24-29 St. Giles, Oxford OX1 3LB, UK

A. B. Givental, Lenin Institute for Physics and Chemistry, Moscow, USSR

Robert E. Gompf, Department of Mathematics, University of Texas, Austin TX, USA

D. H. Hartley, Department of Physics, University of Lancaster, Lancaster, UK

N. J. Hitchin, Mathematical Institute, 24-29 St. Giles, Oxford OX1 3LB, UK

H. Hofer, FB Mathematik, Ruhr Universität Bochum, Universitätstr. 150, D-463 Bochum, FRG

Lisa Jeffrey, Mathematical Institute, 24-29 St. Giles, Oxford OX1 3LB, UK

F. A. E. Johnson, Department of Mathematics, University College, London WC1E 6BT, UK

J. D. S. Jones, Mathematics Institute, University of Warwick, Coventry CV4 7AL, UK

Robion Kirby, Department of Mathematics, University of California, Berkeley CA 94720, USA

Dieter Kotschick, Queen's College, Cambridge CB3 9ET, UK, *and*, The Institute for Advanced Study, Princeton NJ 08540, USA

Matthias Kreck, Max-Planck-Institut für Mathematik, 23 Gottfried Claren Str., Bonn, Germany

N. S. Manton, Department of Applied Mathematics and Mathematical Physics, University of Cambridge, Silver St, Cambridge CB3 9EW, UK

Dusa McDuff, Department of Mathematics, SUNY, Stony Brook NY, USA

Paul Melvin, Department of Mathematics, Bryn Mawr College, Bryn Mawr PA 19010, USA

Christian Okonek, Math Institut der Universität Bonn, Wegelerstr. 10, D-5300 Bonn 1, FRG

J. H. Rubinstein, Department of Mathematics, University of Melbourne, Melbourne, Australia, *and*, The Institute for Advanced Study, Princeton NJ 08540, USA

Ronald J. Stern, Department of Mathematics, University of California, Irvine CA 92717, USA

L. R. Taylor, Department of Mathematics, Notre Dame University , Notre Dame IN 46556, USA

C. B. Thomas, Department of Pure Mathematics and Mathematical Statistics, University of Cambridge, 16, Mill Lane, Cambridge CB3 9EW, UK

K. P. Tod, Mathematical Institute, 24-29 St. Giles, Oxford OX1 3LB, UK

R. W. Tucker, Department of Physics, University of Lancaster, Lancaster, UK

Edward Witten, Institute for Advanced Study, Princeton NJ 08540, USA

John C. Wood, Department of Pure Mathematics, University of Leeds, Leeds, UK

Names of Participants

N. A'Campo (Basel)
M. Atiyah (Oxford)
H. Azcan (Sussex)
M. Batchelor (Cambridge)
S. Bauer (Bonn)
I.M. Benn (Newcastle, NSW)
D. Bennequin (Strasbourg)
W. Browder (Princeton/Bonn)
R. Brussee (Leiden)
P. Bryant (Cambridge)
F. Burstall (Bath)
E. Corrigan (Durham)
S. de Michelis (San Diego)
S. Donaldson (Oxford)
S. Dostoglu (Warwick)
J. Eells (Warwick/Trieste)
Y. Eliashberg (Stanford)
D. Fairlie (Durham)
R. Fintushel (MSU, East Lansing)
A. Floer (Berkeley)
M. Furuta (Tokyo/Oxford)
G. Gibbons (Cambridge)
A. Givental (Moscow)
R. Gompf (Austin, TX)
C. Gordon (Austin, TX)
J-C. Hausmann (Geneva)
N. Hitchin (Warwick)
H. Hofer (Bochum)
J. Hurtebise (Montreal)
D. Husemoller (Haverford/Bonn)
P. Iglesias (Marseille)
L. Jeffreys (Oxford)
F. Johnson (London)
J. Jones (Warwick)
R. Kirby (Berkeley)
D. Kotschick (Oxford)
M. Kreck (Mainz)
R. Lickorish (Cambridge)
J. Mackenzie (Melbourne)
N. Manton (Cambridge)
G. Massbaum (Nantes)
G. Matić (MIT)
D. McDuff (SUNY, Stony Brook)
M. Micallef (Warwick)
C. Okonek (Bonn)
P. Pansu (Paris)
H. Rubinstein (Melbourne)
D. Salamon (Warwick)
G. Segal (Oxford)
R. Stern (Irvine, CA)
C. Thomas (Cambridge)
K. Tod (Oxford)
K. Tsuboi (Tokyo)
R. Tucker (Lancaster)
C.T.C. Wall (Liverpool)
S. Wang (Oxford)
R. Ward (Durham)
P.M.H. Wilson (Cambridge)
E. Witten (IAS, Princeton)
J. Wood (Leeds)

INTRODUCTION

In the past decade there have been a number of exciting new developments in an area lying roughly between *manifold theory* and *geometry*. More specifically, the principal developments concern:

(1) geometric structures on manifolds,
(2) symplectic topology and geometry,
(3) applications of Yang-Mills theory to three- and four-dimensional manifolds,
(4) new invariants of 3-manifolds and knots.

Although they have diverse origins and roots spreading out across a wide range of mathematics and physics, these different developments display many common features—some detailed and precise and some more general. Taken together, these developments have brought about a shift in the emphasis of current research on manifolds, bringing the subject much closer to geometry, in its various guises, and physics.

One unifying feature of these geometrical developments, which contrasts with some geometrical trends in earlier decades, is that in large part they treat phenomena in specific, low, dimensions. This mirrors the distinction, long recognised in topology, between the flavours of "low-dimensional" and "high-dimensional" manifold theory (although a detailed understanding of the connection between the special roles of the dimension in different contexts seems to lie some way off). This feature explains the title of the meeting held in Durham in 1989 and in turn of these volumes of Proceedings, and we hope that it captures some of the spirit of these different developments.

It may be interesting in a general introduction to recall the the emergence of some of these ideas, and some of the papers which seem to us to have been landmarks. (We postpone mathematical technicalities to the specialised introductions to the six separate sections of these volumes.) The developments can be said to have begun with the lectures [T] given in Princeton in 1978-79 by W.Thurston, in which he developed his "geometrisation" programme for 3-manifolds. Apart from the impetus given to old classification problems, Thurston's work was important for the way in which it encouraged mathematicians to look at a manifold in terms of various concomitant geometrical structures. For example, among the ideas exploited in [T] the following were to have perhaps half-suspected fall-out: representations of link groups as discrete subgroups of $PSL_2(\mathbf{C})$, surgery compatible with geometric structure, rigidity, Gromov's norm with values in the real singular homology, and most important of all, use of the theory of Riemann surfaces and Fuchsian groups to develop a feel for what might be true for special classes of manifolds in higher dimensions.

Meanwhile, another important signpost for future developments was Y. Eliashberg's proof in 1981 of "symplectic rigidity"– the fact that the group of symplectic diffeomorphisms of a symplectic manifold is C^0-closed in the full diffeomorphism group.

This is perhaps a rather technical result, but it had been isolated by Gromov in 1970 as the crux of a comprehensive "hard versus soft" alternative in "symplectic topology": Gromov showed that if this rigidity result was not true then any problem in symplectic topology (for example the classification of symplectic structures) would admit a purely algebro-topological solution (in terms of cohomology, characteristic classes, bundle theory etc.) Conversely, the rigidity result shows the need to study deeper and more specifically geometrical phenomena, beyond those of algebraic topology.

Eliashberg's original proof of symplectic rigidity was never fully published but there are now a number of proofs available, each using new phenomena in symplectic geometry as these have been uncovered. The best known of these is the "Arnol'd Conjecture" [A] on fixed points of symplectic diffeomeorphisms. The original form of the conjecture, for a torus, was proved by Conley and Zehnder in 1982 [CZ] and this established rigidity, since it showed that the symplectic hypothesis forced more fixed points than required by ordinary topological considerations. Another demonstration of this rigidity, this time for contact manifolds, was provided in 1982 by Bennequin with his construction [B] of "exotic" contact structures on $\mathbf{R}^3$.

Staying with symplectic geometry, but moving on to 1984, Gromov [G] introduced "pseudo-holomorphic curves" as a new tool, thus bringing into play techniques from algebraic and differential geometry and analysis. He used these techniques to prove many rigidity results, including some extensions of the Arnol'd conjecture and the existence of exotic symplectic structures on Euclidean space. (Our "low-dimensional" theme may appear not to cover these developments in symplectic geometry, which in large part apply to symplectic manifolds of all dimensions: what one should have in mind are the crucial properties of the *two-dimensional* surfaces, or pseudo-holomorphic curves, used in Gromov's theory. Moreover his results seem to be particularly sharp in low dimensions.)

We turn now to 4-manifolds and step back two years. At the Bonner Arbeitstagung in June 1982 Michael Atiyah lectured on Donaldson's work on smooth 4-manifolds with definite intersection form, proving that the intersection form of such a manifold must be "standard". This was the first application of the "instanton" solutions of the Yang-Mills equations as a tool in 4-manifold theory, using the moduli space of solutions to provide a cobordism between such a 4-manifold and a specific union of $\mathbf{CP}^2$'s [D]. This approach again brought a substantial amount of analysis and differential geometry to bear in a new way, using analytical techniques which were developed shortly before. Seminal ideas go back to the 1980 paper [SU] of Sacks and Uhlenbeck. They showed what could be done with non-linear elliptic problems for which, because of conformal invariance, the relevant estimates lie on the borderline of the Sobolev inequalities. These analytical techniques are relevant both in the Yang-Mills theory and also to pseudo-holomorphic curves. Other important and influential analytical techniques, motivated in part by Physics, were developed by C.Taubes [Ta].

Combined with the topological h-cobordism theorem of M. Freedman, proved shortly before, the result on smooth 4-manifolds with definite forms was quickly used to deduce, among other things, that $\mathbf{R}^4$ admits exotic smooth structures. Many different applications of these instantons, leading to strong differential-topological conclusions, were made in the following years by a number of mathematicians; the other main strand in the work being the definition of new invariants for smooth 4-manifolds, and their use to detect distinct differentiable structures on complex algebraic surfaces (thus refuting the smooth h-cobordism theorem in four dimensions).

From an apparently totally different direction the *Jones polynomial* emerged in a series of seminars held at the University of Geneva in the summer of 1984. This was a new invariant of knots and links which, in its original form [J], is defined by the traces of a series of representations of the Braid Groups which had been encountered in the theory of von Neumann algebras, and were previously known in statistical mechanics. For some time, in spite of its obvious power as an invariant of knots and links in ordinary space, the geometric meaning of the Jones invariant remained rather mysterious, although a multitude of connections were discovered with (among other things) combinatorics, exactly soluble models in statistical physics and conformal field theories.

In the spring of the next year, 1985, A. Casson gave a series of lectures in Berkeley on a new integer invariant for homology 3-spheres which he had discovered. This Casson invariant "counts" the number of representations of the fundamental group in $SU(2)$ and has a number of very interesting properties. On the one hand it gives an integer lifting of the well-established Rohlin $\mathbf{Z}/2$ μ-invariant. On the other hand Casson's definition was very geometric, employing the moduli spaces of unitary representations of the fundamental groups of surfaces in an essential way. (These moduli spaces had been extensively studied by algebraic geometers, and from the point of view of Yang-Mills theory in the influential 1982 paper of Atiyah and Bott [AB].) Since such representations correspond to flat connections it was clear that Casson's theory would very likely make contact with the more analytical work on Yang-Mills fields. On the other hand Casson showed, in his study of the behaviour of the invariant under surgery, that there was a rich connection with knot theory and more familiar techniques in geometric topology. For a very readable account of Cassons work see the survey by A. Marin [M].

Around 1986 A. Floer introduced important new ideas which applied both to symplectic geometry and to Yang-Mills theory, providing a prime example of the interaction between these two fields. Floer's theory brought together a number of powerful ingredients; one of the most distinctive was his novel use of ideas from Morse theory. An important motivation for Floer's approach was the 1982 paper by E. Witten [W1] which, among other things, gave a new analytical proof of the Morse inequalities and explained their connection with instantons, as used in Quantum Theory.

In symplectic geometry one of Floer's main acheivements was the proof of a generalised form of the Arnol'd conjecture [F1]. On the Yang-Mills side, Floer defined new invariants of homology 3-spheres, the *instanton homology groups* [F2]. By work of Taubes the Casson invariant equals one half of the Euler characteristic of these homology groups. Their definition uses moduli spaces of instantons over a 4-dimensional tube, asymptotic to flat connections at the ends, and these are interpreted in the Morse theory picture as the gradient flow lines connecting critical points of the Chern-Simons functional.

Even more recently (1988), Witten has provided a quantum field theoretic interpretation of the various Yang-Mills invariants of 4-manifolds and, in the other direction, has used ideas from quantum field theory to give a purely 3-dimensional definition of the Jones link invariants [W2]. Witten's idea is to use a functional integral involving the Chern-Simons invariant and holonomy around loops, over the space of all connections over a 3-manifold. The beauty of this approach is illustrated by the fact that the choices (quantisations) involved in the construction of the representations used by Jones reflect the need to make this integral actually defined. In addition Witten was able to find new invariants for 3-manifolds.

It should be clear, even from this bald historical summary, how fruitful the cros-fertilisation between the various theories has been. When the idea of a Durham conference on this area was first mooted, in the summer of 1984, the organisers certainly intended that it should cover Yang-Mills theory, symplectic geometry and related developments in theoretical physics. However the proposal was left vague enough to allow for unpredictable progress, sudden shifts of interest, new insights, and the travel plans of those invited. We believe that the richness of the contributions in both volumes has justified our approach, but as always the final judgement rests with the reader.

References

[A] Arnold, V.I. *Mathematical Methods of Classical Mechanics* Springer, Graduate Texts in Mathematics, New York (1978)

[AB] Atiyah, M.F. and Bott, R. *The Yang-Mills equations over Riemann surfaces* Phil. Trans. Roy. Soc. London, Ser. A **308** (1982) 523-615

[B] Bennequin, D. *Entrelacements et équations de Pfaff* Astérisque **107-108** 1983) 87-91

[CZ] Conley, C. and Zehnder, E. *The Birkhoff-Lewis fixed-point theorem and a conjecture of V.I. Arnold* Inventiones Math.**73** (1983) 33-49

[D] Donaldson, S.K. *An application of gauge theory to four dimensional topology* Jour. Differential Geometry **18** (1983) 269-316

[F1] Floer, A. *Morse Theory for Lagrangian intersections* Jour. Differential Geometry **28** (1988) 513-547

[F2] Floer, A. *An instanton invariant for 3-manifolds* Commun. Math. Phys. **118** (1988) 215-240

[G] Gromov, M. *Pseudo-holomorphic curves in symplectic manifolds* Inventiones Math. **82** (1985) 307-347

[J] Jones, V.R.F. *A polynomial invariant for links via Von Neumann algebras* Bull. AMS **12** (1985) 103-111

[M] Marin, A. (after A. Casson) *Un nouvel invariant pour les spheres d'homologie de dimension trois* Sem. Bourbaki, no. 693, fevrier 1988 (Astérisque **161-162** (1988) 151-164)

[SU] Sacks, J. and Uhlenbeck, K.K. *The existence of minimal immersions of 2-spheres* Annals of Math. **113** (1981) 1-24

[T] Thurston, W.P. *The Topology and Geometry of 3-manifolds* Princeton University Lecture Notes, 1978

[Ta] Taubes, C.H. *Self-dual connections on non-self-dual four manifolds* Jour. Differential Geometry **17** (1982) 139-170

[W1] Witten, E. *Supersymmetry and Morse Theory* Jour. Differential Geometry **17** (1982) 661-692

[W2] Witten, E. *Some geometrical applications of Quantum Field Theory* Proc. IXth. International Congress on Mathematical Physics, Adam Hilger (Bristol) 1989, pp. 77-110.

Acknowledgements

We should like to take this opportunity to thank the London Mathematical Society and the Science and Engineering Research Council for their generous support of the Symposium in Durham. We thank the members of the Durham Mathematics Department, particularly Professor Philip Higgins, Dr. John Bolton and Dr. Richard Ward, for their work and hospitality in putting on the meeting, and Mrs. S. Nesbitt and Mrs. J. Gibson who provided most efficient organisation. We also thank all those at Grey College who arranged the accommodation for the participants. Finally we should like to thank Dieter Kotschick and Lisa Jeffrey for writing up notes on some of the lectures, which have made an important addition to these volumes.

PART 1

FOUR-MANIFOLDS AND ALGEBRAIC SURFACES

The last few years have seen important advances in our understanding of 4-manifolds: their topology, differential topology and geometry. On the topological side there is a good picture of the full classification, through Freedman's h-cobordism and restricted s-cobordism theorems. In the differential topological category we are now well-aquainted with the special features of 4-manifold theory which are detected by the instanton solutions of the Yang-Mills equations, but the general classification is, for the moment, a matter of speculation. The 4-manifolds underlying complex algebraic surfaces have always provided a particularly interesting stock of examples, and the fascinating problems of understanding the interaction between the complex structure and the differential topology lie at the forefront of current research. One can obtain a good idea of the present position of the subject, and of the progress that has been made in recent years, by reading the two survey articles [M], [FM].

The five articles in this section cover many facets of the subject. The paper of Donaldson contains a general account of the use of Yang-Mills moduli spaces to define 4-manifold invariants, and some discussion of geometrical apects of the theory. In particular it gives a brief summary of the link between Yang-Mills theory over complex surfaces and stable holomorphic bundles, which in large measure accounts for the prominence of algebraic surfaces in the results. The paper of Gompf surveys the general picture of smooth 4-manifolds, especially algebraic surfaces, and presents partial classification results. It also contains wonderfully explicit "Kirby calculus" descriptions of some distinct differentiable structures on a family of open 4-manifolds, and ties these in to the ideas of Floer homology which we consider at greater length in the next section. The paper of Kotschick takes a more algebro-geometric stance, and surveys what is known about the differential topology of a special, but very important, class of complex surfaces. This class includes the "Dolgachev surfaces", which provided some of the first applications of the new techniques from Yang-Mills theory and which are also the starting point for Gompf's examples. The interaction between the complex geometry and the topology is particularly apparent in Kotschick's paper, and leading open problems, of detecting *rationality*, can be traced back to early work on algebraic surfaces.

The Dolgachev surfaces are also the starting point for the work described in the article of Kreck; the general setting is the relative theory, of 2-dimensional surfaces in 4-manifolds, and the Dolgachev manifolds appear as branched covers. Kreck's paper gives us an example of the application of the topological s-cobordism theorem, together with surgery theory, to a very concrete problem.

The paper of Johnson deals with a rather different facet of the topology of algebraic varieties; the structure of the fundamental group. There has been a good deal of activity in the last few years on the problems of describing what groups can occur as the fundamental groups of Kähler manifolds or of complex projective manifolds, with work of Johnson and Rees, Gromov, Toledo, Corlette, Goldman and Millson and others. A wide variety of techniques have been used, ranging from algebra to differential geometry and analysis. These questions are, at least vaguely, related to the techniques applied in defining differentiable invariants of complex surfaces, since the moduli spaces of unitary representations of the fundamental group of a compact Kähler manifold can be interpreted as moduli spaces of stable holomorphic vector bundles (compare, for example, the contribution of Okonek below).

[FM] Friedman, R. and Morgan, J.W. *Algebraic surfaces and 4-manifolds: some conjectures and speculations* Bull. Amer. Math. Soc. (New Series) **18** (1988) 1-18

[M] Mandelbaum,R. *Four-dimensional topology: an introduction* Bull. Amer. Math. Soc. (New Series) **2** (1980) 1-159

Yang-Mills Invariants of Four-manifolds

S.K.DONALDSON

The Mathematical Institute, Oxford.

This article is based on three lectures given at the Symposium in Durham. In the first section we review the well-known analogies between Yang-Mills instantons over 4-manifolds and pseudo-holomorphic curves in almost-Kahler manifolds. The second section contains a rapid summary of the definition of invariants for smooth 4-manifolds using Yang-Mills moduli spaces, and of their main properties. In the third section we outline an extension of this theory, defining new invariants which we hope will have applications to connected sums of complex algebraic surfaces. Finally, in the fourth section, we take the opportunity to make some observations on pseudo-holomorphic curves and discuss the possibility of using linear analysis to construct symplectic submanifolds, in analogy with the Kodaira embedding theorem from complex geometry.

SECTION 1, ELLIPTIC TECHNIQUES IN TOPOLOGICAL PROBLEMS

The last ten years have seen the development and application of new techniques in the two fields of 4-manifold topology and symplectic geometry. There are striking parallels between these developments, both in detail and in general methodology. In the first case one is interested primarily in smooth, oriented 4-manifolds, and the problems of classification up to diffeomorphism. In the second case one is interested in, for example, problems of existence and uniqueness of symplectic structures (closed, nowhere degenerate, 2-forms). In each case the structure considered is locally standard : the only questions are global ones and it is reasonable to describe both subjects as "topological" in an extended sense of the word.

The new developments which we have in mind bring methods of geometry and analysis to bear on these topological questions. One introduces, as an auxiliary tool, some appropriate geometrical structure, which will have local invariants like curvature and torsion. In the case of symplectic manifolds this structure is a Riemannian metric adapted to the symplectic form or, equivalently, a compatible almost complex structure. Such a metric can appropriately be called *almost-Kahler* . In the other case one considers Riemannian metrics on 4-manifolds. With this structure fixed we study associated geometric objects : in the first case these are the *pseudo-holomorphic curves* in an almost complex manifold V (i.e. maps $f : \Sigma \to V$ from a

Riemann surface Σ with complex-linear derivative) ; in the other case the objects are the *Yang-Mills instantons* over a 4-manifold X (i.e. connections A on a principal bundle $P \to X$ with anti-self-dual curvature). In either case the objects can be viewed as the solutions of certain non-linear, elliptic, differential equations. Information about the original topological problem is extracted from properties of the solutions of these equations. In the symplectic case this strategy was first employed by Gromov [**15**], and the developments in both fields are instances of the use of "hard " techniques, in the terminology of Gromov [**16**].
The detailed analogies between these two set-ups are wide ranging. Among the most important are

(1) In each theory there is a "classical " or "integrable " case. On the one hand we can consider Kahler metrics on *complex* manifolds V, and their associated symplectic forms. Then the pseudo-holomorphic curves are the holomorphic curves in the ordinary sense. On the other hand we can consider the 4-manifolds obtained from complex projective surfaces, with Kahler metrics. Then, as we shall describe in Section 2 (c) below, the Yang-Mills instantons can be identified with certain holomorphic bundles over the complex surface. So in either theory our differential geometric objects can be described in algebro-geometric terms in these important cases.
(2) There is a fundamental integral formula in each case. The *area* of a compact pseudo-holomorphic curve equals its topological "degree " (the pairing of its fundamental class with the cohomology class of the symplectic form) ; and the *Yang-Mills energy* (mean- square of the curvature) of an instanton over a compact base manifold is a topological characteristic number of the bundle carrying the conection.
(3) Both theories are conformally invariant ; with regard to the structures on Σ and X respectively.
(4) The non-linear elliptic differential equations which arise in the two cases can have non-zero Fredholm indices. Thus the solutions are typically not isolated but are parametrised by moduli manifolds.
(5) Both theories enjoy strong links with Mathematical Physics (σ - models and gauge theories). A unified treatment of these developments from the point of view of quantum field theory has been given by Witten [**22**].
(6) Both theories exploit exploit special "low-dimensional" features – they are tied to the 2-dimensionality of Σ and the 4-dimensionality of X respectively.

There are many other points of contact between the theories. Notable among these are the developments in the two fields brought about through the magnificent work of Floer (see [**10**], and the articles on Floer's work in these Proceedings). Many of the developments in the two fields bear strongly on the *representation variety* W of conjugacy classes of representations of the fundamental group of a closed Riemann surface, which has a natural Kahler structure. For example the Casson invariant of a 3-manifold can be obtained from the intersection number of a pair of Lagrangian submanifolds in W. In a different setting we will encounter the space W in Section 2 (c) below, in our discussion of instantons over complex algebraic surfaces. It is intriguing that these representation spaces have also come to the fore recently in the Jones/Witten theory of invariants for knots and 3-manifolds (see the contributions of Atiyah, Hitchin, Kirby and Witten in the accompanying volume), and it seems quite likely that this points the way towards the possibility of obtaining some unified understanding of these different developments in Low-Dimensional Topology and Geometry.

SECTION 2, YANG-MILLS INVARIANTS

(a) Definition. We will now describe how the Yang-Mills instantons yield invariants of certain smooth 4-manifolds. For more details see [**8**] or [**9**]. For brevity we will confine our discussion here to the gauge group SU(2), so we fix attention on a principal SU(2) bundle P over a compact, oriented Riemannian 4-manifold X. We will also assume that X is simply connected. The bundle P is determined up to isomorphism by the integer $k = \langle c_2(P), [X] \rangle$, and if P is to support any anti-self-dual connection k must be non-negative, by the integral formula mentioned in (2) of Section 1. For each $k \geq 0$ we have a *moduli space* M_k of anti-self-dual connections on P modulo equivalence, and M_0 consists of a single point, representing the product connection on the trivial bundle.
Let A_0 be a solution of the instanton equations, i.e. $F^+(A_0) = 0$, where $F^+ = (1/2)(F + *F)$ denotes the self-dual part of the curvature. The curvature of another connection $A_0 + a$ can be written

$$F(A_0 + a) = F(A_0) + d_{A_0} a + a \wedge a,$$

where d_{A_0} is the coupled exterior derivative. Taking the self-dual part we get, in standard notation,

$$F^+(A_0 + a) = d^+_{A_0} a + (a \wedge a)^+.$$

The moduli space is obtained by dividing the solutions of this equation by the action of the "gauge group" $\mathcal{G}$ = Aut P. For small deformations a this division can be replaced by imposing the Coulomb gauge condition (provided the connection A_0 is irreducible)

$$d^*_{A_0} a = 0 \ ,$$

which defines a local transversal slice for the action of $\mathcal{G}$. Thus (assuming irreducibility) a neighbourhood of the point $[A_0]$ in the moduli space is given by the solutions of the differential equations

$$d^*_{A_0} a = 0$$
$$d^+_{A_0} a + (a \wedge a)^+ = 0.$$

These are non-linear, first order, equations; the non-linearity coming from the quadratic term $(a \wedge a)^+$. The linearisation about $a = 0$ can be written $\delta_{A_0} a = 0$, where $\delta_{A_0} = d^*_{A_0} \oplus d^+_{A_0}$ is a elliptic operator which plays the role in this four-dimensional situation of the Cauchy-Riemann operator in the theory of pseudo-holomorphic curves. The Fredholm index s = ind δ_{A_0} of this operator is given by the formula:

$$s = 8k - 3(1 + b^+(X) ,$$

in which $b^+(X)$ is the dimension of a maximal positive subspace for the intersection form on $H^2(X)$. The number s is the "virtual dimension" of the moduli space; more precisely, according to a theorem of Freed and Uhlenbeck [**11**], [**9**], for a generic Riemannian metric on X the part of the moduli space consisting of irreducible connections will be a smooth manifold of dimension s.

Let us now assume that $b^+(X)$ is strictly positive. Then it can be shown that for generic metrics and all $k \geq 1$ every instanton is irreducible. It is easy to see why b^+ enters here. A reducible anti-self-dual connection on P corresponds to an element c of $H^2(X; \mathbf{R})$ which is in the intersection of the integer lattice and the subspace $H^- \subset H^2$ consisting of classes represented by anti-self dual forms. The codimension of H^- is b^+, so if $b^+ > 0$ and H^- is in general position there are no non-zero classes in the intersection. On the same lines one can show that if $b^+ > 1$ then for generic 1 -parameter families of Riemannian metrics on X we do not encounter any non-trivial reducible connections.

We can now indicate how to define differential topological invariants of the underlying 4-manifold X. We introduce the space $\mathcal{B}^*$ of <u>all</u> irreducible connections on P, modulo equivalence. It is an infinite dimensional manifold and, under our assumptions the moduli space M_k is a submanifold of $\mathcal{B}^*$, for generic metrics on X. Roughly, the invariants we define are the pairings of the *fundamental homology class* of the moduli space with the cohomology of $\mathcal{B}^*$. To see that this is a reasonable strategy we have to consider the dependence of the definition on the Riemannian metric on X. The moduli space itself certainly depends on the choice of metric, so let us temporarily write $M_k(g)$ for the the moduli space defined with respect to a metric g. Suppose g_0 , g_1 are two generic metrics on X. We join them by a smooth path g_t; $t \in [0,1]$ of metrics. If $b^+ > 1$ then, as explained above, we do not encounter any reducible connections so we can define

$$\mathcal{N} = \{ ([A], t) \in \mathcal{B}^* \times [0,1] \mid [A] \in M_k(g_t)\}.$$

For a generic path g_t the space $\mathcal{N}$ is a manifold- with- boundary, the boundary consisting of the disjoint union of $M_k(g_0)$ and $M_k(g_1)$. Using the obvious projection from $\mathcal{N}$ to $\mathcal{B}^*$, we can regard $\mathcal{N}$ as giving a "homology" between the two moduli spaces.

This idea needs to be amplified in a number of ways. First we need to show that the moduli space is orientable (and to fix signs one must find a rule for choosing a definite orientation). Second we need to construct cohomology classes on $\mathcal{B}^*$. This second step is an exercise in algebraic topology. Fix a base point in X and let $\tilde{\mathcal{B}}$ be the $SO(3)$ bundle over $\mathcal{B}^*$ whose points represent equivalence classes of connections on a bundle which is trivialised over the base point. The space $\tilde{\mathcal{B}}$ is weak-homotopy equivalent to the space $Maps(X, BG)$ of based maps (of "degree" k) from X to the classifying space BG (which can be identified with $\mathbf{HP}^{\infty}$) of the structure group SU(2). One can show then that the rational cohomology of $\tilde{\mathcal{B}}$ is a polynomial algebra on 2-dimensional cohomology classes labelled by a basis for the 2-dimensional homology of X. That is, the cohomology is generated by the image of a natural map

$$\tilde{\mu} : H_2(X; \mathbf{Z}) \to H^2(\tilde{\mathcal{B}}; \mathbf{Z}),$$

which is just the slant product in $Maps(X, BG) \times X$ with the 4-dimensional class pulled back from the generator of $H^4(BG)$ under the evaluation pairing $Maps(X, BG) \times X \to BG$. One can show further that this map $\tilde{\mu}$ descends to a map

$$\mu : H_2(X; \mathbf{Z}) \to H^2(\mathcal{B}^*; \mathbf{Z}),$$

and that the rational cohomology of $\mathcal{B}^*$ is freely generated as a ring by the image of this map and by a 4-dimensional class (the Pontryagin class of the fibration $\tilde{\mathcal{B}} \to \mathcal{B}^*$). The upshot of this algebro-topological excursion is that the *rational* cohomology classes of $\mathcal{B}^*$ are labelled by *polynomials* in the homology of X.
The third and most important step required to define invariants is to understand the compactness properties of the moduli space. If the moduli spaces were compact then they would carry fundamental homology classes in the usual way and there would be little extra to say. However in practice the moduli spaces are scarcely ever compact, but they do have natural compactifications. The compactification $\overline{M}_k$ of M_k is a subset of

$$M_k \cup M_{k-1} \times X \cup M_{k-2} \times s^2(X) \cup \ldots .$$

The topology is defined by a notion of convergence of the following kind. If $(x_1, \ldots, x_l)$ is a point in the symmetric product $s^l(X)$, a sequence $[A_n]$ in M_k converges to a limit $([A], (x_1, \ldots x_l)) \in M_{k-l} \times s^l(X)$ if the connections converge (up to equivalence) away from $x_1, \ldots, x_l$, and the energy densities $|F(A_n)|^2$ converge as measures to

$$|F(A)|^2 + 8\pi^2 \sum_{i=1}^{l} \delta_{x_i}.$$

The statement that the closure $\overline{M}_k$ of M_k in this topology is compact is essentially a handy formulation of analytical results of Uhlenbeck on Yang-Mills fields. This theory enters into our discussion of invariants because it can be used to show that

if the moduli space has even dimension, $s = 2d$ say, then for k such that $4k > (3b^+(X)+3)$ there is a natural pairing between the moduli space M_k and a product of cohomology classes $\mu(\alpha_1) \smile \mu(\alpha_2) \smile \cdots \smile \mu(\alpha_d)$, for any $\alpha_1, \ldots, \alpha_d$ in $H_2(X)$. We will refer to this range of values of as the "stable range" for k.
The cleanest conceptual definition of these pairings proceeds by extending the cohomology classes to the compactified space. For $l > 0$ and $c \in H^2(X)$ let $s^l(c) \in H^2(s^l(X))$ be the natural "symmetric sum " of copies of c. Then for α in $H^2(X)$ we let $\alpha^{(l)}$ be the class

$$\alpha^{(l)} = \pi_1^*(\mu(\alpha)) + \pi_2^*(s^l(c)) \in H^2(M_{k-l} \times s^l(X)\ ),$$

where c is the Poincaré dual of α. One then shows that, for any k, there is an extension $\overline{\mu}(\alpha)$ of $\mu(\alpha)$ to $H^2(\overline{M}_k)$, which agrees with $\alpha^{(l)}$ on $M_{k,l} \equiv \overline{M}_k \cap (M_{k-l} \times s^l(X))$. Consequently, for any $\alpha_1, \ldots, \alpha_d$ there is a class

$$\Pi = \overline{\mu}(\alpha_1) \smile \cdots \smile \overline{\mu}(\alpha_d) \in H^{2d}(\overline{M}_k).$$

Granted this we can define a pairing $< \Pi, [\overline{M}_k] >$ so long as the compactified space carries a fundamental homology class, and this fact follows from standard homology theory provided that the "strata " $M_{k,l}$ making up $\overline{M}_k$ have codimension 2 or more, for $l > 0$. But the dimension of $M_{k,l}$ is certainly bounded by that of $M_{k-l} \times s^l(X)$ which is :

(1) $\dim M_{k-l} + 4l = \dim M_k \ - 8l + 4l = \dim M_k - 4l$, if $l < k$;
(2) $\dim s_k(X) = 4k$, if $l = k$.

Since b^+ is odd the condition for $M_{k,k}$ to have codimension 2 is that $8k - 3(1 + b^+(X)) > 4k$, which is just the stable range condition stated above.
The disadvantage with this approach is that the only definition of the classes $\overline{\mu}(\alpha)$ known to the author is rather complicated (the main points in the definition are given in Chapter 7 of [**9**]). However the same pairing can be defined by a much more elementary, although less perspicuous, procedure. For a generic surface Σ in X the restriction of any ireducible anti-self-dual connection over X to Σ is again irreducible, so we get restriction maps :

$$r : M_j \rightarrow \mathcal{B}_\Sigma^*,$$

where $\mathcal{B}_\Sigma^*$ is the space of irreducible connections over Σ, modulo equivalence. If α is the fundamental class of Σ in $H_2(X)$ the cohomology class $\mu(\alpha)$ is pulled back from $\mathcal{B}_\Sigma^*$ by the restriction map. We choose a generic codimension 2 submanifold in this target space which represents the cohomology class, and let V_Σ be the pre-image of this in the moduli space. By abuse of notation we use the same symbol to denote subsets of all the different moduli spaces M_j (since they are all pulled back from the same representative over Σ). Let now $\Sigma_1, \ldots, \Sigma_d$ be surfaces in X, in general position, and write V_i for representatives V_{Σ_i}, as above. The crux of the matter is to show that, for k in the stable range, the intersection

$$M_k \cap V_1 \cap \cdots \cap V_d$$

is compact. We can then define the pairing to be the corresponding algebraic intersection number ; the number of points, counted with signs. The argument to establish this compactness is elementary, given two basic facts. First we can choose the V_i so that all intersections in all the moduli spaces are transverse (and the product connection is not in the closure of the V_i). Second, if $[A_n]$ is a sequence in $V_i \subset M_k$ which converges to $([A], x_1, \ldots, x_l)$ in the sense considered above, and if none of the points x_j lies in Σ_i then the limit $[A]$ is in $V_i \subset M_j$. One then goes on to show that this intersection number is independent of the choice of Riemannian metric on X by intersecting $\mathcal{N}$ with the V_i. Similar arguments show that the intersection number is independent of the choice of V_i, and of the surfaces Σ_i, within their homology classes.

In sum, we have found new invariants of 4-manifolds which are multi-linear functions in the homology. We introduce the notation

$$\mathrm{Sym}^d_{X,R}$$

for the set of d-linear, symmetric, functions on $H_2(X;\mathbf{Z})$ with values in a ring R. Then we have

THEOREM 1. *Let X be a smooth, oriented, compact and simply connected 4-manifold with* $b^+(X) = 2a + 1$ *for* $a \geq 1$. *For each* k *with* $4k > (3b^+(X) + 3)$ *the map :*

$$q_k = q_{k,X} : ([\Sigma_1], \ldots, [\Sigma_d]) \mapsto \sharp(V_1 \cap \cdots \cap V_d \cap M_k)$$

defines an element of $\mathrm{Sym}^d_{X,\mathbf{Z}}$, *where* $d = 4k - 3(1 + a)$, *which is (up to sign) a differential-topological invariant of X, natural with respect to orientation -preserving diffeomeorphisms.*

We interpose a few remarks here. First, if $b^+ = 1$ one can still define invariants, but these have a more complicated form; see the article by Kotschick in these Proceedings. Second, it should be possible to extend the range of values of k for which invariants are defined. In a simple model case (where $b^+ = k = 1$) one knows how to introduce a boundary term to compensate for a codimension-1 stratum $M_{1,1}$, then one obtains the "Γ-invariant" of a 4-manifold. This approach has been extended in the Oxford D.Phil. thesis of K.C.Mong, and can probably be applied quite generally, although this has yet to be worked out in detail. A simpler procedure has been developed by J.W. Morgan, using components of the invariants for a connected sum $X \sharp\, r\overline{\mathbf{CP}}^2$, to define $q_{k,X}$ for values of k below the "stable range".

As a third remark; it would be good to have a definition of the invariants which was both elementary and conceptually clear. To do this one would need to fully understand the interaction between the topology used to define the compactification and the homotopy theory of the spaces of connections. It is worth emphasising that the anti-self dual equation itself plays no essential part in this discussion. Let $\mathcal{B}_k^*$ denote the space of irreducible connections modulo equivalence on a bundle of Chern class k. We can define a topology on the union:

$$\overline{\mathcal{B}_k^*} = \mathcal{B}_k^* \cup \mathcal{B}_{k-1}^* \times X \cup \mathcal{B}_{k-2}^* \times s^2(X) \cup \ldots$$

in much the same way as before, decreeing that a sequence $[A_n]$ converges to $([A], (x_1, \ldots, x_l)$ if

(1) The connections converge away from the x_i.
(2) The self-dual parts $|F^+(A_n)|^2$ of the energy densities are uniformly bounded.
(3) The Chern-Weil integrands $Tr(F(A_n)^2)$ converge as measures to the limit $Tr(F(A))^2 + 8\pi^2 \sum \delta_{x_i}$.

It would be interesting to identify the homotopy type of $\overline{\mathcal{B}_k^*}$. Similar questions can be posed for the spaces of maps from a Riemann surface, which are relevant to the analogous "weak" convergence encountered in the theory of harmonic maps and holomorphic curves.

(b) Connected sums. One of the main features of the invariants constructed above is that they vanish for a large class of connected sums. We have

THEOREM 2. *Let X be a 4-manifold which satisfies the conditions of Theorem 1. If X can be written as a smooth, oriented, connected sum $X = X_1 \sharp X_2$ and each of the numbers $b^+(X_i)$ is strictly positive, then $q_{k,X}$ is identically zero for all k.*

This strong statement reflects the fact that one can give a rather detailed description of the moduli spaces over a connected sum, in terms of data on each factor. This uses analytical techniques, which go back to work of Taubes [**18**], for "glueing" together anti-self- dual solutions, and the ideas lead on to Floer's instanton homology groups (which appear in the context of "generalised conected sums" across a homology 3-sphere). We will now indicate the kind of analytical techniques involved, and sketch how they lead to Theorem 2.

Let A_1, A_2 be instantons on bundles P_1, P_2 over the manifolds X_1, X_2 respectively. Assume that the connections are irreducible and that the operators $d^+_{A_i}$ appearing in the linearisation of the anti-self-dual equations are surjective (which is true for generic metrics on X_i). We also suppose that the metrics on the X_i are flat in small neighbourhoods of points x_i. We introduce a parameter $\lambda > 0$ and consider a conformal structure on the connected sum based on the "glueing" map given, in local Euclidean co-ordinates about these points, by

$$\xi \mapsto \frac{\lambda}{|\xi|^2} \bar{\xi},$$

where $\xi \mapsto \bar{\xi}$ is a reflection. A suitable metric on the conformal class represents a connected sum with a "neck" of diameter $O(\lambda^{1/2})$ (another, conformally equivalent, model is a connected sum joined by a tube of radius $O(1)$ and length $O(\exp(\lambda^{-1}))$. We want to construct an instanton on X, for a small parameter λ, which is close to A_i away from the neck region in the connected sum. As an approximation to what we want we fix an identification of the fibres :

$$\rho : (P_1)_{x_1} \to (P_2)_{x_2}.$$

We construct a connection A_0 on a bundle P over X by flattening the connections A_i near x_i, and glueing together the bundles using the identification ρ, spread out over balls around the x_i using the flat structures. We want to find an anti-self-dual connection $A_0 + a$ near to A_0. This is rather similar to our discussion above of the local behaviour of the moduli space about a solution, the difference is that now A_0 is not itself a solution. We want to solve the equation

$$d^+_{A_0} a = -F^+(A_0) + (a \wedge a)^+,$$

with a small. Suppose that S is a right inverse to $d^+_{A_0}$ i.e. $d^+_{A_0} S\omega = \omega$, and that we have a uniform bound on the operator norm of S, mapping from L^2 to L^4, that is

$$\|S(\omega)\|_{L^4} \leq C\|\omega\|_{L^2},$$

with a constant C independent of λ (which should be regarded as a parameter throughout the discussion). Note that the L^2 norm on 2-forms and L^4 norm on 1-forms are conformally invariant, so we need only specify the conformal structure on X. We will come back to the construction of S in a moment, but first we show how it leads to a solution of our problem. We seek a solution in the form $a = S(\omega)$, so the equation becomes :

$$\omega = -(S(\omega) \wedge S(\omega))^+ - F^+(A_0).$$

We use the Cauchy- Schwartz inequality to estimate the quadratic term, or rather the corresponding bilinear form:

$$\|S(\omega_1) \wedge S(\omega_2)\|_{L^2} \leq (1/\sqrt{2})\|S(\omega_1)\|_{L^4}\|S(\omega_2)\|_{L^4} \leq (1/\sqrt{2})C^2\|\omega_1\|_{L^2}\|\omega_2\|_{L^2}.$$

This means that we can write our equation in the form $\omega = T(\omega)$, where $T(\omega) = -(S(\omega) \wedge S(\omega))^+ - F^+(A_0)$, and we have :

$$\|T(\omega_1) - T(\omega_2)\|_{L^2} \leq (1/2)\|\omega_1 - \omega_2\|_{L^2},$$

say, if the L^2-norms of ω_1, ω_2 are smaller than some fixed constant, independent of λ. On the other hand

$$T(0) = -F^+(A_0)$$

and it is easy to see that this can be made arbitrarily small by making λ small (since one neeeds to flatten the connections over corespondingly small neighbourhoods of the points x_i). It follows easily then from the contraction mapping principle that, for small λ, there is a solution to our problem in the form

$$\omega = \lim_{n \to \infty} T^n(0).$$

We now come back to explain how to construct the right inverse S, obeying the crucial uniform estimate. By standard elliptic theory there are right inverses S_i to the operators $d^+_{A_i}$ over the compact manifolds X_i which are bounded as maps from L^2 to L^4. To save notation (and an additional, unimportant, term in the estimates) let us at this stage ignore the distinction between A_i and the slightly flattened connection over X_i used to form A_0. Let ϕ_1, ϕ_2 be cut-off functions on X whose derivatives are supported in the neck region, with ϕ_i equal to 1 on the "X_i side " and to 0 on the other side, and with $\phi_1^2 + \phi_2^2 = 1$ on X. The function ϕ_i can be regarded in an obvious way as a function on X_i, and we can choose the functions so that (for small λ) the L^4 norm of $d\phi_i$ is as small as we please. (This is essentially the failure of the Sobolev embedding $L^p_1 \to C^0$, when $p = 4$.) By the conformal invariance it does not matter whether we measure this L^4 norm in X or in X_i.

Now, as a first approximation to the desired inverse S, we set

$$N(\omega) = \phi_1 S_1(\phi_1 \omega_1) + \phi_2 S_2(\phi_2 \omega_2),$$

for any self-dual 2-form ω over X. The cut-off functions allow us to make sense of this formula, over X, even though the S_i are defined over X_i, using obvious identifications. Moreover we have

$$d^+_{A_0} N\omega = \omega + \sum_i \phi_i (d\phi_i \wedge S_i(\phi_i \omega))^+.$$

This gives, much as before, that

$$\|d^+_{A_0} N\omega - \omega\|_{L^2} \leq \text{const.}(\sum_i \|d\phi_i\|_{L^4}) \|\omega\|_{L^2}.$$

When λ is small we can choose ϕ_i with derivative small in L^4, then this inequality says that $d^+_{A_0} N - 1$ is a contraction ; hence $d^+_{A_0} N$ is invertible and we can put

$$S = N \circ (d^+_{A_0} N)^{-1}.$$

This completes our brief excursion into the analytical aspects of the theory. Taking the ideas further one shows that, with A_i fixed and λ small, one constructs a family of solutions parametrised by a copy of SO(3), the choice of gluing parameter ρ. Letting the A_i vary we construct open sets in the moduli space $M_{k,X}$ which are fibre bundles over open sets in $M_{k_1,X_1} \times M_{k_2,X_2}$, for $k = k_1 + k_2$ and each $k_i > 0$. When one of the k_i is zero, say k_2, the picture is different, since the operator $d^+_{A_2}$ then has a cokernel of dimension $3b^+(X_2)$. One obtains another open set in $M_{k,X}$ which is modelled on a subset Z of M_{k,X_1}, this being the zero set of a section of a vector bundle E of rank $3b^+(X_2)$ over M_{k,X_1}. The bundle E is the direct sum of $b^+(X_2)$ copies of a canonical 3-plane bundle over the moduli space : the vector bundle associated by the adjoint representation to the principal SO(3) bundle $\tilde{\mathcal{B}} \rightarrow \mathcal{B}^*$ mentioned in (a) above. In particular the rational *Euler class* of E is zero.

The relation between these different open sets in the manifold $M_{k,X}$, and their dependence on the parameter λ, is rather complicated but to sketch the ideas involved in the proof of the "vanishing theorem" (Theorem 2) we can proced by imagining that the moduli space $M_{k,X}$ is actually decomposed into compact components in this way, labelled by (k_1, k_2). We then invoke two mechanisms. First, for components with neither k_i equal to zero, the SO(3) fibre in the description of the component fibering over $M_{k_1,X_1} \times M_{k_2,X_2}$ is trivial as far as the cohomology classes $\mu(\alpha)$ are concerned. These classes are all lifted up from the base in the fibration (think of restricting to surfaces in X_1, X_2) and so their cup-product must obviously vanish on the fundamental class. The second mechanism applies when one of the k_i is zero, k_2 say. We can then think (under our unrealistic hypotheses) of the corresponding component of $M_{k,X}$ as being identified with the zero set Z. Under this identification the cohomology classes $\mu(\alpha)$ are all obtained by restricting the corresponding classes over M_{k,X_1}. On the other hand the fundamental class of Z in the homology of M_{k,X_1} is Poincaré dual to the Euler class of E, and hence is zero in rational homology, so the contribution from this component to all the homology pairings gives zero.

We emphasise again that all we have tried to do here is to give the main ideas in the proof of Theorem 2, since we will take up these ideas again in Section 3 ; the detailed proof is long and complicated and we refer to [8] for this.

(c) Instantons and holomorphic bundles. We will now consider the "integrable case " mentioned in Section 1. We suppose that our base manifold is endowed with a compatible complex structure : then we will see that any instanton naturally defines a *holomorphic bundle.* In this discussion it is simplest to work with vector bundles, so we identify our connections with covariant derivatives on the complex vector bundle associated to the fundamental representation of SU(2). The relation with holomorphic structures can been seen most simply if we consider first the case when the base space is $\mathbf{C}^2$, with the standard flat metric, and choose complex co-ordinates $z = x_1 + ix_2, w = x_3 + ix_4$. A covariant derivative has components

$$\nabla_i = \frac{\partial}{\partial x_i} + A_i,$$

and its' curvature has components

$$F_{ij} = [\nabla_i, \nabla_j].$$

The anti-self-dual condition becomes the three equations :

$$F_{12} + F_{34} = 0$$
$$F_{13} + F_{42} = 0$$
$$F_{14} + F_{23} = 0.$$

Now write $D_z = (1/2)(\nabla_1 + i\nabla_2), D_w = (1/2)(\nabla_3 + i\nabla_4)$; these are the coupled Cauchy-Riemann operators in the two complex directions. Then the second and third of the three anti-self-dual equations can be expressed in the tidy form

$$[D_z, D_w] = 0.$$

This is the *integrability condition* which is necessary and sufficient for the existence of a map g from $\mathbf{C}^2$ to $GL(2, \mathbf{C})$ such that :

$$gD_z g^{-1} = \frac{\partial}{\partial \overline{z}} \quad , \quad gD_w g^{-1} = \frac{\partial}{\partial \overline{w}}.$$

So, in the presence of the complex structure on the base, we can write our three anti-self-dual equations as the integrability condition plus the remaining equation, which can be written :

$$[D_z, D_z^*] + [D_w, D_w^*] = 0.$$

For a global formulation of this we suppose X is a complex Kahler surface and ω is the metric 2-form on X. The anti-self dual forms are just the "primitive" (1,1) forms: the forms of type (1,1) which are orthogonal to the Kahler form. The covariant derivative of a connection over X can be decomposed into $(1,0)$ and $(0,1)$ parts :

$$\nabla_A = \nabla'_A + \nabla''_A$$

We extend the operators to coupled exterior derivatives ∂_A , $\overline{\partial}_A$ on bundle-valued forms , equal to ∇'_A, ∇''_A respectively on the 0-forms. If the curvature has type (1,1) then $\overline{\partial}_A^2 = 0$ and the connection defines a holomorphic bundle, whose local holomorphic sections are the solutions of the equation $\overline{\partial}_A s = 0$. So an anti-self-dual connection defines a holomorphic bundle. Conversely, given a holomorphic bundle, we get an anti-self dual connection from any compatible unitary connection which satisfies the remaining equation $F(A).\omega = 0$. This relation can be taken much further.It has been shown ([6], [19]) that it induces a (1,1) correspondence between:

(1) The equivalence classes of irreducible anti-self-dual connections (with structure group SU(2) in the present discussion).
(2) Equivalence classes of holomorphic $SL(2, \mathbf{C})$ bundles E over X which satisfy the condition of "stability " with respect to the polarisation $[\omega] \in H^2(X)$. This stability condition is the requirement that for every line bundle bundle L which admits a holomorphic map to E we must have $c_1(L) \smile [\omega] < 0$.

The substance of this assertion is an existence theorem : for any stable bundle we can find a compatible connection A which satisfies the differential equation $F(A).\omega = 0$. The effect is that, as far as discussion of moduli questions go (and in particular for the purposes of defining invariants), we can shift our focus from the differential geometry of anti-self-dual connections to the algebraic geometry of holomorphic bundles.

These ideas have been used in two ways. On the one hand we can, in favourable cases, apply algebraic techniques to describe the moduli spaces explicitly and then calculate invariants. Two standard techniques are available for constructing rank-2 holomorphic bundles over surfaces. In one we consider a bundle V of rank 2, with a holomorphic section s which vanishes on a set of points $\{x_i\}$ in X, with multiplicity one at each point. Then we have an exact sequence

$$0 \to \mathcal{O} \to V \to \Lambda \otimes \mathcal{I} \to 0,$$

where Λ is the line bundle $\Lambda^2 V$ and $\mathcal{I}$ is the ideal sheaf of functions vanishing on the points x_i. These extensions are classified by a group $Ext \equiv Ext^1(\Lambda \otimes \mathcal{I}, \mathcal{O})$, which fits into an exact sequence:

$$H^1(\Lambda^*) \to Ext \to \bigoplus_i (K_X \otimes \Lambda)^*_{x_i} \to H^0(K_X \otimes \Lambda)^*,$$

where the last map is the transpose of the evaluation map at the points x_i. So we can read off complete information about these extensions if we have sufficient knowledge of the cohomology groups of the line bundles over X. In principle this approach can be used to describe all bundles over X since, if E is any rank-2 bundle we can always find a line bundle L such that $V = E \otimes L$ has a section vanishing at an isolated set of points (for a complete theory one needs also to consider zeros with higher multiplicity).

For the second construction technique we consider a double branched-cover $\pi : \tilde{X} \to X$. If J is a line bundle over $\tilde{X}$ the direct image $\pi_*(J)$ is a rank-2 vector bundle (locally free sheaf) over X. Conversely, starting from a rank-2 bundle E over X, if we have a trace-free section s of the bundle $EndE \otimes L$ for some line bundle L over X, which has distinct eigenvalues at the generic point of X, and whose determinant vanishes with multiplicity one on a curve C in X then we can construct a double cover $\tilde{X} \to X$, branched over C. The points of $\tilde{X}$ represent choices of eigenvalue of s. The associated eigenspace defines a line bundle J' over $\tilde{X}$, and E is the direct image of $J = J' \otimes [C]$, where C is regarded as a divisor in $\tilde{X}$. This theory can be extended to cases where $\tilde{X}$ is singular and J is a rank-1 sheaf of a suitable kind.

The other application of these ideas is more general. The Yang-Mills invariants give strong information about the differential topology of complex surfaces even in cases where one cannot, at present, calculate the invariants explicitly. This comes about through a general positivity property of the invariants. Let $\alpha \in H_2(X)$ be Poincaré dual to the Kahler class $[\omega]$ over a surface X, as above. Then we have :

THEOREM 3.
For all large enough k the invariant $q_{k,X}$ satisfies $q_{k,X}(\alpha, \alpha, \ldots, \alpha) > 0$.

For the proof of this one considers the restriction of holomorphic bundles over X to a hyperplane section Σ – a complex curve representing α. We have a moduli space W_Σ of stable bundles over Σ, just as considered in Witten's interpretation of the Jones invariants.(See the account of Witten's lectures in these Proceedings). Let us suppose for simplicity that stable bundles over X remain stable when restricted to C (the technical difficultyies that arise here can be overcome by replacing α by $p\alpha$ for $p >> 0$, and considering restriction to a finite collection of curves.) Then we have a restriction map

$$r : M_k \to W_\Sigma.$$

Over W_Σ we have a basic holomorphic line bundle $\mathcal{L}$, again just as considered in Witten's theory. It is easy enough to show that $\mu(\alpha)$ is the pull-back by r of the first Chern class of $\mathcal{L}$. On the other hand $\mathcal{L}$ is an ample line bundle over W_Σ; for large N the sections of $\mathcal{L}^N$ define a holomorphic embedding $j : W_\Sigma \to \mathbf{CP}^m$. Furthermore one can easily see that r is an embedding, so the composite $j \circ r$ gives a projective embedding of M_k, and $N\mu(\alpha)$ is the restriction of the hyperplane class over projective space. In this way one shows that,*under one important hypothesis*, the pairing $N^d q_{k,X}(\alpha, \ldots, \alpha)$ is the degree of the closure of the image of M_k in $\mathbf{CP}^m$ (a projective variety). The degree of a non-empty projective variety is positive, and this gives the result.

The vital hypothesis we require for this argument to work is the condition that, at least over a dense set of points in M_k, the Kahler metric ω behaves like a generic Riemannian metric, i.e. that for a dense set of anti-self-dual connections A the cokernel of the operator d_A^+ is zero. In algebro-geometric terms we require that for a dense set of stable bundles E the cohomology group $H^2(\mathrm{End}_0 E)$ should be zero. (Here End_0 denotes the trace -free endomorphisms).

To complete the proof of (3) then we must show that this hypothesis is satisfied, and this is where the condition that k be large enters. What one proves is that the subset S_k of the moduli space representing bundles E with $H^2(End_0 E)$ non-zero has complex dimension bounded by

$$\dim_C S_k \leq 3k + Ak^{1/2} + B, \tag{4}$$

for some constants A,B. This grows more more slowly than the virtual (complex) dimension $d = 4k - (3/2)(1 + b^+(X))$ of the moduli space M_k , and it follows that $M_k \backslash S_k$ is dense in M_k, for large k. To establish the bound (4) one uses the fact that $H^2(End_0 E)$ is Serre- dual to $H^0(End_0 E \otimes K_X)$. So if E represents a point in S_k there is a non-trivial section s of the bundle $End_0 E \otimes K_X$. Two cases arise according to whether the determinant of s is identically zero or not. If the determinant is zero the kernel of s defines a line bundle L^* and a section of $E \otimes L$. Then we can fit into the first construction described above and estimate the number of parameters available in the group Ext in terms of k. If the determinant is non-zero we fit into the second construction, using a branched cover, and we again estimate the number of parameters which determine the branched cover $\tilde{X}$ and rank-1 sheaf J.

(d) Remarks.
One can roughly summarise the first results which are obtained from these Yang-Mills invariants by saying that they show that there are at least two distinct classes of such manifolds (up to diffeomorphism), which are not detected by classical methods. On the one hand we have the connected sums of elementary building blocks, for example the manifolds :

$$X_{\alpha,\beta} = \underbrace{(\sharp \mathbf{CP}^2 \sharp \ldots \sharp \mathbf{CP}^2)}_{\alpha \text{ copies}} \ \sharp \ \underbrace{\overline{\mathbf{CP}}^2 \sharp \ldots \sharp \overline{\mathbf{CP}}^2}_{\beta \text{ copies}}$$

for which the invariants are trivial. Any (simply connected) 4-manifold X with odd intersection form is homotopy equivalent to one of the $X_{\alpha,\beta}$. On the other hand we have complex algebraic surfaces, where the invariants are non-trivial. (There is some overlap between these classes in the case when $\alpha = 1$.) For a more extensive discussion of this general picture see the article by Gompf in these proceedings. More refined results show that the second class of 4-manifolds itself contains many distinct manifolds with the same classical invariants (that is, homotopy equivalent but non-diffeomorphic, simply connected smooth 4-manifolds.) The strongest results of this kind have been obtained by Friedman and Morgan in their work on elliptic surfaces.

Recall that a K3 surface is a compact, simply connected complex surface with trivial canonical bundle. All K3 surfaces are diffeomorphic, but not necessarily bi-holomorphically equivalent. Some K3 surfaces are "elliptic surfaces", that is they admit a holomorphic map $\pi : S \to \mathbf{CP}^1$ whose generic fibre is an elliptic curve (2-dimensional torus). Starting with such a K3 surface one can construct a family of complex surfaces $S_{p,q}$ $p,q \geq 1$ by performing logarithmic transformations to a pair of fibres of π, with multiplicities p and q. From a differentiable point of view a logarithmic transform of multiplicity r can be effected by removing a tubular neighbourhood of a fibre, with boundary a 3-dimensional torus, and glueing it back using the automorphism of the 3-torus specified in a standard way by the matrix:

$$\begin{pmatrix} 1 & r-1 & 0 \\ 0 & 1 & 0 \\ 0 & 0 & 1 \end{pmatrix}$$

in $SL(3, \mathbf{Z}) \cong Aut\ T^3$. The $S_{p,q}$ are again elliptic surfaces and, as was pointed out by Kodaira [**17**], are homotopy equivalent to S if $p+q$ is even. In particular $b^+(S_{p,q}) = 3$ and, for $k \geq 4$, Theorem 1 gives an invariant $q_{k,S_{p,q}}$, which is a multilinear function of degree $d = 4k-6$. Friedman and Morgan have announced a partial evaluation of these invariants, for $k > 4$ using a description due to Friedman of the moduli spaces of stable holomorphic bundles over $S_{p,q}$, for a suitable Hodge metric [**12**],[**13**]. Friedman's description of the moduli spaces starts from an analysis of the restriction of bundles on $S_{p,q}$ to the fibres of the elliptic fibration. On a fibre the bundle is either decomposable into a connected sum, or an extension of the trivial bundle by itself. The first condition is open and for a bundle which decomposes on the generic fibre the choice of a factor in such a decomposition defines a double branched cover of the surface; then the bundle can be recovered using the second construction mentioned above. Friedman also analyses the other bundles by the first technique, using extensions, and these turn out to have fewer moduli. In this way Friedman is able to obtain a very general and quite detailed description of the moduli spaces.
To state the result of Friedman and Morgan we regard the invariants as polynomial functions on the homology of the 4-manifolds. There are two basic such functions: the intersection form Q - viewed as a quadratic polynomial, and the linear function:

$$\kappa_{p,q} : H_2(S_{p,q}) \to \mathbf{Z},$$

given by pairing with the cohomology class $-c_1(S_{p,q})$. The Yang-Mills invariants can be expressed as polynomials in Q and $\kappa_{p,q}$ and have the form:

(5) $$q_{k,S_{p,q}} = (pq)\ Q^{[l]} + \sum_{i=1}^{l} a_i Q^{[l-i]} \kappa_{p,q}^{2i}.$$

Here $l = (d/2) = (1/4)\dim M_k$, and we have written $Q^{[l]}$ for the "divided power" $(1/l!)Q^l$. The formulae depend a little on ones choice of conventions for multiplication in the ring $\mathrm{Sym}^*_{X,\mathbf{Z}}$: explicitly we have, for example :

$$Q^{[2]}(\alpha_1, \alpha_2 \alpha_3, \alpha_4) = (\alpha_1.\alpha_2)(\alpha_3.\alpha_4) + (\alpha_1.\alpha_3)(\alpha_4.\alpha_2) + (\alpha_1.\alpha_4)(\alpha_2.\alpha_3).$$

In the formula (5) the a_i are unknown integers. Friedman and Morgan deduce from this partial calculation that the product (pq) is a differentiable invariant of the 4-manifold $S_{p,q}$. In particular there are infinitely many diffeomorphism types realised within the one homotopy (or homeomorphism) class. (Indeed Friedman and Morgan work much more generally, considering all simply connected elliptic surfaces with $b^+ \geq 3$.)

A natural question to ask is to what extent such information can be derived without recourse to algebraic geometry and explicit descriptions of the moduli spaces. One obvious approach to this is to think about the differentiable description of the logarithmic transformation in terms of cutting and pasting along a 3-torus. One can get a good theoretical understanding of the effect on the Yang-Mills invariants of such cutting and pasting operations across a homology 3-sphere using Floer's instanton homology groups. One would like to have an extension theory of this to more general 3-manifolds, like the 3-torus. The author has been told that analytical results in this direction have been obtained by T. Mokwra.

Another general problem is to find if there are any other simply connected 4-manifolds, beyond the connected sums and algebraic surfaces noted above. We call attention to a very interesting family of 4-manifolds, which provides at present many candidates for such examples. These candidates are obtained by starting with a complex algebraic surface X defined over the real numbers, so there is an anti-holomorphic involution $\sigma : X \to X$ with fixed point set a real form $X_{\mathbf{R}}$ of X – a real algebraic surface. We let Y be the quotient space X/σ, which naturally has the structure of a smooth manifold (since the fixed point set has real codimension 2). If X is simply connected and $X_{\mathbf{R}}$ is non-empty the quotient Y is also simply connected and its' classical numerical invariants can be found from the formulae

$$b^+(Y) = p_g(X) \quad , \quad 2\chi(Y) = \chi(X) + \chi(X_{\mathbf{R}}) \quad .$$

Going backwards, the manifold X can be recovered as the double cover of Y, branched over the obvious copy of $X_{\mathbf{R}}$.

This construction has been used by Finashin, Kreck and Viro in the case when X is a Dolgachev surface (with $p_g = 0$). In this case the quotient Y does not give a new differentiable structure – for a suitable choice of σ it is diffeomorphic to the 4-sphere. Instead they show that the branch surfaces give new exotic knottings in S^4 –see Kreck's article in these Proceedings. A similar picture holds if we take X to be one of the manifolds $S_{p,q}$ considered above. First, if (S, σ) is a K3 surface with anti-holomorphic involution then S/σ is one of the standard manifolds : $S^2 \times S^2$ or $X_{1,\beta}$. Indeed, Yau's solution of the Calabi conjecture gives a $\sigma-$ invariant *hyperkahler* metric on S, compatible with a family of complex structures. It is easy to see that with respect to one such complex structure, J say, the map σ is a *holomorphic* involution of S. (This complex structure is orthogonal to the original one present in our explicit complex description of S.) Then J induces a complex structure on the quotient space $T = (S/\sigma)$, such that the projection map is a holomorphic branched cover. But it is a simple fact from complex surface theory that if a K3 surface is a branched cover of a surface T then T is a rational surface; hence S/σ is rational and so diffeomeorphic to $S^2 \times S^2$ or some $X_{1,\beta}$.

Now the argument of Finashin, Kreck and Viro shows that the quotient of a logarithmic transform $S_{p,q}$ by an anti-holomorphic involution is again diffemorphic to one of these standard manifolds. By this means one can get "knotted complex curves" in, for example, $S^2 \times S^2$, i.e. embedded surfaces homologous to a complex curve of the same genus, but not isotopic to a complex curve.

While we do not obtain any new manifolds by this quotient construction in the two cases considered above, in more general cases the problem of understanding the diffeomorphism type of the quotient seems to be quite open. An attractive feature of this class of manifolds is that one can still hope to get some explicit geometrical information about the Yang-Mills solutions. The anti-holomorphic involution σ of X induces an anti-holomorphic involution $\tilde{\sigma}$ of the moduli spaces $M_{k,X}$. Recently S-G.Wang has shown that the moduli space $M_{j,Y}$ can essentially be identified with a component of the fixed-point set of $\tilde{\sigma}$ in $M_{2j,X}$ (the "real" bundles over X). On the other hand these real bundles can, in principle, be analysed algebro-geometrically.

Section Three, Torsion invariants

(a) More cohomology classes. The theory outlined in Section Two can be extended in a number of directions. In this Section we will consider one such extension; where we define additional invariants which exploit the torsion in the homology of the space of connections. This extension was greatly stimulated by conversations with R.Gompf during the Durham Symposium, and for aditional background we refer again to Gompf's article in these Proceedings.

Our starting point is the following question : does the connected sum of a pair of algebraic surfaces decompose into "elementary " factors ? For example, can we split off an $S^2 \times S^2$ summand ? The invariants we have defined so far are not at all useful for these problems, since they are trivial on such connected sums. So we will now look for finer invariants, which will not have such drastic "vanishing" properties. These invariants use more subtle topological features.

As we explained in Section Two the rational cohomology of the space $\mathcal{B}^*$ of equivalence classes of irreducible connections on an SU(2) bundle over a compact 4-manifold X is very simple. The *integral* cohomology of $\mathcal{B}^*$ the other hand, is much more complicated. For example, consider the case when $X = S^4$ and, as in Section 2, let $\tilde{\mathcal{B}}$ be the space of "framed " connections - homotopy equivalent to an SO(3) bundle over $\mathcal{B}^*$. This basic example was discussed in detail by Atiyah and Jones [1]. The space $\tilde{\mathcal{B}}$ is homotopy equivalent to $\Omega^3 S^3$ – the third loop space of S^3. The rational cohomology is trivial, but the cohomology with finite co-efficient groups is very rich. Many non-zero homology classes are detected by a virtual bundle which corresponds, in the framework of connections, to the index of the family of coupled

Dirac operators parametrised by $\tilde{\mathcal{B}}$. In general on an arbitrary spin 4-manifold X we can use the Dirac family to construct corresponding classes, as in [7]. One can then go on to consider the problem of pushing these classes down to $\mathcal{B}^*$. For our application below we want a certain class $u \in H^1(\mathcal{B}^*; \mathbf{Z}/2)$, or equivalently a real line bundle η over $\mathcal{B}^*$. This is defined when the Chern class k of the bundle P over X we are considering is *even* . We recall the construction from [7]. Over $\mathcal{A}$ there is a determinant line bundle $\tilde{\eta}$ with fibres

$$\tilde{\eta}_A = \Lambda^{max.} ker D_A \otimes \Lambda^{max.} ker D_A^*,$$

where D_A is the Dirac operator coupled to A via the fundamental representation of SU(2), and regarded as a *real* operator. This admits a natural action of the gauge group $\mathcal{G}$; and the element -1 in the centre of $\mathcal{G}$ (which acts trivially on $\mathcal{A}$) acts as $(-1)^{ind D_A}$ on the fibres of $\tilde{\eta}$. On the other hand the numerical index of the coupled operator compares with that of the ordinary Dirac operator D by

$$\text{ind } D_A = k + 2 \text{ ind } D.$$

(The factor 2 appears here as the dimension of the fundamental representation.) It follows that $(-1) \in \mathcal{G}$ acts trivially on $\tilde{\eta}$ precisely when k is even, and in this case the bundle descends to a line bundle $\eta \to \mathcal{B}^*$. We then put $u = w_1(\eta)$.

(b) Additional invariants.

Consider first a general case where we have a cohomology class $\theta \in H^s(\mathcal{B}^*; R)$, for some co-efficient group R, and a Yang-Mills moduli space $M_k \subset \mathcal{B}^*$ of dimension s. If we can construct a natural pairing between a fundamental class of the moduli space and θ we obtain a numerical invariant of X. We recall that in Section 2 such a pairing could be obtained, when θ is a product of classes $\mu(\alpha)$, by extending the cohomology classes to the compactified space $\overline{M}_k$ which carries a fundamental homology class once k is large. It seems that this approach cannot be extended, without reservation, to *all* the cohomology of $\mathcal{B}^*$.(It is certainly not true that all the cohomology of $\mathcal{B}^*$ extends to $\overline{M}_k$.) However, as we shall now show, it can be carried through when the class θ contains a large enough number of factors of the form $\mu(\alpha)$. Suppose then that the virtual dimension s of the moduli space under consideration has the form $s = 2d + r$, where $r = 1$ or 2 and θ is a cohomology class of the shape :

$$\theta = \mu(\alpha_1) \smile \cdots \smile \mu(\alpha_d) \smile \phi,$$

where $\phi \in H^r(\mathcal{B}^*; R)$. To construct a pairing between θ and the moduli space we proceed as follows. As in the second construction of Section 2 we let $V_1, \ldots, V_d$ be codimension 2 representatives for the $\mu(\alpha_i)$, based on surfaces Σ_i in X, and chosen so that all multiple intersections are transverse to all moduli spaces. Then the intersection :

$$I = M_k \cap V_1 \cap \cdots \cap V_d$$

is an r-dimensional, oriented, submanifold of $M_k \subset \mathcal{B}^*$. If I is compact we can evaluate the remaining factor ϕ on I to obtain an invariant in the co-efficient group R.

The argument to show that I is compact, when k is large, is just the same as that used in the basic case (when $r = 0$) considered in Section 1. In terms of the compactified space, we exploit here the fact that the lower strata have codimension at least 4, and this is where the hypothesis $r \leq 2$ enters. In fact, at this point we only need $r \leq 3$. In detail ; suppose that $[A_\alpha]$ is an infinite sequence in I. Taking a subsequence we may assume that it converges to $([A], (x_1, \ldots, x_l)$ in $\overline{M}_k$. There are at most $2l$ of the surfaces which contain one of the points x_i, so $[A]$ must lie in at least $d - 2l$ of the V_j. If $l = k$, so A is flat, $[A]$ does not lie in any of the V_j, so in this case we must have $d \leq 2k$ i.e.$4k \leq 3(1 + b^+(X)) + r$. So if we assume that

$$4k > 3(1 + b^+(X)) + r$$

this case does not occur. On the other hand if $l < k$ the dimension of M_{k-l} is $2d + r - 8l$ and this must be at least $2(d - 2l)$, since A lies in $d - 2l$ of the V_j. Hence $r \geq 4l$, and since $r \leq 3$ we must have $l = 0$. So A is a limit point of the sequence in I.

Now a similar argument involving families shows that for any two generic metrics on X, or choices of V_j, the intersections are cobordant in $\mathcal{B}^*$. This is where we need to use the assumption that $r \leq 2$, since we introduce an extra parameter into our "dimension counting".It follows then that the pairings are the same. Finally let us note that the group of orientation- preserving self- homotopy equivalences of X acts naturally on the cohomology of $\mathcal{B}^*$. For simplicity we suppose that the class ϕ is fixed by this action, we just call such a class an *invariant class* . Then to sum up we obtain

THEOREM 7. *Let X be a compact, smooth, oriented,and simply connected 4-manifold with $b^+(X) > 1$. Let ϕ be an invariant class in $H^r(\mathcal{B}^*, R)$ for $r \leq 2$. If $4k > 3(1 + b^+(X) + r$ and the dimension $s = 8k - 3(1 + b^+(X)$ equals $2d + r$ then the map*

$$q_{k,\phi,X} : H_2(X; \mathbf{Z}) \times \cdots \times H^2(X : \mathbf{Z}) \to R,$$

given by $q_{k,\phi,X}([\Sigma_1], \ldots, [\Sigma_d]) =< \phi\, , M_k \cap V_1 \cap V_2 \cap \cdots \cap V_d >$ defines an element of $Sym^d_{X,R}$ which is (up to sign) a differential-topological invariant of X, natural with respect to orientation preserving diffeomorphisms.

(c) Loss of compactness. Unfortunately, the author does not know any interesting potential applications for the invariants of Theorem 7. So we now go further and see what can be done if we take $r = 3$ in the set-up above. For definiteness we now fix the class $\phi \in H^3(\mathcal{B}^*; \mathbf{Z}/2)$ to be u^3, the cup-cube of the class u described in (a). Thus we should assume that X is spin and that the Chern class k is even, the dimension formula shows that we must then have $b^+(X)$ *even* . The essential fact about this class u^3 is that it can detect the "glueing parameter" which appears when we join together instantons over two different regions, after the fashion of our connected sum construction in 2(b). We shall use this fact twice below so we will recall the main point now. Consider a pair of irreducible connections A_1, A_2 on bundles P_1, P_2 over spin manifolds X_1, X_2. Let the Chern classes of the bundles be k_1, k_2, with $k_1 + k_2$ even. Flattening the connections in small balls we construct a connection $A_0(\rho)$ for each gluing parameter ρ, and in this way we obtain a family of gauge -equivalence classes of connections over the connected sum parametrised by $SO(3)$. Up to homotopy this family is independent of the particular connections A_i, or the particular flattening procedure. We can restrict our determinant line bundle to this family, getting a real line bundle over $SO(3)$. A simple application of the Atiyah-Singer "Excision Axiom " shows that this bundle is

$$\eta = (-1)^{k_1}\xi, \tag{8}$$

where ξ is the Hopf line bundle over $SO(3)$, viewed as projective 3-space.(See [7].) Note that there is no loss in symmetry in this formula, since $k_1 + k_2$ is even. It follows then that the pairing of u^3 with the fundamental class of $SO(3)$ is $(-1)^{k_1}$. With this fact at hand we will now go back to our discussion of invariants. Let the dimension of the moduli space $M_k(g)$ be $2d+3$ and let $I(g)$ be the intersection of the moduli space with $V_1, \dots, V_d$. As we noted above, $I(g)$ is still compact for generic metrics g (so long as $4k > 6 + 3b^+$) and we can form the pairing of $[I(g)]$ with u^3. The argument to show that this is independent of the choices of surfaces and codimension-2 representatives V_i goes through just as before, and we obtain a multilinear function

$$\beta_g : H_2(X) \times \cdots \times H_2(X) \to \mathbf{Z}/2,$$

by setting $\beta_g([\Sigma_1], \dots, [\Sigma_d]) = < u^3, I(g) >$.

The new feature that we encounter is that β_g is not now independent of the generic metric g. The problem comes from the next stratum $M_{k,1} = M_{k-1} \times X$ in the compactified space. The moduli space M_{k-1} has dimension $2(d-2)-1$ so in a typical 1-parameter family of metrics g_t we should expect there to be some isolated times when $M_{k-1}(g_t)$ meets $d-2$ of the V_i, say $V_1, \dots, V_{d-2}$. If A is a connection in such an intersection and x is a point in the intersection $V_{d-1} \cap V_d$ then the pair $([A], x)$ can lie in the closure of $\mathcal{I} \equiv \mathcal{N} \cap V_1 \cdots \cap V_d$, and in that case $\mathcal{I}$ does not give a compact cobordism from $I(g_0)$ to $I(g_1)$.

All is not lost, however, through this failure of compactness. The same analytical techniques used for connected sums allow one to model quite precisely the behaviour

of the compactified moduli space around $([A], x)$, see [**7**]. The link L of the stratum $M_{k,1}$ in the compactified space is a copy of $SO(3)$, representing the gluing parameter which attachs a highly concentrated instanton to the background connection A. It follows from the discussion in the previous paragraphs that the pairing of u^3 with L is 1. On the other hand a simple topological argument shows that a suitable truncation of the space $\mathcal{I}$ is a compact manifold–with–boundary, whose boundary has, in addition to $I(g_0)$ and $I(g_1)$, a component for each pair $([A], x)$, and this component is a small perturbation of the link L. So the difference $(\beta_{g_0} - \beta_{g_1}([\Sigma_1], \dots, [\Sigma_d]))$ is exactly the total number of pairs $([A], x)$. (Here we have, of course, to allow all partitions of $\{1, \dots, d\}$ of type $(d-2, 2)$ when counting the pairs $([A], x)$.)
To understand this better we consider briefly another kind of generalisation of the Yang-Mills invariants. Suppose we have a situation where the moduli space M_j has virtual dimension -1, and so is empty for generic metrics. We define an invariant for a path of metrics g_t by counting the number of points in the associated moduli space $\mathcal{N}$. If $b^+ \geq 3$ this number depends on the path only through its' homotopy class, with fixed (generic) end points. This collection of invariants of paths give a class in $H^1(\mathcal{R}^*)$, where $\mathcal{R}^*$ is the space of Riemannian metrics on X with trivial isometry group, modulo diffeomorphism. More generally, if we have a moduli space of dimension $(2d-1)$ we can take the intersection with subvarieties V_i to obtain a multi-linear invariant of paths of metrics :

$$\sigma([g_t])([\Sigma_1], \dots, [\Sigma_d]) = \sharp(\mathcal{N} \cap V_1 \cdots \cap V_d).$$

This is independent of the representatives V_i, and yields a homotopy invariant of paths. It naturally defines a class in the twisted cohomology $H^1(\mathcal{R}^*; \Pi)$, where Π is the local co-efficient system over $\mathcal{R}^*$ corresponding the representation of the diffeomorphism group on the multilinear, $\mathbf{Z}/2$ -valued functions in the homology of X. (We can, of course, go further in this direction to define higher cohomology classes over $\mathcal{R}^*$.)
Our analysis of the ends of the manifold $\mathcal{I}$ now leads immediately to the formula

$$\beta_{g_1} - \beta_{g_0} = Q \cdot \sigma([g_t]), \tag{9}$$

where σ is the $(d-2)$ -linear invariant of paths defined by the moduli space M_{k-1}, as described above. In (9) we take any path from g_0 to g_1, and on the right hand side we use the multiplication in the ring of multilinear functions $\mathrm{Sym}^*_{X,\mathbf{Z}/2}$ with the intersection form Q of X. In particular the functions β_{g_t} are equal modulo the ideal generated by the intersection form, and we obtain an intrinsic invariant in the quotient graded- ring

$$\mathrm{Sym}^*_{X,\mathbf{Z}/2} / < Q > .$$

(Notice that another consequence of (9) is that the product of the cohomology class defined by σ with Q is zero in $H^1(\mathcal{R}^*; \Pi)$.)
To sum up then we have :

THEOREM 10. *Let X be a compact, simply connected, oriented, spin 4-manifold with $b^+(X) > 1$. Suppose k is even and is such that $8k - 3(1 + b^+(X) = 2d + 3$ and $4k > 3(1 + b^+(X) + 3$. Then the pairing*

$$\beta_{k,X}([\Sigma_1], \dots, [\Sigma_d]) = < u^3, M_k \cap V_1 \cap \dots \cap V_d >$$

defines a differential- topological invariant $\beta_{k,X}$ in $Sym^d_{X,\mathbf{Z}/2}/ < Q >$.

(Invariants with this kind of ambiguity have appeared in a slightly different context in the works of Kotschick and Mong. The identification of the precise correction factor arising from the failure of compactness has been discussed, in this other context, by Kotschick.)

(d)Invariants for connected sums. We suppose now that the manifold X appearing in Theorem 9 is a smooth, oriented, connected sum $X_1 \sharp X_2$ and that each of $b^+(X_1), b^+(X_2)$ is odd. We shall use the analytical techniques described in Section 2 (b) to partially calculate the new invariant of X in terms of the factors in the sum. The arguments involved are very similar to those in the Vanishing theorem for the rational invariants - but we shall see by contrast that these torsion invariants for X need not be trivial, due to the fact that they detect the glueing parameter which appeared in our description of the moduli space. The discussion here is very similar to that in [7] for the complementary problem of the existence of 4-manifolds: it is also very similar to Furuta's use of such torsion classes in his generalisation of Floer's cohomology groups ; see Furuta's article in these proceedings.
To analyse the invariant $\beta_{k,X}$ we fix a partition $d = d_1 + d_2$ and homology classes $[\Sigma_1], \dots, [\Sigma_{d_1}]$ in X_1, represented by surfaces Σ_j in the obvious way, and classes $[\Sigma'_1], \dots, [\Sigma'_{d_2}]$ in X_2. Recall that $8k - 3(1 + b^+(X)) = 2d + 3$, where

$$4k > 6 + 3b^+(X).$$

We shall evaluate the pairing $\beta = \beta_{k,X}([\Sigma_1], \dots, [\Sigma_{d_1}], [\Sigma'_1, \dots, [\Sigma'_{d_2}])$ assuming that

$$d_i > (3/2)(1 + b^+(X_i)). \tag{11}$$

The point of this condition is that if we define k_i by

$$8k_i - 3(1 + b^+(X_i) = 2d_i,$$

both of k_1, k_2 are in the range where the polynomial invariants q_{k_i,X_i} developed in Section 2(a) are defined. Let us write $q_1, q_2 \in \mathbf{Z}$ for the evaluation of these invariants on the classes $[\Sigma_j], [\Sigma'_j]$, in $H_2(X_1), H_2(X_2)$ respectively.
We now proceed in the familiar fashion, considering a family of metrics $g(\lambda)$ on X, with the neck diameter $O(\lambda^{1/2})$, and "converging" to given, sufficiently generic,

metrics g_1, g_2 on X_1, X_2. We let $I(\lambda) \subset M_k, X(g(\lambda))$ be the intersection of moduli space with all of the V_j and V'_j. We will show that, for small λ, $I(\lambda)$ is a disjoint union of copies of SO(3). In one direction, suppose that A_1 is a connection over X_1 which represents a point of

$$I_1 = M_{k_1,X_1} \cap V_1 \cap \cdots \cap V_{d_1},$$

and similarly that A_2 is a connection over X_2 which represents a point of the intersection I_2 of M_{k_2,X_2} with the V'_j. Then the glueing theory sketched in 2(b) shows that, for small enough λ, there is a family of ASD connections over X parametrised by a the product of a copy of SO(3) (the gluing parameter), and neighbourhoods of the points $[A_i]$ in their respective moduli spaces. Taking the intersection with the V_j and V'_j is effectively the same as removing these two latter sets of parameters in the family; so we obtain a copy $I([A_1],[A_2])$ of SO(3) in the intersection, which clearly forms a complete connected component of $I(\lambda)$.
Now, under the condition (10) the sets I_1, I_2 are finite, so for small λ we find $|I_1|.|I_2|$ copies of SO(3) in $I(\lambda)$. We will now show that these make up all of $I(\lambda)$. Again the argument takes a familiar form : suppose we have a sequence $\lambda_n \to 0$ and connections A_n in $I(\lambda_n)$. After taking a subsequence we can suppose that the connections converge to limits B_1,B_2 over the complement of sets of sizes l_1, l_2 in the two punctured manifolds; where B_i is an anti-self-dual connection on a bundle with Chern class κ_i over X_i. We have an "energy" inequality

$$\kappa_1 + \kappa_2 + l_1 + l_2 \leq k. \tag{12}$$

Now the argument is the usual dimension counting. First note that at least one of the κ_i must be strictly positive, by (11). Suppose next that κ_2, say, is zero, so B_2 is the product connection. Then each surface Σ'_j must contain one of the l_2 exceptional points in X_2, so:

$$d_2 \leq 2\, l_2.$$

Over the other piece : at least $d_1 - 2\, l_1$ of the V_j must meet the moduli space for Chern class κ_1, so :

$$2d_1 \leq 4l_1 + 8\kappa_1 - 3(1 + b^+(X_1).$$

Combining these inequalities with (11) we obtain a contradiction. Similarly, in the case when neither κ_1 nor κ_2 is zero one deduces that in fact $l_1 = l_2 = 0$ and $k = \kappa_1 + \kappa_2$. It follows then that for large n the point $[A(n)]$ lies in $I([B_1],[B_2])$, and hence that $I(\lambda)$ is indeed the union of these components, for small λ.
We can now use the relation (8) between the class u and the gluing construction to evaluate β. The copies $I([A_1],[A_2])$ of $SO(3)$ are small perturbations of those

obtained by flattening the connections, so the cohomoloogy classes restrict in just the same way. We obtain the formula

$$\begin{aligned}\beta &= 0 \ \text{ if } k_1, k_2 \text{ even} \\ &= q_1 . q_2 \ \text{ if } k_1, k_2 \text{ odd.}\end{aligned}$$

We can sum up in the following theorem

THEOREM 13. *Let X be a simply connected, spin, 4-manifold with $b^+(X)$ even and k be even with $4k > 6 + 3b^+(X)$. If X can be written as a connected sum $X = X_1 \sharp X_2$, with each of $b^+(X_i)$ odd, the invariant $\beta_{k,X}$ has the form :*

$$\beta_{k,X} = \Big(\sum_{k_1, k_2 \text{ odd}, k_1+k_2=k} q_{k_1,X_1} . q_{k_2,X_2} \Big) + \epsilon_1 + \epsilon_2 \mod 2,$$

where ϵ_i contains terms of degree at most $(3/2)(1 + b^+(X_i))$ in $H_2(X_i)$.

We see then that the torsion invariants are more sensitive than those defined by the rational cohomology: the latter are killed by connected sums, since the glueing parameter is rationally trivial, but the torsion classes can detect the gluing parameter and give potentially non-trivial invariants. Moreover, if X_1 and X_2 are complex algebraic surfaces we can hope to calculate some components of the new invariant for the conected sum, using Theorem 13, and hence show, for example, that the manifold does not split off an $S^2 \times S^2$ summand. In this direction, one can use a theorem of Wall [**20**], which tells us that if $X = Y \sharp (S^2 \times S^2)$ and $b^+(Y) \geq 1$ then the automorphism group of X realises all symmetries of the intersection form. The invariants for such a manifold must be preserved by the automorphism group, and this gives strong restrictions. The author has, however, not yet found any examples where this scheme can be applied: in a few simple examples various arithmetical factors seem to conspire against the success of the method. Perhaps more elaborate examples will be successful, or perhaps there is some deeper phenomenon in play which makes these torsion invariants also vanish on connected sums.

SECTION 4, REMARKS ON HOLOMORPHIC AND PSEUDO-HOLOMORPHIC CURVES

(a) Curves, Line bundles and Linear systems.

We will now change tack and make make two remarks on the relation between holomorphic and pseudo-holomorphic curves. These remarks can be motivated by the analogies with Yang-Mills theory on 4-manifolds described in Section 1. We have seen in Section 2 that, at present, the utility of the Yang-Mills invariants is derived largely from the link with algebraic geometry in the case when the 4-manifold is a complex algebraic surface. One might hope that, in a similar way, the space of pseudo-holomorphic curves in a general almost Kahler manifold captures information which depends only on the symplectic structure and which reduces, in the special case of Kahler manifolds, to well-known facts about complex curves. This is certainly true to some extent : for example Gromov proved in [**15**] that a symplectic 4- manifold which has the homotopy type of $\mathbf{CP}^2$ and contains a complex line of self-intersection 1 (for a suitable, compatible, almost complex structure) is symplectomorphic to the standard $\mathbf{CP}^2$. Gromov's argument reduces in the integrable case to classical geometry, effectively a step in the Enriques-Castuelnuevo theorem [**14**]. For many more general results on these lines see the contribution of Mac Duff to the Proceedings. Observe, by the way, that an embedded (real) surface Σ in a symplectic manifold (V, ω) is pseudo-holomorphic with respect to some compatible almost complex structure if and only if it is a *symplectic submanifold* , i.e. if ω restricts to a symplectic form on Σ. So Gromov's Theorem asserts that a symplectic homotopy $\mathbf{CP}^2$ in which a generator of the homology can be represented by a symplectic 2-sphere is standard.

With this in mind we consider what can be said about moduli spaces of holomorphic and pseudo-holomorphic curves. In the integrable case one can of course apply a great deal of existing algebro-geometric theory. First, in a general way, the moduli spaces of holomorphic curves will be quasi-projective varieties – and the coresponding hyperplane class has a simple toploogical description, much as in the Yang-Mills case. While it is a difficult problem to find holomorphic curves in general there is one class of examples which are easy to find and describe – the curves given by *complete intersections* of hypersurfaces in a complex manifold. We recall that in a hypersurface W in a complex manifold V can be identified with a line bundle ξ over V and a holomorphic section s of ξ. If V is simply connected then the line bundle ξ is in turn specified uniquely by its' first Chern class - an integral class of type (1,1). Having fixed ξ the corresponding hypersurfaces form a *linear system* , parametrised by the projective space $\mathbf{P}(\Gamma(\xi)^*)$. Thus the study of complete intersection curves, and in particular of all holomomorphic curves in a complex surface, reduces to questions about line bundles and their holomorphic sections.

This familiar theory has a number of simple consequences. We will concentrate on the case of symplectic 4-manifolds and complex surfaces, although some of our remarks apply in higher dimensions. First, the existence of *any* holomorphic curves at all in some complex manifolds is a very unstable phenomenon. Take for example a generic (Kahler) metric on a K3 surface. The integer lattice in H^2 only meets the subspace of (1,1) classes in the origin, so there are no non-trivial holomorphic curves. (Note that the ideas here are very close to those we encountered when discussing how to avoid reducible instantons). The same phenomenon applies more generally, and we shall now see how it can be understood in the framework of the local deformation theory of solutions to the holomorphic equation, and cohomology.

Let us now go back for a while to review some of the general theory of pseudo-holomorphic curves, of a given genus g and a given homotopy class, in an almost complex manifold V. We can define two moduli spaces $\mathcal{M}_\Sigma$ and $\mathcal{M}$; the first being the space of holomorphic maps from a fixed Riemann surface Σ, and the second being the space of all pseudo holomorphic curves of the given topological type, in which the induced complex structure is allowed to vary. (Thus $\mathcal{M}_\Sigma$ is a fibre of the natural map from $\mathcal{M}$ to the moduli space of Riemann surfaces of genus g.) The linearisation of the equation defining $\mathcal{M}$ about a given solution $f : \Sigma \to V$ (which we take, for simplicity, to be an embedding) is given by a linear elliptic operator δ_f, acting on sections of the normal bundle. The Fredholm index s of δ_f is easily calculated to be

$$s = 2(\Sigma.\Sigma + 1 - g)$$

and this index is the virtual dimension of the moduli space $\mathcal{M}$ of all the pseudo-holomorphic curves. To describe $\mathcal{M}_\Sigma$ locally we introduce a similar operator $\delta_{f,0}$, acting on the pull back of the tangent bundle of V. The Fredholm index of $\delta_{f,0}$ is $s - (6g - 6)$, and this is the virtual dimension of $\mathcal{M}_\Sigma$.

Now for a generic *almost-Kahler* structure on V one can show that the operators δ_f and $\delta_{f,0}$ are surjective, for all embedded pseudo-holomorphic curves (the analogue of the Freed-Uhlenbeck result in the Yang-Mills case), see [**15**]. This means that the moduli spaces $\mathcal{M}$, $\mathcal{M}_\Sigma$ are smooth manifolds whose dimension agrees with their virtual dimension. We will now see that the picture for *Kahler* metrics is quite different. Consider a holomorphic curve $f : \Sigma \to V$ in a complex Kahler surface V ; for simplicity we assume f is an embedding. The cokernels of δ_f and $\delta_{f,0}$ can be identified with the sheaf cohomology groups $H^1(\Sigma;\nu), H^1(\Sigma, TV|_\Sigma)$ respectively, where ν is the normal bundle of Σ in V. We have then :

PROPOSITION 14.

If V is a compact complex surface with $p_g(V) > 0$ and Σ is an embedded curve in V then $H^1(\Sigma;\nu)$ and $H^1(\Sigma; TV|_\Sigma)$ are both non-zero, except for the cases

(1) *$p_g(V) = 1$ and Σ (or some multiple thereof) is cut out by the section of K_V.*
(2) *Σ is an exceptional curve in V (i.e. an embedded 2-sphere with self-intersection -1).*
(3) *V is an elliptic surface, and Σ is a multiple fibre in V - a 2-torus whose normal bundle is a holomorphic root of the trivial bundle.*

To prove this it suffices to consider the normal bundle, since the holomorphic map from $TV|_\Sigma$ to ν induces a surjection on H^1. Now if ξ is the line bundle over V corresponding to Σ the normal bundle ν is the restiction of ξ to Σ and we have an exact sequence

$$0 \to \mathcal{O}_V \to \xi \to \xi|_\Sigma = \nu \to 0.$$

This induces a long exact cohomology sequence, the relevant part of which is:

$$H^1(\Sigma;\nu) \to H^2(V;\mathcal{O}) \to H^2(V;\xi).$$

The space $H^2(V;\mathcal{O})$ has dimension $p_g(V)$ which is positive by hypothesis. To show that $H^1(\nu)$ is non-zero it suffices to show that the map to $H^2(V;\xi)$ is not injective. By Serre duality this is equivalent to showing that the map between the duals

$$m_s : H^0(V;\xi^* \otimes K_V) \to H^0(V;K_V)$$

is not surjective. Here m_s is multiplication by the section s of ξ cutting out Σ. So m_s is surjective if and only if all sections of the canonical line bundle K_V vanish on Σ. Thus we have established that $H^1(\Sigma;\nu)$ *is non-zero if* Σ *is not a fixed component of* $|K_V|$.

To complete the proof we examine the case when Σ is a fixed component of $|K_V|$. Thus we can write $K_V = [\Sigma + C]$, where C is another curve in V. Now if E is an exceptional curve in V then certainly all sections of K_V must vanish on E, so we get a fixed component this way, as allowed for in case (i) of the Proposition. Conversely, leaving aside exceptional curves, we may as well replace V by its minimal model. So we assume now that V is itself minimal. We now appeal to the classification of surfaces, as on p.188 of [**2**]. The only cases that can occur are when V is an minimal elliptic surface or a minimal surface of general type. In the first case the curve Σ must be a fibre of the elliptic fibration. If it is an ordinary fibre the normal bundle is trivial and $H^1(\nu)$ is non-zero, so the only curves that occur in this way are the multiple fibres allowed for in part(3) of the Proposition. In the second case, when V is of general type, we can assume C is non-empty, otherwise we fall into category (i) of the proposition. Then we must have $\Sigma.C > 0$, since $|K_V|$ is connected ([**2**], page 218). On the other hand some routine manipulation using the adjunction formula shows that the holomorphic Euler characteristic $\chi(\nu)$ is given by $-\Sigma.C$, so $H^1(\nu)$ must be non-zero.

We see then that for most purposes the Kahler metrics are quite *unlike* the generic almost-Kahler metrics as far as the pseudo-holomorphic curves which they defined are concerned. Again, one should contrast this discussion with that for holomorphic bundles and instantons where the key result, obtained from the estimate (4), was the fact that the Kahler metrics behave quite *like* the generic metrics.

A partial remedy for the degeneracy we have noted above can be achieved by allowing the symplectic form to vary. Fix a conformal structure on V such that the symplectic form ω is self-dual. Then ω is an element of the space $\mathcal{H}^+$ of self-dual harmonic forms, which has dimension $b^+(V)$. There is an open set $U \subset \mathcal{H}^+$ containing ω such that any $\omega' \in U$ is a non-degenerate 2-form, defining a symplectic structure on V. Also there is a unique metric in the conformal class which is almost-Kahler with metric form ω'. Thus we have a natural family of almost- Kahler structures on V parametrised by U. Fixing the volume of V we get a $b^+ - 1$ dimensional family parametrised by the subset $S(U)$ of the sphere $S(\mathcal{H}^+)$. We can then consider an enlarged moduli space $\mathcal{M}^+$ whose points consist of pairs (ω', f), where ω' is in $S(U)$ and $f : \Sigma \to V$ is pseudo-holomorphic with respect to the

corresponding structure. Thus the space $\mathcal{M}$ considered before is a fibre of the natural map from $\mathcal{M}^+$ to $S(U)$.
The point of this construction is that the space $\mathcal{M}^+$ is in many respects the more appropriate generalisation of the moduli space of holomorphic curves in a Kahler surface. The Hodge decomposition of the cohomology shows that if the original metric is Kahler then $\mathcal{M} = \mathcal{M}^+$ - i.e. there are no pseudo-holomorphic curves, in the given homology class, for the perturbed structures. Moreover the dimension of M^+ in the Kahler case is then typically equal to the virtual dimension $s + b^+(V) - 1$. From this point of view the degeneracy detected by Proposition 14 in the Kahler case appears as the degeneracy of the map from $\mathcal{M}^+$ to $\mathcal{M}$. The great drawback with this approach, as far as applications go, is the fact that in general we will lose the basic compactness properties in the manifold $\mathcal{M}^+$, since the symplectic structure itself will break down at the boundary of U.(In the special case of a K3 surface, with a hyperkahler metric, the set $S(U)$ is the whole 2-sphere and this breakdown does not occur : the importance of considering all the complex structures in this case was pointed out to the writer by Mario Micallef.)

(b) Harmonic theory on almost Kahler manifolds.
We will now focus on a specific question – the existence of symplectic submanifolds in a general almost Kahler manifold. To be quite precise we will consider the following problem :

PROBLEM. *Let V be a compact 2n-dimensional manifold and ω a symplectic form on V with integral periods, i.e. $[\omega] \in H^2(V;\mathbf{R})$ is the reduction of a class α in $H^2(X;\mathbf{Z})$. Is there a positive integer k such that the Poincaré dual of $k\alpha$ is represented by a symplectic submanifold of V ?*

(We have mentioned in (i) that, when $n = 2$, such a submanifold would be pseudo-holomorphic for a suitable almost-Kahler structure on V.)
The reason for phrasing the problem in this way is that there is a simple, familiar, answer in the case when the form is compatible with an integrable complex structure. There is then a holomorphic line bundle $\xi \to V$ with $c_1(\xi) = \alpha$, and a compatible unitary connection on ξ having curvature form $-2\pi i\omega$. Thus ξ is a *positive line bundle* and, according to the Kodaira embdding theorem, ξ is *ample*, i.e the sections of some power ξ^k , $k >> 0$, define a projective embedding of V. Given this we obtain many holomorphic curves, Poincaré dual to $k\alpha$, as hyperplane sections in the projective space, or equivalently as the zero sets of holomorphic sections of ξ^k. These holomorphic curves are a fortiori symplectic submanifolds, so we answer the problem affirmatively in the "classical" case.
It seems that there is esentially only one proof of the Kodaira theorem known; using vanishing theorems and harmonic theory over Kahler manifolds (see for example **[14]** , **[21]**). Thus it is natural to ask whether this kind of proof can be adapted to answer the problem in the general, non-integrable case. In the remainder of this section we will make some first moves in this direction. Let us consider then an

almost- Kahler manifold V, of any dimension and mimic, as far as possible, the usual differential geometric theory from the Kahler case.

First, we can decompose the differential forms on V into bi-type (since this is a purely algebraic operation) and define operators :

$$\partial : \Omega_V^{p,q} \to \Omega_V^{p+1,q} \ , \ \overline{\partial} : \Omega_V^{p,q} \to \Omega_V^{p,q+1}$$

by taking the relevant components of the exterior derivative d. In general we do not have $d = \partial + \overline{\partial}$, and $\overline{\partial}^2$ is not zero. Instead we have,

$$d = \partial + \overline{\partial} + N + \overline{N},$$

and

$$\overline{\partial}^2 = N \circ \partial : \Omega_V^{p,q} \to \Omega_V^{p,q+2},$$

where N is the *Nijenhius tensor* in $T^{1,0} \otimes \Lambda^{0,2}$, which defines bundle maps from $\Lambda^{p+1,q}$ to $\Lambda^{p,q+2}$ (see [4]).
We will now go on to consider vector bundles over V. Let $E \to V$ be a complex Hermitian bundle with a connection having covariant derivative ∇_E and curvature F_E. We decompose ∇_E, much as in Section 2(c), to write:

$$\nabla_E = \nabla_E' + \nabla_E'' : \Omega^0(E) \to \Omega^{1,0}(E) \oplus \Omega^{0,1}(E).$$

We can extend these operators to $E-$ valued differential forms, getting operators $\partial_E, \overline{\partial}_E$ with :

$$d_E = \overline{\partial}_E + \overline{\partial}_E + N + \overline{N},$$

such that $\partial_E = \nabla_E', \overline{\partial}_E = \nabla_E''$ on sections of E. The important case for us will be when E is a complex line bundle ξ^k. Let s be a smooth section of ξ^k, with zero set $Z \subset V$. It is a simple exercise in linear algebra to show that Z will be a symplectic submanifold of V if the section s satisfies the condition:

$$(15) \qquad |\overline{\partial}_{\xi^k}(s)| < |\partial_{\xi^k}(s)| \ \text{ on } \ Z.$$

This condition is thus a natural generalisation, from the point of view of the zero set, of the notion of a holomorphic section in the integrable case. We can interpret the problem of finding a symplectic submanifold as the problem of finding "approximately holomorphic" sections in this sense. To see why it is plausible that such sections should exist we will go on to consider the interaction between curvature and the coupled $\overline{\partial}$-operator , in the almost-Kahler case.

First, turning back to the general differential geometric theory for the bundle E over V, we have:

$$\partial_E^2 = F_E^{2,0} + \overline{N}\overline{\partial}_E \ , \ \overline{\partial}_E^2 = F_E^{0,2} + N\overline{\partial}_E \ , \ \partial_E\overline{\partial}_E + \overline{\partial}_E\partial_E = F_E^{1,1} + N\overline{N} + \overline{N}N,$$

where $F_E^{p,q}$ is the (p,q) component of the curvature. We wish to combine these formulae with the "Kahler identities". To state these we introduce the algebraic map

$$\Lambda : \Omega_V^{p,q} \to \Omega_V^{p-1,q-1},$$

which is the adjoint of multiplication by the metric form ω, with respect to the standard inner product on the forms. The basic Kahler identities extend to the almost-Kahler case, in that we have:

PROPOSITION 16. *On any almost Kahler manifold the formal adjoints of the operators $\partial, \overline{\partial}$ are :*

$$\partial^* = i[\Lambda, \overline{\partial}] \quad , \quad \overline{\partial}^* = -i[\Lambda, \partial].$$

These identities can be verified by checking that the usual proof for the Kahler case, as for example in [**21**], does not use the equations $\overline{\partial}^2 = \partial^2 = 0$. Consider for example the formula for the operator $\overline{\partial}^*$ on forms of type $(0,q)$. This is very easy to prove. We need to show that any $\theta \in \Omega^{0,q}, \phi \in \Omega^{0,q-1}$ with compact supports satisfy :

$$< \theta, \overline{\partial}\phi >=< i\Lambda\partial\theta, \phi >=< i\partial\theta, \omega \wedge \phi > .$$

This follows from the algebraic identity:

$$\rho \wedge \overline{\rho} \wedge \omega^{2n-2r} = |\rho|^2 d\mu$$

for $\rho \in \Omega^{0,r}$. We have:

$$< \theta, \overline{\partial}\phi >= \int \theta \wedge \partial\overline{\phi} \wedge \omega^{n-q} = \int \theta \wedge d\overline{\phi} \wedge \omega^{n-r},$$

which equals

$$\int \partial\theta \wedge \overline{\phi} \wedge \omega^{n-r} = \int \partial\theta \wedge \overline{\phi} \wedge \omega^{n-r},$$

using Stokes' theorem and the fact that ω is closed. Now the same algebraic identity shows that the last expression is just $< i\Lambda\partial\theta, \omega \wedge \phi >$, as required. The Kahler identities extend to bundle-valued forms (since the connection is trivial to first order). We apply this first to the two Laplacians on sections of E to get:

$$(\nabla_E')^*\nabla_E' - (\nabla_E'')^*\nabla_E'' = i\Lambda(F_E^{1,1} + N\overline{N} + \overline{N}N). \tag{17}$$

Turning now to the E -valued $(0,q)$ forms, we can consider three Laplace- type operators. First we have the "$\overline{\partial}$ -Laplacian "

$$\Delta = \overline{\partial}_E^* \overline{\partial}_E + \overline{\partial}_E \overline{\partial}_E^*.$$

Second we have the ∂-Laplacian on $\Omega_E^{0,q}$:

$$\Box' = \partial_E^* \partial_E.$$

The Kahler identities give :

$$\Delta = \Box' + i\Lambda(F_E^{1,1} + N\overline{N} + \overline{N}N). \tag{18}$$

On the other hand there is a unique connection on the bundle $\Lambda^{0,q} \otimes E$ compatible with the metric and having ∂_E for its $(0,1)$ component. We write the covariant derivative of this connection as $\nabla = \nabla' + \nabla''$, so $\nabla' = \partial_E$. We can form the third operator

$$\Box'' = \nabla''^* \nabla''.$$

Applying (17) to the bundle $E \otimes \Lambda^{0,q}$ we get :

$$\Box' - \Box'' = i\Lambda\Phi, \tag{19}$$

where Φ is the (1,1) part of the curvature of the connection ∇. Combining (18) with (19) we obtain :

$$\Delta = \Box'' + i\Lambda F_E^{1,1} + i\Lambda(N\overline{N} + \overline{N}N\Phi). \tag{20}$$

Now suppose that E is the line bundle ξ^k, where ξ has curvature $2\pi i\omega$. Then it is easy to see that the operator $i\Lambda F_E^{1,1}$ is multiplication by $2\pi k(n-q)$ on $(0,q)$ forms. The curvature Φ has two components – one from ξ^k and one from the bundle $\Lambda^{0,q}$. The former gives a contribution $2\pi kn$ to $i\Lambda\Phi$ and the latter is independent of k. Similarly the term $\overline{N}N + N\overline{N}$ does not vary with k. In sum then we get the two formulae

$$\Delta = \Box'' + 2\pi kq + C_1 \quad \text{on } \Omega^{0,q}(\xi^k) \tag{21}$$

and

$$\Delta = \Box' - 2\pi k(n-q) + C_2 \quad \text{on } \Omega^{0,q}(\xi^k), \tag{22}$$

where C_1, C_2 are tensors which depend only on the geometry of the base manifold. The most important of these formulae for us is the first, which shows that, for large k, Δ is a very positive operator on $\Omega^{0,q}(\xi^k)$, for all $q > 0$.

Now it is easy to see, in various ways, that when k is large the operator $D = \partial_E^* + \overline{\partial}_E^*$ taking $\Omega^{0,\text{even}}(\xi^k)$ to $\Omega^{0,\text{odd}}(\xi^k)$ has a large kernel. The quickest route is to use the Atiyah-Singer index theorem, which shows that the index of this operator is given by the familiar "Riemann-Roch " formula

$$\text{index } D = < \text{ch } (\xi^k) \text{ Td } (V), [V] >,$$

where the Todd class is defined in the usual way by the almost complex structure on V. This formula represents a polynomial of degree n in k and the leading term is k^n Volume(V), thus we have:

$$\dim \ker D \geq \text{index} D = k^n \text{Volume}(V) + O(k^{n-1}).$$

Consider now an element of the kernel of D, which we write in the form $(s+\sigma)$, where s is a section of ξ^k and σ contains the terms in $\Omega^{0,2r}(\xi^k)$ for $r \geq 1$. We write:

$$0 = D^* D(s+\sigma) = \Delta(s+\sigma) + (\overline{\partial}^2 + (\overline{\partial}^*)^2)(s+\sigma).$$

Here we have, for simplicity, written $\overline{\partial}, \overline{\partial}^*$ for the operators coupled to ξ^k, using the fixed connection. We obtain then:

$$\Delta\sigma = \partial^* N^* \sigma + \partial^* N^* s + N\partial s \tag{23}$$

and

$$\Delta s = \partial^* N^* \sigma. \tag{24}$$

Taking the inner product of equation (23) with σ and using the formula (21) above we get:

$$\|\nabla''\sigma\|^2 + 2\pi k \|\sigma\|^2 = < C_1\sigma, \sigma > -2 < \partial^* N^* \sigma, \sigma > - < \partial^* N^*, s > .$$

Here all norms are L^2. The key observation now is that the term $\partial^* N^* \sigma$ is bounded pointwise by a multiple of $|\nabla''\sigma| + |\sigma|$, where the multiple is independent of k. This is plain when one expands out the terms by the Leibnitz rule, in local co-ordinates, and observes that ∂^* only involves the derivatives in the "antiholomorphic " directions. We get then, for large enough k, an estimate of the form:

$$\|\nabla''\sigma\|^2 + (2\pi k - C)\|\sigma\|^2 \leq \text{const.}(\|\sigma\| + \|s\|)(\|\sigma\| + \|s\|).$$

It is elementary to deduce from this that

$$\|\sigma\| \leq \text{const. } k^{-1/2}\|s\| \text{ , } \|\nabla''\sigma\| \leq \text{const. } \|s\|,$$

with constants independent of k.Now use (19), and the fact that the curvature is of order k, to obtain

$$\|\nabla'\sigma\|^2 \leq \|\nabla''\sigma\|^2 + \text{ const. } k\|\sigma\|^2,$$

which gives a bound on the "holomorphic" derivative:

$$\|\nabla'\sigma\| \leq \text{ const. } \|s\|.$$

Using the first order relation $\overline{\partial}s = -\partial^*\sigma$, we deduce that

$$\|\overline{\partial}s\| \leq \text{ const. } \|s\|.$$

On the other hand, returning to the equation (24), and using (22), we get:

$$\|\partial s\|^2 - 2\pi kn\|s\|^2 =< C_2 s, s > - < \partial^* N^* \sigma, s >,$$

so

$$\|\partial s\|^2 \geq 2\pi kn\|s\|^2 - \text{const. } \|s\|^2 - \text{const.}\|\nabla''\sigma\|\|s\|,$$

from which we deduce that

$$\|\partial s\| \geq \text{const. } k^{1/2}\|s\|.$$

We see then that each element of the kernel of D gives rise, when k is large, to an "approximately holomorphic" section s of ξ^k, in that the L^2 norm of $\overline{\partial}s$ is much less than that of ∂s. Precisely, we have a bound

$$\|\overline{\partial}s\| \leq \text{ const. } k^{-1/2} \|\partial s\|.$$

We sum up in the following proposition.

PROPOSITION 25. *Let ξ be a line bundle with a Hermitian connection over an almost- Kahler manifold (V, ω) whose curvature is $2\pi i\omega$ There is a linear space H_k of sections of ξ^k, with*

$$\textit{dim } H_k \backsim \textit{Volume } V \ k^n \ ,$$

and a constant C, independent of k, such that

$$\|\overline{\partial}s\|_{L^2} \leq Ck^{-1/2}\|\partial s\|_{L^2}$$

for all s in H_k.

Here we have identified the space H_k, as defined above, with a space of sections of ξ^k by taking the $\Omega^0(\xi^k)$ component.

This L^2 result falls, of course, a long way short of proving the existence of a section satisfying the pointwise inequality (15) which would give a symplectic submanifold, but it seems possible that a more sophisticated analysis of the kernel of D, for large k, would show that suitable elements of this kernel do indeed satisfy (15). (One can compare here the work of Demailly [**5**] and Bismut [**3**] in the integrable case.) In a similar vein, one can show that for large k the sections H_k generate ξ^k, so they define a smooth map :

$$j_k : V \to \mathbf{P}(H_k^*).$$

It is interesting to investigate in what sense j_k is, for large k, an "approximately holomorphic" map.

REFERENCES

1. M.F.Atiyah and J.D.S.Jones, *Topological aspects of Yang-Mills Theory*, Commun. Math. Phys. **61** (1978), 97 -118.
2. W.Barth, C.Peters and A. Van de Ven, "Compact Complex Surfaces," Springer, Berlin, 1984.
3. J.M.Bismut, *Demailly's Asymptotic Morse Inequalities : A Heat equation proof*, Jour. of Functional Analysis **72** (1987), 263-278.
4. S.S.Chern, "Complex manifolds without potential theory," Springer, Berlin, 1979.
5. J.P. Demailly, *Champs magnétiques et inégalités de Morse pour la d''-cohomologie*, Ann. Inst. Fourier **35(4)** (1985), 189-229.
6. S.K. Donaldson, *Anti-self-dual Yang-Mills connections on complex algebraic surfaces and stable vector bundles*, Proc. London Mathematical Society **3** (1985), 1-26.
7. S.K.Donaldson, *Connections, cohomology and the intersection forms of 4-manifolds*, Jour. Differential Geometry **24** (1986), 275-341.
8. S.K.Donaldson, *Polynomial invariants for smooth 4-manifolds*, Topology (In press).
9. S.K.Donaldson and P.B.Kronheimer, "The Geometry of Four Manifolds," Oxford University Press, 1990.
10. A.Floer, *An instanton invariant for 3-manifolds*, Commun. Math. Phys. **118** (1989), 215-240.
11. D.S.Freed and K.K.Uhlenbeck, "Instantons and Four-Manifolds," Springer, New York, 1984.
12. R.Friedman, *Rank two vector bundles over regular elliptic surfaces*, Invent. Math. **96** (1989), 283-332.
13. R.Friedman and J.W.Morgan, *Complex versus differentiable classification of algebraic surfaces*, Columbia University Preprint.
14. P.Griffiths and J.Harris, "Principles of Algebraic Geometry," John Wiley, New York, 1978.
15. M.Gromov, *Pseudo-holomorphic curves in Symplectic manifolds*, Invent. Math. **82** (1885), 307-347.
16. M.Gromov, *Soft and Hard Symplectic Geometry*, in "Proc. International Congress of Mathematicians, Berkeley,U.S.A., 1986," American Mathematical society, Providence, Rhode Island, 1987, pp. 81-98.
17. K.Kodaira, *On homotopy K3 surfaces*, in "Essays on topology and related topics (dedicated to G. de Rham)," Springer, New York, 1970, pp. 58-96.
18. C.H. Taubes, *Self-dual Yang-Mills connections over non-self-dual 4-manifolds*, Jour. Differential Geometry **17** (1982), 139-170.
19. K.K. Uhlenbeck and S-T.Yau, *On the existence of Hermitian Yang-Mills connections on stable bundles over compact Kahler manifolds*, Commun. Pure and Applied Maths. **39** (1986), 257-293.
20. C.T.C.Wall, *Diffeomorphisms of 4-manifolds*, Jour. Lond. Math. Soc. **39** (1964), 131-140.

21. R.O.Wells,Jr., "Differential analysis on Complex manifolds," Springer, New York, 1980.
22. E. Witten, *Topological Quantum Field Theory*, Commun.Math.Phys. **117** (1988), 353-386.

On the topology of algebraic surfaces

ROBERT E. GOMPF

The University of Texas at Austin

1 INTRODUCTION

A major focus of current research in topology is the classification problem for smooth, closed, simply connected 4-manifolds. While we are still a long way from a complete solution of this problem, progress is now being made, driven mainly by the new gauge-theoretic tools introduced by Donaldson. In this article, we will break the problem into several smaller problems, and discuss what is known about each.

The most basic approach to a classification problem is to list examples. The simplest example of a smooth, closed, simply connected 4-manifold is the 4-sphere S^4. After this, we may think of $\mathbb{C}P^2$, $S^2 \times S^2$ and perhaps the famous $K3$ surface. The latter example has second Betti number $b_2 = 22$ and signature $\sigma = -16$, and it is the simplest known example which is spin and has nonzero signature. We see that except for S^4, these examples are all *algebraic surfaces*, or 4-manifolds obtained as zero loci in $\mathbb{C}P^n$ of collections of homogeneous polynomials. In fact, they are *hypersurfaces* in $\mathbb{C}P^3$, cut out by a single polynomial. For each $d \geq 1$, there is a unique diffeomorphism type of smooth hypersurface in $\mathbb{C}P^3$ obtained from a degree d polynomial. For $d = 1, 2, 4$ we obtain $\mathbb{C}P^2$, $S^2 \times S^2$ and $K3$. The degree 3 hypersurface is the connected sum $\mathbb{C}P^2 \#_6 \overline{\mathbb{C}P^2}$ of seven $\mathbb{C}P^2$'s, six of which have reversed orientation. (Algebraically, it is $\mathbb{C}P^2$ blown up at 6 points.) For each $d \geq 5$ we obtain a new example of a smooth, closed, simply connected 4-manifold. In general, the simply connected algebraic surfaces provide a rich class of examples.

Next, we ask what simply connected 4-manifolds are not algebraic surfaces. Connected sums of algebraic surfaces provide many examples. For example, $M = \mathbb{C}P^2 \# \mathbb{C}P^2$ is not algebraic, since its tangent bundle admits no complex structure. (If it did, we would have the characteristic class identity $c_1^2[M] = (2c_2 + p_1)[M] = 2\chi(M) + 3\sigma(M) = 2\cdot 4 + 3\cdot 2 = 14$. But it is easily checked that no element of $H^2(M;\mathbf{Z})$ has square 14.) In fact, Donaldson's invariants show that simply connected algebraic surfaces can never split as connected sums of pieces with $b_2^+ > 0$ [**6**], so we obtain many new examples this way.

What simply connected 4-manifolds are not connected sums of algebraic surfaces? Here we are stymied. Although many attempts have been made to construct such examples, none has been proven successful. In fact, many constructions have yielded only known examples such as connected sums $\# \pm \mathbf{C}P^2$ of $\mathbf{C}P^2$'s with both orientations. It is possible that all simply connected 4-manifolds are sums of algebraic surfaces.

We now see that the classification problem for smooth, closed, simply connected 4-manifolds splits into three subproblems:

Problem A. Classify simply connected algebraic surfaces up to diffeomorphism.

Problem B. Understand how algebraic surfaces behave under connected sum.

Problem C. Is every smooth, closed, simply connected 4-manifold a connected sum of algebraic surfaces?

As previously indicated, Problem C is a major open problem, about which nothing is known. We will address Problems A and B in the next two sections.

2 ALGEBRAIC SURFACES

There is a partial classification of algebraic surfaces (or, more generally, compact, complex surfaces) due to Kodaira [**19**]. (See also [**3**].) In the simply connected setting, algebraic surfaces (up to diffeomorphism) fall into three types: rational, elliptic, and general type. The rational surfaces include $\mathbf{C}P^2$ and its blow-ups $\mathbf{C}P^2 \#_n \overline{\mathbf{C}P^2}$, as well as $S^2 \times S^2$. This provides a complete list of diffeomorphism types of rational surfaces, although there are many algebraic structures on these manifolds. The elliptic surfaces represent many more diffeomorphism types, including the $K3$ surface, and will be described in detail below. The simply connected surfaces of general type include everything which is neither rational nor elliptic, such as the hypersurfaces in $\mathbf{C}P^3$ of degree $d \geq 5$. This collection is poorly understood. Fortunately, it is fairly "small" in the following sense: For any fixed integer b, there are only finitely many simply connected surfaces of general type (up to diffeomorphism) with second Betti number $b_2 \leq b$ [**11**].

The "largest" class of simply connected algebraic surfaces is the simply connected elliptic surfaces, which will now be described. (See also [**3**], [**15**], [**16**], [**20**].) Without loss of generality, we will restrict attention to *minimal* elliptic surfaces. (Arbitrary elliptic surfaces are obtained from these by blow-ups, *i.e.*, connected sums with $\overline{\mathbb{C}P^2}$.) In general, an elliptic surface is a compact, complex surface with an *elliptic fibration*, a holomorphic map $\pi : V \to C$ onto a complex curve (*i.e.*, Riemann surface), such that the generic fibers of π are "elliptic curves," *i.e.*, complex tori. By elementary arguments, π will have only finitely many critical values, whose preimages are called *singular fibers*, and away from these π will be a fiber bundle projection with torus fibers (called *regular* fibers). Note that if V is simply connected then C must be the Riemann sphere $\mathbb{C}P^1 = S^2$, since any homotopically nontrivial loop in C would lift to a nontrivial loop in V. Note also that for any simply connected, closed 4-manifold, the Euler characteristic $\chi \geq 2$, since $b_1 = b_3 = 0$.

The simplest example of a simply connected elliptic surface is the *rational elliptic surface*, which we denote by V_1. To construct this surface, we begin with a generic pencil of cubic curves on $\mathbb{C}P^2$. That is, we let P_0 and P_1 denote a generic pair of homogeneous cubic polynomials on $\mathbb{C}^3$. For each $t = [t_0, t_1] \in \mathbb{C}P^1$, we let F_t denote the zero locus of $t_0P_0 + t_1P_1$ in $\mathbb{C}P^2$. For all but finitely many values of t, F_t will be a smoothly embedded torus in $\mathbb{C}P^2$. The *base locus* B of the pencil $\{F_t | t \in \mathbb{C}P^1\}$ is the set of points in $\mathbb{C}P^2$ where P_0 and P_1 simultaneously vanish. Since this is a generic pair of cubics, the base locus consists of exactly 9 points. Clearly, $B \subset F_t$ for any $t \in \mathbb{C}P^1$. For any $z \in \mathbb{C}P^2 - B$, however, there is a unique F_t containing z, as can be seen by solving $t_0P_0(z) + t_1P_1(z) = 0$ uniquely (up to scale) for t. Now we blow up each point $p \in B$. (See, for example [**15**].) That is, we delete p from $\mathbb{C}P^2$ and replace it by $\mathbb{C}P^1$, thought of as the set of complex tangent lines through p in $\mathbb{C}P^2$. If the new $\mathbb{C}P^1$ is suitably parametrized, each F_t will intersect it precisely in the point t. Thus, the 9 blow-ups will make the sets F_t disjoint so that they fiber the new ambient manifold $\mathbb{C}P^2 \#_9 \overline{\mathbb{C}P^2}$. This is our rational elliptic surface V_1, with the elliptic fibration $\pi : V_1 \to \mathbb{C}P^1$ given by $F_t \mapsto t$.

We obtain other elliptic surfaces by a procedure called *fiber sum*. Suppose V and W are elliptic surfaces. Let $N \subset V$ be the preimage under π of a closed

2-disk containing no critical values of π. (Thus, N is diffeomorphic to $T^2 \times D^2$.) Let $\varphi : N \hookrightarrow W$ be a fiber-preserving, orientation reversing embedding onto a similar neighborhood in W, and let M be obtained by gluing $V - \text{int}\, N$ to $W - \text{int}\, \varphi(N)$ along their boundaries via the map $\varphi|\partial N$. We call M the *fiber sum* of V and W along φ. We construct manifolds V_n, $n \geq 2$, by taking the fiber sum of n copies of V_1. This turns out to be independent of all choices (except n). In particular, the map φ may be changed by self-diffeomorphisms of N, using the monodromy of the bundle part of V_1. Although this fiber sum construction is not holomorphic in nature, the manifold V_n actually admits an (algebraic) elliptic surface structure whose fibration is the obvious one. Note that since V_1 has Euler characteristic $\chi = 12$ and signature $\sigma = -8$, we have $\chi(V_n) = 12n$ and $\sigma(V_n) = -8n$. As an example, V_2 is diffeomorphic to the $K3$ surface.

To construct further examples, we introduce an operation called *logarithmic transform*. Let V be an elliptic surface, with $N \subset V$ as before, a closed tubular neighborhood of a regular fiber. In the smooth category, a logarithmic transform is performed by deleting $\text{int}\, N$ and gluing $N \approx T^2 \times D^2$ back in, by some diffeomorphism $\psi : T^2 \times S^1 \to \partial N$. The *multiplicity* is defined to be the absolute value of the winding number of $\pi \circ \psi|\,\text{point} \times S^1$ as a map into $\pi(\partial N) \approx S^1$. A logarithmic transform of multiplicity zero destroys the fibration π and the complex structure, but any positive multiplicity p can be realized by a holomorphic logarithmic transform. This changes the fibration by the addition of a singular fiber called a *smooth multiple fiber*, a smoothly embedded torus which is p-fold covered by nearby regular fibers.

Let $V_n(p_1, \ldots, p_k)$ denote the manifold obtained from V_n by logarithmic transforms of multiplicities $p_1, \ldots, p_k$. The diffeomorphism type of this manifold is completely determined by n and the unordered k-tuple $\{p_1, \ldots, p_k\}$. (This is due to the monodromy of the bundle part of V_n and the symmetries of $T^2 \times D^2$. See, for example, [**13**].) In particular, we may add or delete p_i's equal to one without disturbing the diffeomorphism type, since the trivial logarithmic transform (regluing N by the identity map) has multiplicity one. If no p_i equals zero, $V_n(p_1, \ldots, p_k)$ will admit algebraic surface structures which are elliptic. Furthermore, any minimal elliptic surface over S^2 with nonzero Euler characteristic will be diffeomorphic to some $V_n(p_1, \ldots, p_k)$, $(p_1, \ldots, p_k \geq 2)$ [**17**], [**23**]. The manifolds $V_n(p_1, \ldots, p_k)$ which are simply connected are precisely those which can be put in the form $V_n(p, q)$, p, q relatively prime (including $p = 0$, $q = 1$) by adding or deleting p_i's equal to one. (Note that this includes V_n and $V_n(p)$.)

The homeomorphism classification of the manifolds $V_n(p,q)$ (p,q relatively prime) is a corollary of Freedman's Classification Theorem for simply connected, closed, topological 4-manifolds [**8**]. For a fixed odd n, the manifolds $V_n(p,q)$ all fall into one homeomorphism type, that of $\#_{2n-1} \mathbf{C}P^2 \#_{10n-1} \overline{\mathbf{C}P^2}$. For fixed even n, there will be two homeomorphism types, distinguished by the existence of a spin structure. If p and q are both odd (and n is even) then $V_n(p,q)$ will admit a spin structure, and it will be homeomorphic to $\#_{\frac{1}{2}n} K3 \#_{\frac{1}{2}n-1} S^2 \times S^2$. Otherwise it will not admit a spin structure and be homeomorphic to $\#_{2n-1} \mathbf{C}P^2 \#_{10n-1} \overline{\mathbf{C}P^2}$.

The diffeomorphism classification of the manifolds $V_n(p,q)$ is much more complex, and only partially understood. It is well-known that the algebraic surfaces $V_1(p)$ ($p \geq 1$) are rational, and hence diffeomorphic to $\mathbf{C}P^2 \#_9 \overline{\mathbf{C}P^2}$. (A topological proof of this appears in [**13**].) Each $V_n(0)$ is diffeomorphic to $\#_{2n-1} \mathbf{C}P^2 \#_{10n-1} \overline{\mathbf{C}P^2}$ [**13**]. Further results require Donaldson's invariants from gauge theory [**5**], [**6**]. In the $n = 1$ case, Friedman and Morgan [**9**] and Okonek and Van de Ven [**24**] showed that each diffeomorphism type is realized by only finitely many $V_1(p,q)$ ($2 \leq p < q$), and in particular, no two of the manifolds $V_1(2,q)$ $q = 1,3,5,7,\ldots$ are diffeomorphic. For $n \geq 2$ ($p,q \geq 1$), Friedman and Morgan [**10**] showed that the product pq is a smooth invariant, implying a similar finiteness result, and showing that no two of the manifolds $V_n(p)$ $p = 0,1,2,\ldots$ are diffeomorphic. (The $p = 0$ case follows from the decomposition $V_n(0) \approx \# \pm \mathbf{C}P^2$, together with Donaldson's theorem [**6**] that a simply connected algebraic surface cannot be decomposed into two pieces with $b_2^+ > 0$.)

These results about elliptic surfaces are quite surprising from a topologist's viewpoint. Observe that we have many families (one for each odd n and two for each even n), each of which contains only one homeomorphism type, but infinitely many diffeomorphism types. We may interpret each family as a single topological manifold which admits infinitely many nondiffeomorphic smooth structures. This contrasts strikingly with topology in dimensions $\neq 4$, where a compact manifold admits only finitely many diffeomorphism types of smooth structures. (In fact, high dimensional smoothing theory would predict that a simply connected, closed 4-manifold should admit no more than one diffeomorphism type of smooth structure.) Furthermore, a classical result of Wall [**26**] implies that all members of a given family will be smoothly h-cobordant, so we have infinite families of counterexamples to the smooth h-Cobordism Conjecture for 4-manifolds. In Section 4, we will further analyze the topology of elliptic surfaces, and use these to construct other surprising examples.

3 CONNECTED SUMS OF ALGEBRAIC SURFACES

We begin by considering connected sums of algebraic surfaces with rational surfaces. Observe that connected sum with $\overline{CP^2}$ is the same as blowing-up, which keeps us within the category of algebraic surfaces. Thus, we should not expect too much information to be lost during this procedure. In fact, Donaldson's invariants are stable under blow-ups **[9]**, **[10]**, so that our infinite families of distinct elliptic surfaces remain distinct after sum with any number of $\overline{CP^2}$'s. In contrast, sum with $+CP^2$ is much more damaging. Mandelbaum and Moishezon **[20]**, **[23]** showed that if M is a simply connected elliptic surface, then $M \# CP^2$ always decomposes as a connected sum of $\pm CP^2$'s. They obtained similar results for many other algebraic surfaces, including hypersurfaces in CP^3 and *complete intersections*, which are those algebraic surfaces obtained as transverse intersections of N hypersurfaces in CP^{N+2}. Mandelbaum has conjectured that for any simply connected algebraic surface M, $M \# CP^2$ should decompose as $\# \pm CP^2$.

Sum with $S^2 \times S^2$ is similarly damaging. In fact, Wall **[25]** showed that if M is a simply connected non-spin 4-manifold, then $M \# S^2 \times S^2$ is diffeomorphic to $M \# S^2 \tilde{\times} S^2$, where $S^2 \tilde{\times} S^2$ denotes the twisted S^2-bundle over S^2, which is diffeomorphic to $CP^2 \# \overline{CP^2}$. (An analogous phenomenon occurs in dimension 2: If M^2 is nonorientable, then $M \# S^1 \times S^1$ is diffeomorphic to $M \# S^1 \tilde{\times} S^1$ where $S^1 \tilde{\times} S^1$, the Klein bottle, is diffeomorphic to $RP^2 \# RP^2$.) It follows immediately that if M is a simply connected elliptic surface or complete intersection, and if M is nonspin, then $M \# S^2 \times S^2 \approx \# \pm CP^2$. If M is spin, such a decomposition cannot occur (since $M \# S^2 \times S^2$ will be spin and $\# \pm CP^2$ will not), but we might expect a similar decomposition into simple spin manifolds. In fact, Mandelbaum showed that for M a simply connected, spin elliptic surface, $M \# S^2 \times S^2$ decomposes as a connected sum of $K3$ surfaces (with their usual orientations) and $S^2 \times S^2$'s. This suggests the following:

Definition. A 4-manifold M *dissolves* if it is diffeomorphic to either $\#_k CP^2 \#_\ell \overline{CP^2}$ or $\pm(\#_k K3 \#_\ell S^2 \times S^2)$ for some $k, \ell \geq 0$.

Note that for any given M, at most one of the two possibilities can occur, and this, as well as k, ℓ and the sign ($\pm$) are determined by the (oriented) homotopy type of M. (In fact, the intersection form suffices.) We can now state the results of Mandelbaum and Moishezon for elliptic surfaces concisely: If M is a simply connected elliptic surface, then $M \# \mathbb{C}P^2$ and $M \# S^2 \times S^2$ dissolve.

We now turn to more general connected sums. It can be shown that for M, N simply connected elliptic surfaces, $M \# \overline{N}$ dissolves [**14**]. This result still holds if M and N are also allowed to be complete intersections other than $\mathbb{C}P^2$, provided that at least one of M, N is not spin [**12**]. One is free to conjecture that $M \# \overline{N}$ dissolves for M, N any simply connected algebraic surfaces except $\mathbb{C}P^2$. These results, as well as those of Mandelbaum and Moishezon, are proven essentially by elementary cut-and-paste techniques. Ultimately, they rely on various versions of a lemma of Mandelbaum [**21**] which shows how to decompose fiber sums and related objects into ordinary connected sums in the presence of an $S^2 \times S^2$ or $S^2 \tilde{\times} S^2$. A unified discussion of the results for elliptic surfaces appears in [**14**].

The case of connected sums with compatible orientations seems harder. For example, there is no known example of irrational algebraic surfaces $M_1, \ldots, M_k$ such that $\#_{i=1}^k M_i$ dissolves. It is conceivable that such sums never dissolve, and perhaps such connected sum decompositions (for simply connected, minimal irrational surfaces) are even unique. However, the usual Donaldson invariants will vanish for these sums, making analysis of this situation difficult.

The crucial dependence of the topology on orientations is even more graphically illustrated by the following result. Suppose M is made as a fiber sum of two elliptic surfaces with nonzero Euler characteristic, but assume that the sum reverses orientation (*i.e.*, the gluing map φ *preserves* orientation). If M is simply connected, then it dissolves [**14**]. Of course, fiber sums with the usual choice of orientation are elliptic, so they never dissolve (except for the rational case and $V_2 = K3$). In practice, this difference arises from the "negativity" of most irrational algebraic surfaces. Most (and perhaps all) simply connected, irrational algebraic surfaces contain embedded spheres with negative normal Euler number, but few (and perhaps no) such manifolds contain embedded spheres with positive normal Euler number. (In fact, such spheres cannot exist if the algebraic surface has $b_2^+ > 1$. Otherwise, by blowing up we could obtain a sphere with normal Euler number one. A tubular neighborhood of this would be diffeomorphic to $\mathbb{C}P^2 - \{\text{point}\}$, and we would have a connected sum decomposition of an algebraic surface into two pieces (one of which is $\mathbb{C}P^2$), both of which have $b_2^+ > 0$. This contradicts a theorem of Donaldson [**6**].) When we connected sum or fiber sum an algebraic surface with one of reversed orientation, this typically introduces spheres of positive normal Euler number. It is the interaction of positive spheres with negative spheres which provides the $S^2 \times S^2$ summand required for the application of Mandelbaum's lemma. This also explains why $\overline{\mathbf{C}P^2}$ behaves more like a typical algebraic surface than $\mathbf{C}P^2$ does: $\mathbf{C}P^2$ contains embedded spheres with normal Euler numbers +1 and +4, but no negative spheres. One further example of this phenomenon is the following: If V is a simply connected elliptic surface and M is a nonorientable 4-manifold, then $V \# M$ is diffeomorphic to $W \# M$ for some W which dissolves [**14**]. (Roughly, this is because positive and negative are indistinguishable in a nonorientable manifold.)

4 NUCLEI OF ELLIPTIC SURFACES

The new theory of Floer and Donaldson (for example, [**2**]) motivates the study of homology 3-spheres in 4-manifolds. Specifically, if a 4-manifold is split into two pieces along a homology 3-sphere, we can understand Donaldson's invariants of the 4-manifold in terms of certain invariants of the pieces which take values in the "instanton homology" of the boundary homology sphere. In this section, we will show how to split an elliptic surface along a homology 3-sphere with known instanton homology, in such a way that one piece is very small but still contains all of the topological information of the elliptic surface. This "nucleus" is itself an interesting 4-manifold with boundary. (For more detail, see [**13**].)

We begin with any $V_n(p_1, \ldots, p_k)$. We will find embedded in this the Brieskorn homology sphere $\Sigma(2, 3, 6n-1)$, whose instanton homology was computed by Fintushel and Stern [**7**]. This splits the manifold into two pieces. The small piece, or *nucleus* $N_n(p_1, \ldots, p_k)$ has Euler characteristic 3 (compared with $12n$ for the ambient space). Its diffeomorphism type depends only on n and $p_1, \ldots, p_k$, and doesn't change if we add or delete p_i's equal to 1 (just as with $V_n(p_1, \ldots, p_k)$). The large piece, with $\chi = 12n - 3$, is independent of $p_1, \ldots, p_k$, and will be denoted Φ_n. It can be shown that Φ_n is actually diffeomorphic to a well-known manifold, namely the *Milnor fiber* of $\Sigma(2, 3, 6n-1)$. This is the locus of $x^2 + y^3 + z^{6n-1} = \varepsilon$ in the closed unit ball in $\mathbf{C}^3$. The boundary, $\Sigma(2, 3, 6n-1)$, has at most two self-diffeomorphisms (up to isotopy) [**4**], and the nontrivial one (when it exists) extends over the Milnor fiber as complex conjugation. Thus, there is a canonical procedure for reconstructing $V_n(p_1, \ldots, p_k)$ from its nucleus

$N_n(p_1, \ldots, p_k)$. (Simply glue on the Milnor fiber by any diffeomorphism of the boundaries.) In particular, $V_n(p_1, \ldots, p_k)$ and $V_n(q_1, \ldots, q_\ell)$ will be diffeomorphic if their nuclei are diffeomorphic.

We consider the nuclei in detail. It is not hard to verify from the construction we will give that the inclusion $N_n(p_1, \ldots, p_k) \hookrightarrow V_n(p_1, \ldots, p_k)$ induces an isomorphism of fundamental groups. We will restrict attention to the simply connected case: $N_n(p, q)$, p, q relatively prime. In this case, the nucleus has $b_1 = 0$ and $b_2 = 2$. It follows from Freedman theory that two of the manifolds $V_n(p, q)$ will be homeomorphic if and only if their nuclei are. We have already seen that nondiffeomorphic $V_n(p, q)$'s have nondiffeomorphic nuclei. (The converse also holds in the cases where two $V_n(p, q)$'s are known to be the same. That is, the nuclei $N_1(p)$ are all diffeomorphic.) It follows that our families of homeomorphic but nondiffeomorphic elliptic surfaces yield families of homeomorphic but nondiffeomorphic nuclei. Thus, we may trim away much useless complexity from the elliptic surfaces, and see their most important topological properties captured in the much simpler nuclei. As a measure of the simplicity of these nuclei, consider the ones $N_n(p)$ with a single multiple fiber. (Note that these are all nondiffeomorphic if $n > 1$.) Each $N_n(p)$ has a handle decomposition with only three handles: One 0-handle and two 2-handles. (Equivalently, it has a perfect Morse function with only three critical points.) These handle decompositions can be drawn explicitly.

The art of drawing pictures of 4-manifolds as handlebodies is called *Kirby calculus*. (See, for example, [**18**].) Suppose, for example, that H is a 4-dimensional handlebody built with one 0-handle and k 2-handles. The 0-handle is just a 4-ball B^4, and each 2-handle is a copy of $D^2 \times D^2$, glued onto the boundary $\partial B^4 = S^3$ along $S^1 \times D^2$, by some embedding $f_i : S^1 \times D^2 \to S^3$ $(1 \leq i \leq k)$. Figure 1 is a schematic picture of this, which is literally a pair of 2-dimensional 1-handles glued to a 0-handle (D^2) to yield a punctured torus. To specify f_i up to isotopy, we need two pieces of information. First, we specify $f_i | S^1 \times \{0\}$ up to isotopy, which is just a knot in S^3. Then, we specify the twisting of the normal vectors as follows: If $p \neq 0$ is a point in D^2, then $f_i(S^1 \times \{0\})$ and $f_i(S^1 \times \{p\})$ are a pair of disjoint knots in S^3. Their linking number, an integer called the *framing* of the 2-handle, specifies the normal twisting as required. To completely determine H, we must simultaneously specify all gluing maps of 2-handles. This is achieved by a *framed link*, or a link $f : \coprod_k S^1 \hookrightarrow S^3$ with an integer attached to each component. Identifying S^3 with $\mathbf{R}^3 \cup \{\infty\}$, we may draw the framed link to obtain a complete picture of H. (For comparison, the 2-dimensional handlebody drawn literally in Figure 1 is specified by the nontrivial link of two 0-spheres in S^1, together with a framing in $\mathbf{Z}_2$ for each 1-handle. If we change one framing, *i.e.*, put a half-twist in one 1-handle, the manifold is changed to a punctured Klein bottle.)

Now we can see the manifolds $N_n(p)$. The simplest pictures, $N_n(0)$ and $N_n = N_n(1)$ are shown in Figure 2. The boxes indicate twists in the link. In the picture for $N_n(0)$, for example, the box indicates n 360° left-handed twists. Our previous discussion implies

Theorem.

a) *For any fixed odd integer $n \geq 3$, the two handlebodies shown in Figure 2 are homeomorphic but not diffeomorphic.*

b) *For each homeomorphism type shown in Figure 2 ($n \geq 1$ arbitrary) there are infinitely many diffeomorphism types.*

Figure 2 is undoubtably the most visualizable example known of homeomorphic but nondiffeomorphic manifolds, with the possible exception of Akbulut's example [**1**] of two such manifolds, each built with a single 2-handle. (His proof depends on a calculation in Donaldson-Floer theory by Fintushel and Stern and, of course, Freedman theory for the topological part.)

Figure 3 shows a general $N_n(p)$. The ribbon at the top represents a spiral with p loops.

Theorem.

Fix $n \geq 2$. If n is even, fix the $\bmod 2$ residue of p. Then the manifolds in the infinite family given by Figure 3 are all homeomorphic, but no two are diffeomorphic.

It remains to sketch the construction of the nuclei. Details (and further properties of the nuclei) can be found in [**13**]. First, consider V_n. By perturbing the elliptic fibration if necessary, we may assume that V_n has a singular fiber of a type called a *cusp fiber*. This is a 2-sphere in V_n which is smoothly embedded except at one non-locally flat point where it is locally a cone on a trefoil knot (*i.e.*, the zero locus of $x^2 + y^3$ in $\mathbf{C}^2$). A regular neighborhood of this will be a handlebody made by attaching a 2-handle to a 0-framed trefoil knot. We may also find a *section* for V_n, *i.e.*, a smoothly embedded 2-sphere intersecting each fiber transversely in a single point. (In V_1, any of the nine $\mathbb{C}P^1$'s created by the blow-ups will work. V_n is made by fiber sum from n copies of V_1, and a section is obtained by splicing together sections of V_1.) The section will have normal Euler number $-n$. Let $N_n \subset V_n$ denote a regular neighborhood of the section union the cusp fiber. The reader can check that this is obtained from a handlebody on a 0-framed trefoil by adding a 2-handle along a $-n$-framed meridian. This is seen explicitly in Figure 2. An easier task is to verify that the intersection form of N_n is $\left[\begin{smallmatrix} 0 & 1 \\ 1 & -n \end{smallmatrix}\right]$ which is unimodular, proving that ∂N_n is indeed a homology sphere. A routine computation with Kirby calculus shows that ∂N_n, as seen in Figure 2, is $-\Sigma(2, 3, 6n - 1)$. More work shows that the closed complement Φ_n is the Milnor fiber. (*Question*: Can this be seen directly by algebraic geometry?)

To construct nuclei in general, observe that by construction N_n contains a neighborhood of a singular fiber. Such a neighborhood contains a continuous family of regular fibers. Change V_n to $V_n(p_1, \ldots, p_k)$ by performing logarithmic transforms on k regular fibers in the interior of N_n. This will not disturb the homology sphere or Φ_n, but it will change N_n to a new manifold $N_n(p_1, \ldots, p_k)$, which is the nucleus of $V_n(p_1, \ldots, p_k)$. Figure 3 may now be derived by Kirby calculus. Details appear in [**13**].

References

[1] Akbulut, S. An exotic 4-manifold. Preprint.

[2] Atiyah, M. (1988). New invariants of 3 and 4 dimensional manifolds. In *The Mathematical Heritage of Herman Weyl.* Proc. Symp. Pure Math., 48. Amer. Math. Soc.

[3] Barth, W., Peters, C. & Van de Ven, A. (1984). Compact complex surfaces. In *Ergebnisse der Mathematik*, Series 3, vol. 4, Springer.

[4] Boileau, M. & Otal, J. (1986). Groupe des difféotopies de certaines variétés de Seifert. *C.R. Acad. Sc.*, (Series I) 303, 19–22.

[5] Donaldson, S. (1987). Irrationality and the h-cobordism conjecture. *J. Diff. Geom.*, 26, 141–168.

[6] Donaldson, S. Polynomial invariants for smooth four-manifolds. Preprint.

[7] Fintushel, R. & Stern, R. Instanton homology of Seifert fibered homology three spheres. Preprint.

[8] Freedman, M. (1982). The topology of four-dimensional manifolds. *J. Diff. Geom.*, 17, 357–453.

[9] Friedman, R. & Morgan, J. (1988). On the diffeomorphism types of certain algebraic surfaces, I and II. *J. Diff. Geom.*, 27, 297–398.

[10] Friedman, R. & Morgan, J. To appear.

[11] Gieseker, D. (1977). Global moduli for surfaces of general type. *Invent. Math.*, 43, 233–282.

[12] Gompf, R. (1988). On sums of algebraic surfaces. *Invent. Math.*, 94, 171–174.

[13] Gompf, R. Nuclei of elliptic surfaces. To appear.

[14] Gompf, R. Sums of elliptic surfaces. To appear.

[15] Griffiths, P. & Harris, J. (1978). *Principles of algebraic geometry.* New York: John Wiley & Sons.

[16] Harer, J., Kas, A. & Kirby, R. (1986). Handlebody decompositions of complex surfaces. *Memoirs, Amer. Math. Soc.*, 62, (350).

[17] Kas, A. (1977). On the deformation types of regular elliptic surfaces. In *Complex Analysis and Algebraic Geometry*, pp. 107–111. Cambridge: Cambridge University Press.

[18] Kirby, R. (1989). The topology of 4-manifolds. *Lecture Notes in Math.*, 1374. Springer.

[19] Kodaira, K. (1969). On the structure of compact complex analytic surfaces IV. *Am. J. Math.*, 90, 1048–1066.

[20] Mandelbaum, R. (1979a). Decomposing analytic surfaces. In *Geometric Topology*, Proc. 1977 Georgia Topology Conference, 147–218.

[21] Mandelbaum, R. (1979b). Irrational connected sums. *Trans. Amer. Math. Soc.*, 247, 137–156.

[22] Mandelbaum, R. (1980). Four-dimensional topology: an introduction. *Bull. Amer. Math. Soc.*, 2, 1–159.

[23] Moishezon, B. (1977). Complex surfaces and connected sums of complex projective planes. *Lecture Notes in Math.*, 603, Springer.

[24] Okonek, C. & Van de Ven, A. (1986). Stable bundles and differentiable structures on certain elliptic surfaces. *Inventiones Math.*, 86, 357–370.

[25] Wall, C.T.C. (1964a). Diffeomorphisms of 4-manifolds. *J. London Math. Soc.*, 39, 131–140.

[26] Wall, C.T.C. (1964b). On simply-connected 4-manifolds. *J. London Math. Soc.*, 39, 141–149.

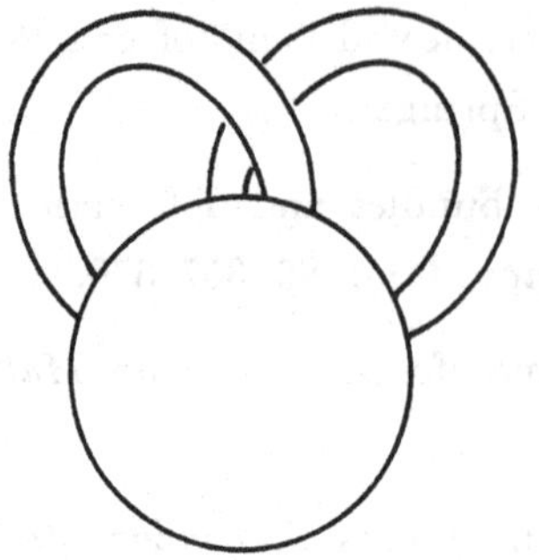

Figure 1

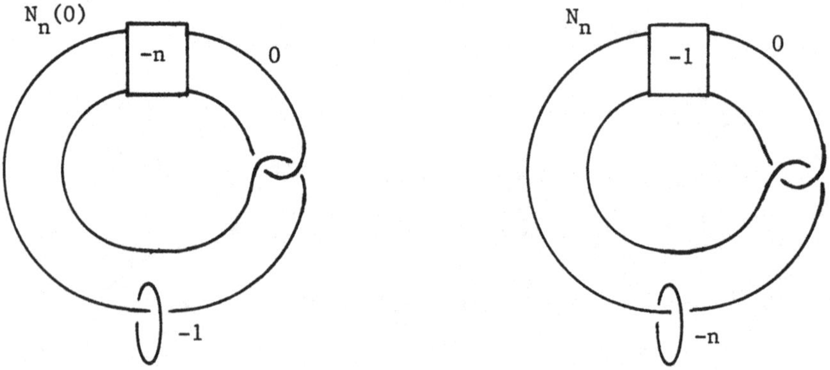

Figure 2

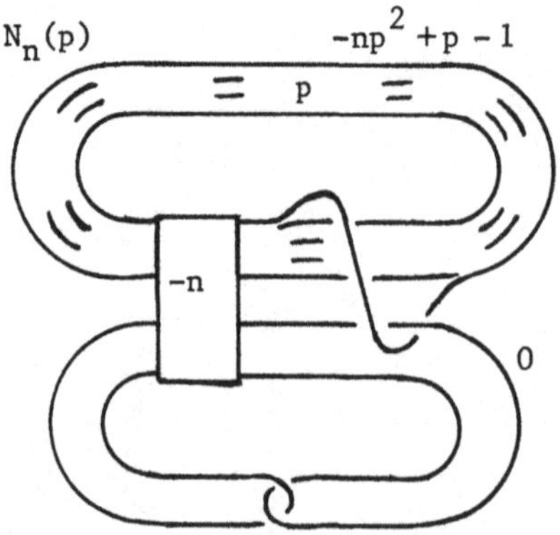

Figure 3

The topology of algebraic surfaces with irregularity and geometric genus zero

DIETER KOTSCHICK[†)]

Queens' College, Cambridge, CB3 9ET, England, *and*
The Institute for Advanced Study, Princeton, NJ 08540 USA

Interest in algebraic surfaces with $p_g = h^0(\mathcal{O}(K)) = 0$ goes back to the work of Enriques and Castelnuovo in the 19th Century. After Clebsch had proved that curves with $p_g = 0$ are rational, these authors considered the analogous question for surfaces. It was clear to them that in this case the irregularity $q = h^1(\mathcal{O}(K))$ has to be controlled as well.

In 1894 Enriques constructed his now famous surface, which is irrational with $q = p_g = 0$, disproving the most obvious rationality criterion. Two years later Castelnuovo proved that the modified conditions $q = P_2 = 0$ do imply rationality. Thus he substituted the second plurigenus from $P_k = h^0(\mathcal{O}(K^k))$ for the first. (For the Enriques surface K is a 2-torsion bundle, so the bigenus is one.)

Over the next forty years more examples of irrational surfaces with $q = p_q = 0$ were constructed. Like the Enriques surface they were all elliptic. Only in 1931 did Godeaux [G] find a surface of general type with these invariants. His construction was disarmingly simple: divide the Fermat quintic in $\mathbf{C}P^3$ by the standard free $\mathbf{Z}_5$-action on the coordinates. Campedelli also gave an example of a surface of general type, introducing his "double plane" construction. This has $H_1(X, \mathbf{Z}) = \mathbf{Z}_2^3$.

On the basis of these examples Severi conjectured in 1949 that the two conditions $H_1(X, \mathbf{Z}) = 0$ and $p_g = 0$ should imply rationality. (Recall that $q = 0$ implies $H_1(X, \mathbf{Q}) = 0$.) This was disproved by Dolgachev, who in 1966 gave examples of simply connected irrational elliptic surfaces with vanishing geometric genus, cf. [**Dv**]. Then, in her 1982 Warwick thesis, R. Barlow constructed a simply connected surface of general type with $p_g = 0$, cf. [**B1**]. To this day it is the only such example.

In this paper we want to summarize the progress made in understanding algebraic surfaces with $p_g = 0$ since the survey of Dolgachev [**Dv**]. In view of the spectacular advances in 4-manifold topology initiated by Freedman and Donaldson we extend the classification programme from the algebraic or analytic to the smooth and topological categories. Although the results are not yet complete, a coherent picture emerges. Namely, if the topology of a surface is sufficiently complicated (e.g. large fundamental group), then it determines the smooth structure. On the other hand, if the topology is simple (e.g. rational surfaces) then there are homeomorphic surfaces which are not diffeomorphic to the given one.

[†)]Supported by NSF Grant Number DMS-8610730

The reader should be warned that nothing is proved in this article. For background for the papers referred to in the third section consult the forthcoming book by S. K. Donaldson and P. B. Kronheimer (Oxford UP). The philosophy is well explained in [**FM2**].

I would like to thank R. Gompf for pointing out an erroneous claim about (non-algebraic) 4-manifolds with 2-torsion in $H_1(X, \mathbf{Z})$ made in the oral version of this lecture.

1. Algebraic classification

The following theorem is part of the classification of surfaces [**BPV**], Chapter VI.

THEOREM 1 (Enriques-Kodaira). *Let X be a smooth minimal compact complex algebraic surface with $q = p_g = 0$. Then X is one of the following:*

A) *a minimal rational surface*
B) *an Enriques surface*
C) *a minimal properly elliptic surface*
D) *a minimal surface of general type.*

We can describe these surfaces in more detail:
A) The minimal rational surfaces are $\mathbf{P}^2$ and the Hirzebruch surfaces Σ^n, $n = 0, 2, 3, 4, \ldots$. Here Σ^n is the $\mathbf{P}^1$-bundle $\mathbf{P}(\mathcal{O} \oplus \mathcal{O}(n))$ over $\mathbf{P}^1$. Thus $\Sigma^0 = \mathbf{P}^1 \times \mathbf{P}^1$. The surface Σ^1 is not minimal, it is $\mathbf{P}^2$ blown up once. Note that $K^2 = 9$ for $\mathbf{P}^2$ and $K^2 = 8$ for Σ^n.
B), C) Let X_9 be the rational elliptic surface obtained from $\mathbf{P}^2$ by blowing up the nine base points of a generic cubic pencil. Then Dolgachev [**Dv**] proved that the surfaces in B), C) are precisely those obtained from X_9 by performing logarithmic transformations on at least two different smooth fibers. We denote the surface obtained in this way by $X(p_1, \ldots, p_\ell)$, where the p_i are the multiplicities of the logarithmic transformations, and call it a Dolgachev surface. (Some authors reserve this name for the case when there are only two multiple fibers and their multiplicities are relatively prime.) Up to the deformation $X(p_1, \ldots, p_\ell)$ does not depend on the choices involved in the construction. Now $X(2, 2)$ is the Enriques surface. It is distinguished from the other Dolgachev surfaces by its Kodaira dimension ($= 0$, rather than 1). Note second $K^2 = 0$ for all of these surfaces, just like for X_9.
D) For a minimal surface of general type we have $c_1^2 > 0$ and $c_2 > 0$. If in addition $q = p_g = 0$, then $c_1^2 + c_2 = 12$ by Noether's formula, and $K^2 = c_1^2 \in \{1, 2, \ldots, 9\}$. Examples of surfaces realizing all these values of K^2 are contained in the table on page 237 of [**BPV**]. The reader can also find there an overview of different methods used in those constructions. Among them are the classical Godeaux and Campedelli constructions mentioned in the introduction. For a modern discussion of these see [**R3**]. The author knows of only three new constructions which have appeared since [**BPV**]. One is J. H. Keum's method using branched double covers of

Enriques surfaces [**Ke**]. He uses this to give examples of surfaces with $K^2 = 2, 3, 4$. Some of them are known to coincide with surfaces constructed earlier by different methods and contained in the table of [**BPV**], whereas others are new. Another recent method of construction is due to M. Oka [**O2**], who uses singularity theory and toroidal embeddings. Unfortunately this method has not yet produced new examples. Finally, a new surface with $K^2 = 2$ is contained in Xiao Gang's book on genus 2 fibrations [**X**] , Example 4.11, cf. also [**Be**].

Having long lists of examples is interesting, but it cannot be the ultimate aim of the theory. Rather the aim is to give complete constructions of *all* such surfaces. To this end one introduces another invariant, besides q, p_g and K^2. This is the group $\operatorname{Tors} X$ of torsion divisors. If $q = 0$ then $\operatorname{Tors} X \cong H_1(X, \mathbf{Z})$. For given q, p_g and K^2 one tries to pin down the possible groups for $\operatorname{Tors} X$, and then gives an exhaustive construction for the whole moduli scheme of surfaces with a fixed torsion group.

Let us look at the case $q = p_g = 0$, $K^2 = 1$ in detail. These surfaces are called numerical Godeaux surfaces, in honor of Lucién Godeaux who constructed the first such surface [**G**]. In this case a result of Deligne and Bombieri [**Bo**] implies $|\operatorname{Tors} X| \leq 6$. This has been refined by Reid [**R1**], who has also determined the moduli spaces in some cases, by using explicit presentations of the canonical ring. His result is:

THEOREM 2 (Reid). *For a numerical Godeaux surface* $\operatorname{Tors} X$ *is one of* 0, $\mathbf{Z}_2$, $\mathbf{Z}_3$, $\mathbf{Z}_4$ *or* $\mathbf{Z}_5$. *In the last three cases the moduli schemes are irreducible 8-dimensional varieties.*

It is natural to conjecture that the irreducibility result will extend to the cases of torsion 0 or $\mathbf{Z}_2$. However, the problem becomes harder for smaller groups, and two recent attempts on the case of $\mathbf{Z}_2$ by Reid [**R2**] and by Catanese-Debarre [**CD**] seem to have failed. At least the existence of such surfaces is known. Examples have been constructed by Barlow [**B1**] for the case of no torsion and by Oort-Peters [**OP**] and by Barlow [**B2**] for the case $\operatorname{Tors} X = \mathbf{Z}_2$. (The Oort-Peters surface is a close cousin of Xiao's surface mentioned above. Both of them are genus 2 fibrations [**X**].)

Of course one would like to have an analogue of Theorem 2 for $K^2 > 1$ as well. However, except for some unpublished work of Reid on the case $K^2 = 2$, not much seems to be known.

2. HOMEOMORPHISM TYPES

In order to identify the homeomorphism types of some of the surfaces described in 1, we have to determine their fundamental groups. The rational surfaces are, of course, simply connected. For the elliptic surfaces we have the result of Dolgachev [**Dv**]: $\pi_1(X(p,q)) = \mathbf{Z}_k$, with $k = g.c.d.(p,q)$. In particular $X(p,q)$ is simply connected if p and q are relatively prime. The fundamental group is non-abelian if there are more than two multiple fibers.

For surfaces of general type the answers are less coherent. In general it is very hard to control π_1 in complicated constructions. Only for Oka's method is there a general theorem [**O1**], which says that π_1 is always finite cyclic. This should encourage attempts to use his methods to find new surfaces (say simply connected ones). Here are some other results. If $K^2 = 9$, then Yau's uniformization result implies that the fundamental group is infinite. If $K^2 = 8$, then the only known examples have infinite π_1, and for $K^2 = 7$, the fundamental group is unknown. For $K^2 = 6$ the example cited in [**BPV**] has infinite π_1. For $1 \leq K^2 < 6$ surfaces with various finite fundamental groups are known. For $K^2 = 4$ there is also an example of J. H. Keum [**Ke**] with fundamental group $\mathbf{Z}^4 \rtimes (\mathbf{Z}_2)^2$.

As in the previous section, more is known in the case of numerical Godeaux surfaces, i.e. $K^2 = 1$. Here $H_1(X, \mathbf{Z}) = 0, \mathbf{Z}_2, \mathbf{Z}_3, \mathbf{Z}_4$ or $\mathbf{Z}_5$ by Theorem 2. Moreover, $\pi_1 = H_1$ for all surfaces with $H_1 = \mathbf{Z}_3, \mathbf{Z}_4$ or $\mathbf{Z}_5$, and for the other known surfaces except possibly the Oort-Peters example. This is because the Barlow surface [**B1**] is simply connected by construction, and the construction in [**B2**] gives $\pi_1 = \mathbf{Z}_2$. In the cases of torsion $\mathbf{Z}_3$, $\mathbf{Z}_4$ or $\mathbf{Z}_5$ all surfaces have the same topological type because of the irreducibility of moduli (Theorem 2). Thus it is enough to exhibit one example with $\pi_1 = H_1$ for each case. This is done in [**Be**][1] for $\mathbf{Z}_3$ and in [**M**] for $\mathbf{Z}_4$. For $\mathbf{Z}_5$ it is obvious in view of the classical Godeaux construction.

We now give the homeomorphism classification in simple cases, which are the only ones where it is known. For the simply connected case we call on Freedman's work [**F1**], who proved that smooth 4-manifolds are classified by their intersection forms.

THEOREM 3 (Freedman). *The simply connected Dolgachev surfaces are homeomorphic to X_9, and the Barlow surface is homeomorphic to X_8. Moreover any simply connected minimal surface of general type with $p_g = 0$ is homeomorphic to X_{9-K^2} or to $S^2 \times S^2$.*

In the non-simply connected case we can only deal with the case $\pi_1 = \mathbf{Z}_k$, due to the work of Hambleton-Kreck. Building on the fact that surgery works [**F2**] they extend Freedman's classification to the case $\pi_1 = \mathbf{Z}_k$ with k odd [**HK1**]. They have also dealt with the case of even k [**HK2**]. Here the classification is more complicated, because it involves the so-called w_2-type of the manifold. Some consequences of their results are:

THEOREM 4 (Hambleton & Kreck). *A numerical Godeaux surface with $\pi_1 = \mathbf{Z}_k$ is homeomorphic to $X_8 \# \Sigma_k$, where Σ_k is any rational homology sphere with $\pi_1 = \mathbf{Z}_k$. Similarly the Dolgachev surfaces $X(p,q)$ and $X(q',q')$ are homeomorphic if and only if g.c.d.$(p,q) =$ g.c.d.$(p',q') = k$ and, when k is even, $\frac{p}{k} + \frac{q}{k} \equiv \frac{p'}{k} + \frac{q'}{k} \bmod 2$.*

---

[1] Beware, the construction of a simply connected surface suggested in [Be] does not work. On the other hand, the constructions using double covers of Enriques surfaces are essentially contained in [Ke].

Here the complication in the case of even k comes precisely from the w_2-type. If k is odd and in the case $\frac{p}{k} + \frac{q}{k} \equiv 1 \bmod 2$ for even k the intersection form of $X(p,q)$ is diagonal and $X(p,q)$ is homeomorphic to $X_g \# \Sigma_k$. In the remaining case the intersection form is $H \oplus E_8$, where $H = \begin{pmatrix} 0 & 1 \\ 1 & 0 \end{pmatrix}$ is the standard hyperbolic, and E_8 is taken to be negative definite. This case includes the Enriques surface.

3. Differentiable structures

In the previous section we did not discuss the homeomorphism classification for those manifolds for which it coincides with the diffeomorphism classification. This is the case for the Hirzebruch surfaces Σ^n, which by construction are diffeomorphic to S^2-bundles over S^2. It is a trivial consequence of Wall's work **[W]** that Σ^n is $S^2 \times S^2$ for n even and $\mathbf{P}^2 \# \overline{\mathbf{P}}^2$ for n odd. A much deeper result is the following consequence of a theorem of Ue **[U]**:

THEOREM 5 (Ue). *The diffeomorphism type of a Dolgachev surface with three or more multiple fibers is determined by its fundamental group.*

Although this result is far from trivial, its proof is in a sense elementary. On the other hand, Theorems 3 and 4 given in the previous section are proved using Freedman's surgery **[F1]**, **[F2]**. This means that they leave room for non-diffeomorphic surfaces in the same homeomorphism type. Indeed, in many cases such non-diffeomorphic pairs can be found using gauge theory. (This of course disproves the 4-dimensional h-cobordism conjecture **[D1]**, **[D2]**.)

The first result of this type was obtained by Donaldson **[D1]**, **[D2]**, who proved that $X(2,3)$ is not diffeomorphic to X_9. His argument was extended by Friedman and Morgan **[FM1]** to show, among other results, the following:

THEOREM 6 (Friedman & Morgan). *No Dolgachev surface is diffeomorphic to X_9. Moreover the map from deformation types to diffeomorphism types of simply connected Dolgachev surfaces is finite-to-one.*

Friedman and Morgan **[FM1]** also showed that this is still true after an arbitrary number of blowups. As a corollary one finds that X_k has infinitely many smooth structures for $k \geq 9$. We have given a simpler proof of Theorem 6 in **[K3]**. This uses a new invariant ϕ_1 defined using gauge theory on SO(3)-bundles with non-trivial Stiefel-Whitney class, instead of Donaldson's Γ-invariant coming from SU(2)-bundles **[D1]**, **[D2]**. Using this type of argument we have also proved **[K2]**:

THEOREM 7 (Kotschick). *The Barlow surface is not diffeomorphic to X_8.*

The generalization of this to arbitrary blowups is very complicated, and has not been completed yet. Thus, to prove the optimal Theorem 8 below, we have reverted in **[K4]**, see also **[K1]**, Chapter V, to the case of the Donaldson invariant instead of our own invariants from **[K3]**.

THEOREM 8 (Kotschick). *For all k the k-fold blowup of the Barlow surface is not diffeomorphic to a rational or (blown up) Dolgachev surface.*

The proof of Theorem 8 is rather interesting, because it uses naturally the geometry of surfaces of general type, and suggests a way of extending the proof to arbitrary such surfaces. If this is successful it will prove the following folk conjecture in the case $p_g = 0$ and $\pi_1 = 0$:

CONJECTURE. *Surfaces of different Kodaira dimension are not diffeomorphic.*

Now let us look at the case of non-simply connected surfaces. In view of Theorems 4 and 5 the interesting cases are those of finite (cyclic) fundamental groups. For these one can obtain results by going over to the universal cover and applying the theory of Donaldson polynomials [**D3**]. This is done for elliptic surfaces in [**FM3**] and for the Godeaux surface in [**HK1**]. However, direct arguments are possible as well.

Thus Maier [**Ma**] extended the work of Donaldson [**D1**], [**D2**] and of Friedman and Morgan [**FM1**] to non-simply connected Dolgachev surfaces, proving that they give infinitely many smooth structures on every $X_9\#\Sigma_k$. Similarly Okonek [**Ok**] treated the case of elliptic surfaces homeomorphic to the Enriques surface. Technically this is the simplest possible case, because the Γ-invariant [**D1**], [**D2**] takes a simple form, allowing arguments of the type used for SO(3)-invariants in [**K3**]. Finally, our own method of proof using SO(3)-invariants [**K2**], [**K3**] works uniformly for any fundamental group, as long as the intersection form is odd. Thus we can deal with non-simply connected numerical Godeaux surfaces and with those Dolgachev surfaces which have diagonal intersection forms. Our method does not apply to the Enriques surface (at least not in the naive form given in [**K3**] cf. the discussion in §4 of [**K3**]).

Constructions of algebraic surfaces:

[B1] R. N. Barlow, *A simply connected surface of general type with $p_g = 0$*, Invent. math. **79**, 293-301 (1985).

[B2] R. N. Barlow, *Some new surfaces with $p_g = 0$*, Duke Math. Journal **51**, No. 4 (1984), 889-904.

[BPV] W. Barth, C. Peters and A. Van de Ven, *Compact Complex Surfaces*, Springer-Verlag, Berlin 1984.

[Be] A. Beauville, *A few more surfaces with $p_g = 0$*, Letter dated Sept. 3, 1984.

[Bo] E. Bombieri, *Canonical models for surfaces of general type*, Publ. Math. IHES **42** (1973), 171-219.

[CD] F. Catanese and O. Debarre, *Surfaces with $K^2 = 1$, $p_g = 1$, $q = 0$*, J. reine angew. Math. **395** (1989), 1-55.

[Dv] I. Dolgachev, *Algebraic surfaces with $q = p_g = 0$*, in "Algebraic Surfaces", CIME 1977, Liguore Editore Napoli (1981), 97-215.

[G] L. Godeaux, *Sur une surface algebrique de genre zero et de bigenre deux*, Atti Acad. Naz. Lincei **14** (1931), 479-481.

[Ke] J. H. Keum, *Some new surfaces of general type with $p_g = 0$*, (Univ. of Utah preprint).

[M] Y. Miyaoka, *Tricanonical Maps of Numerical Godeaux Surfaces*, Invent. math. **34**, 99-111 (1976).

[O1] M. Oka, *On the resolution of Hypersurface Singularities*, Advanced Study in Pure Math. **8** (1986), 405-436.

[O2] M. Oka, *Examples of algebraic surfaces with $q = 0$ and $p_q \leq 1$ which are locally hypersurfaces*, in "A Fete of Topology", ed. Y. Matsumoto et. al., Academic Press, Boston, 1987.

[OP] F. Oort and C. Peters, *A Campedelli surface with torsion group* $\mathbf{Z}/2$, Indagationes Math. **43**, Fasc. 4 (1981), 399-407.

[R1] M. Reid, *Surfaces with $p_g = 0$, $K^2 = 1$*, Journal of Faculty of Science, Univ. of Tokyo, Sec. 1A, Vol. 25, No. 1 (1978), 75-92.

[R2] M. Reid, *Infinitesimal view of extending a hyperplane section – deformation theory and computer algebra*, in "Hyperplane sections and Related Topics", (L'Aquila, May 1988), ed. L. Livorni, Springer LNM (to appear).

[R3] M. Reid, *Campedelli versus Godeaux*, Proc. of the 1988 Cortona conf. on Algebraic Surfaces, INdAM and Academic Press (to appear).

[X] G. Xiao, *Surfaces fibres en courbes de genre deux*, Springer LNM 1137, 1985.

Homeomorphism types:

[F1] M. H. Freedman, *The topology of four-dimensional manifolds*, J. Differential Geo. **17** (1982), 357-453.

[F2] M. H. Freedman, *The disk theorem for four-manifolds*, Proc. ICM Warsaw (1984), 647-663.

[HK1] I. Hambleton and M. Kreck, *On the Classification of Topological 4-Manifolds with Finite Fundamental Group*, Math. Ann. **280**, 85-105, (1988).

[HK2] I. Hambleton and M. Kreck, *Smooth structures on algebraic surfaces with cyclic fundamental group*, Invent. math. **91**, 53-59 (1988).

Differentiable structures:

[D1] S. K. Donaldson, *La topologie differentielle des surfaces complexes*, C. R. Acad. Sci. Paris Sér. I Math. **301**, (1985), 371-320.

[D2] S. K. Donaldson, *Irrationality and the h-cobordism conjecture*, J. Differential Geo. **26** (1987), 141-168.

[D3] S. K. Donaldson, *Polynomial invariants for smooth four-manifolds*, Topology (to appear).

[FM1] R. Friedman and J. W. Morgan, *On the diffeomorphism types of certain algebraic surfaces I,* J. Differential Geo. **27** (1988), 297-369.

[FM2] R. Friedman and J. W. Morgan, *Algebraic surfaces and 4-manifolds: some conjectures and speculations,* Bull. AMS **18** (1) (1988), 1-19.

[FM3] R. Friedman and J. W. Morgan, *Complex versus differentiable classification of algebraic surfaces,* (Columbia Univ. preprint).

[K1] D. Kotschick, *On the geometry of certain 4-manifolds,* Oxford thesis, 1989.

[K2] D. Kotschick, *On manifolds homeomorphic to* $\mathbf{C}P^2 \# 8\overline{\mathbf{C}P^2}$, Invent. math. **95**, 591-600 (1989).

[K3] D. Kotschick, SO(3)-*invariants for 4-manifolds with* $b_2^+ = 1$, (Cambridge and IAS preprint).

[K4] D. Kotschick, *Positivity versus rationality of algebraic surfaces,* (to appear).

[Ma] F. Maier, *On the diffeomorphism type of elliptic surfaces with cyclic fundamental group,* Tulane thesis, 1987.

[Ok] C. Okonek, *Fake Enriques surfaces,* Topology Vol. **27**, No. 4 (1988), 415-427.

[U] M. Ue, *On the diffeomorphism types of elliptic surfaces with multiple fibres,* Invent. math. **84**, 633-643 (1986).

[W] C. T. C. Wall, *Diffeomorphisms of 4-manifolds,* J. London Math. Soc. **39** (1964), 131-140.

ON THE HOMEOMORPHISM CLASSIFICATION OF SMOOTH KNOTTED SURFACES IN THE 4–SPHERE

Matthias KRECK
Max–Planck–Institut für Mathematik, Bonn

1. In [FKV] an infinite family of smooth (real) surfaces $F(k)$ embedded in S^4 was constructed which has the following properties:

i) The knottings $(S^4,F(k))$ and $(S^4,F(\ell))$ are not diffeomorphic for $k \neq \ell$.
ii) $F(k) = \#10(\mathbb{R}P^2)$
iii) $\pi_1(S^4 - F(k)) = \mathbb{Z}_2$
iv) The normal Euler number (with local coefficients) of $F(k)$ in S^4 is 16.

The knottings $(S^4,F(k))$ are constructed from the Dolgachev surfaces $D(2,2k+1)$. There are antiholomorphic involutions c on $D(2,2k+1)$ with fixed point set $F(k) = \#10(\mathbb{R}P^2)$ and orbit space $D(2,2k+1)/c$ diffeomorphic to S^4. Thus the diffeomorphism type of $D(2,2k+1)$, the ramified covering along the knotting, is an invariant and one can distinguish these Dolgachev surfaces by Donaldson's Γ–type invariants [D], [FM], [OV]. It was also proved in [FKV] that the number of homeomorphism types of these knottings is finite and it was conjectured that they are all homeomorphic to the standard embedding (S^4,F) with normal Euler number 16. The main result of this note is an affirmative answer to this conjecture.

More precisely consider the standard embedding of $\mathbb{R}P^2$ into S^4 with normal Euler class -2. This can be considered as the fixed point set of the standard antiholomorphic involution c on $\mathbb{C}P^2$ embedded into $\mathbb{C}P^2/c \cong S^4$. Then the standard embedding (S^4,F) with normal Euler class 16 is obtained by taking the connected sum $(S^4,\mathbb{R}P^2)\#9(-S^4,\mathbb{R}P^2)$.

<u>Theorem:</u> Let $S = \#10(\mathbb{R}P^2)$ be embedded into S^4 with normal Euler number

16 and $\pi_1(S^4-S) = \mathbb{Z}_2$. Then (S^4,S) is homeomorphic to (S^4,F), the standard embedding with normal Euler number 16. The homeomorphism can be chosen as a diffeomorphism on a neighborhood of S and F.

Corollary: The knottings $(S^4,F(k))$ are all homeomorphic to (S^4,F) implying that the standard knotting (S^4,F) has infinitely many smooth structures.

Remark: Recently R. Gompf [G] constructed non–diffeomorphic embeddings of a punctured Klein bottle K (= Klein bottle minus open 2–ball) into D^4 with $\pi_1(D^4 - K) = \mathbb{Z}_2$ and intersection form of the 2–fold ramified covering along K equal to $\langle 1\rangle \oplus \langle -1\rangle$. The same methods as used for the proof of our Theorem show that they are pairwise homeomorphic if they have same relative normal Euler number and the knots ∂K in S^3 are equal. We will comment the necessary modifications of the proof in section 5. I was informed by O. Viro that he has similar knottings of K in D^4 which are related to the construction in [V].

2. Proof: Since F and S have isomorphic normal bundles we can choose a linear identification of open tubular neighborhoods and denote the complements by C and C'. We identify the boundaries, so that $\partial C = \partial C' =: M$. We want to extend the identity on M to a homeomorphism from C to C'. Since C and C' are Spin–manifolds a necessary condition for this is that we can choose Spin–structures on C and C' which agree on the common boundary. Another necessary condition is that the diagram

$$\begin{array}{ccc} \pi_1(M) & \longrightarrow & \pi_1(C) \\ \downarrow & & \downarrow \\ \pi_1(C') & \longrightarrow & \mathbb{Z}/2 \end{array} \tag{1}$$

commutes. One can show that by choosing the linear identification of the tubular neighborhoods appropriately one can achieve these two necessary conditions. I am indepted to O. Viro for this information. To obtain condition (1), choose sections s and s' from F^o resp. S^o (delete an open 2–disk) to M such that the composition with the inclusion to C and C' resp. are trivial on π_1. Since the normal Euler numbers of the knottings are equal one can choose the linear identification of the tubular neighborhoods such that they commute with s and s' resp. yielding (1). To obtain the compatibility of Spin structures on M it is enough to control them on the image of s and s'. Note that for each embedded circle α in F°, $s(\alpha)$ bounds an immersed disk D in C. The normal bundle of α determines a 1–dimensional subbundle of $\nu(D)|_{\partial D}$. The Spin structure on the image of s is characterized by the obstruction mod 4 to extending this subbundle to $\nu(D)$ and gives a quadratic form $q : H_1(F^{\circ}) \longrightarrow \mathbb{Z}/4\mathbb{Z}$ [GM]. Thus we have to control that the identification of F and S respects this form or equivalently that the Brown invariants in $\mathbb{Z}/8\mathbb{Z}$ agree. But this follows from the generalized Rochlin formula [GM].

In the following we will assume that the Spin–structures on $\partial C = \partial C' = M$ agree and the diagram (1) commutes. There is another obvious invariant to be controlled, the intersection form on the universal covering. For this we assign to our knotted surface the 2–fold ramified covering along F denoted by X. A simple calculation shows that X is 1–connected, $e(X) = 12$ and $\operatorname{sign}(X) = -8$. Thus the intersection form on X is indefinite and odd (since otherwise the signature were divisible by 16 by Rochlin's Theorem). By the classification of indefinite forms, the intersection form on X is $\langle 1\rangle \oplus 9\langle -1\rangle$. The long exact homology sequence combined with excision and Poincaré duality leads to an exact sequence

$$0 \longrightarrow H^1(F) \longrightarrow H_2(\tilde{C}) \longrightarrow H_2(X) \longrightarrow \underset{\mathbb{Z}/2}{\underset{\|}{H^2(F)}} \longrightarrow 0$$

and the map $H_2(X) \longrightarrow H^2(F) = \mathbb{Z}_2$ is $\alpha \longmapsto \alpha \circ [F]$, the mod 2 intersection number of α with F. Since $\tilde{C}$ is Spin and X is not Spin (see above) the map $\alpha \longmapsto \alpha \circ [F]$ is given by $w_2(X) : \alpha \circ [F] = \langle w_2(X), \alpha\rangle$. Since the image of $H^1(F)$ is contained in the radical of the intersection form on $H_2(\tilde{C})$ and the form on $H_2(X)$ restricted to the kernel of w_2 is non–singular, the image of $H^1(F)$ is the radical of the form on $H_2(\tilde{C})$. The form on $H_2(\tilde{C})/\mathrm{rad}$ is the restriction of $\langle 1\rangle \oplus 9\langle -1\rangle \cong E_8 \oplus \langle 1\rangle \oplus \langle -1\rangle$ to the kernel of $x \longmapsto x \circ x$ which is $E_8 \oplus \begin{bmatrix} 4 & 2 \\ 2 & 0 \end{bmatrix} \cong E_8 + 2\begin{bmatrix} 0 & 1 \\ 1 & 0 \end{bmatrix}$.

We know that the covering transformation τ acts trivially on $H^1(F)$ and by -1 on $H_2(X)$ (since $X/\tau = S^4$). Thus, if we take the $\Lambda = \mathbb{Z}[\mathbb{Z}_2]$ module structure given by τ on $H_2(\tilde{C})$ into account we have an exact sequence

$$0 \longrightarrow \mathbb{Z}^9_+ \longrightarrow H_2(\tilde{C}) \longrightarrow \mathbb{Z}^{10}_- \longrightarrow 0$$

where $+$ or $-$ indicates the trivial or non–trivial Λ–action. Moreover one can show that $H_2(\tilde{C}) = \mathbb{Z}_- \oplus \Lambda^9$ ([FKV], Lemma 5.2A). We can summarize these considerations as follows:

$H_2(\tilde{C}) \cong \mathbb{Z}_- \oplus \Lambda^9$;

(2) the radical of the intersection form is $H_2(\tilde{C})_+$, the $+1$ eigenspace;

the form on $H_2(\tilde{C})/_{\text{rad}}$ is $E_8 \oplus 2\begin{bmatrix} 0 & 1 \\ 1 & 0 \end{bmatrix}$.

The proof is finished by the following proposition which is the main step.

Proposition: Let C and C' be 4–dimensional Spin manifolds with fundamental group $\mathbb{Z}_2$, $\partial C = \partial C' = M$ and inducing same Spin–structure on M such that the conditions (1) and (2) are fulfilled. Then there is a homeomorphism from C to C' inducing on M the identity.

3. Proof of the Proposition. We use the method of [K]. The normal 1–type of C is the fibration $p : B = \mathbb{RP}^\infty \times B\,\mathrm{Spin} \xrightarrow{p_2} BO$ and a normal smoothing of X in (B,p) is given by the non–trivial map $C \longrightarrow \mathbb{RP}^\infty$ and a Spin–structure on C (given by a lift of the normal Gauß map to B Spin). Thus it is uniquely determined by a Spin–structure. By assumption there exist normal smoothings of C and C' in (B,p) which agree on the common boundary. Thus we can form $C \cup (-C')$, a closed manifold with (B,p)–structure. An easy computation with the Atiyah-Hirzebruch spectral sequence shows that $\Omega_4(B,p) \cong \mathbb{Z}$, detected by the signature. Since sign C = sing C', $C \cup -C'$ is zero bordant in (B,p).

Let W be a zero bordism. Then there exists an obstruction $\Theta(W,C) \in \ell_5(\mathbb{Z}/2)$ such that C is h–cobordant to C′ rel. boundary if and only if $\Theta(W,C)$ is zero bordant [K]. This implies our statement using the topological h–cobordism Theorem [F].

We will not repeat the definition of $\Theta(W,C)$. Instead we formulate some elementary properties which are enough to show that in our situation $\Theta(W,C)$ is zero bordant. Elements in $\ell_5(\mathbb{Z}/2)$ are represented by pairs $(H(\Lambda^r),U)$, where $H(\Lambda^r)$ is the hyperbolic form on $\Lambda^r \times \Lambda^r$ and $U \subset \Lambda^r \times \Lambda^r$ is a half rank free direct summand. Note that the difference to the ordinary Wall groups is, that there U is an addition self annihilating (a hamiltonian). Note also that we can forget here the quadratic refinement of the form since it is determined by it. Since the ordinary Wall group $L_5(\mathbb{Z}_2)$ vanishes one can characterize zero bordant elements in $\ell_5(\mathbb{Z}_2)$ as follows:

(3) $[H(\Lambda^r),u] \in \ell_5(\mathbb{Z}/2)$ is zero bordant if U has a hamiltonian complement V.

By construction of $\Theta(W,C)$ and some elementary considerations it has the following properties:

(4) If $(H(\Lambda^r),U)$ represents $\Theta(W,C)$ then $(H(\Lambda^r),U^{\perp})$ represents $\Theta(W,C')$.

(5) There exists a surjective homomorphism $d : U \twoheadrightarrow H_2(\tilde{C})$ inducing an isometry of the form on U with the intersection form on $H_2(\tilde{C})$.

(6) If $V = \Lambda^s \xrightarrow{f} H_2(\tilde{C})$ is a free Λ–resolution, $\Theta(W,C)$ has a representative $(H(\Lambda^s),V)$ such that d occurring in (5) is equal to f.

Since $H_2(\tilde{C}) = \mathbb{Z}_- \oplus \Lambda^9$ we can take $V = \Lambda^{10}$ with the obvious map $f : V \twoheadrightarrow H_2(\tilde{C})$.

The natural thing for showing that $\Theta(W,C)$ is zero bordant is to prove that in the restriction of $(H(\Lambda^s),V)$ to the ±1–eigenspaces, $V_\pm$ have hamiltonian complements and then to construct from them a hamiltonian complement for V.

The restriction of the hyperbolic form b on $H(\Lambda^s)$ to the ±1 eigenspaces is twice the hyperbolic form on $H(\mathbb{Z}^s)$. In particular the restriction to $V_\pm$ is divisible by two. After dividing by 2 we call this form $b_\pm$ and $V_\pm$ sits isometrically in $H(\mathbb{Z}^s)$.

By assumption (2) the form b_+ vanishes identically on V_+ and thus $(H(\Lambda)_+,V_+)$ represents an element in the ordinary L–group $L_5 = \{0\}$.

We have $V \cong \Lambda^{10} \xrightarrow{f} H_2(\tilde{C}) = \mathbb{Z}_- \oplus \Lambda^9 \longrightarrow H_2(\tilde{C})/\mathrm{rad} = \mathbb{Z}_-^{10}$ and $f|V_-$ maps onto $2\mathbb{Z}_-^{10}$. Thus the form b_- on V_- is

$$4(E_8 \oplus 2 \cdot \begin{bmatrix} 0 & 1 \\ 1 & 0 \end{bmatrix})/2 = 2 \cdot E_8 \oplus 4 \begin{bmatrix} 0 & 1 \\ 1 & 0 \end{bmatrix}.$$

Since by (4), $(H(\Lambda^{10}),V^\perp)$ represents $\Theta(W,C')$ and the form on $H_2(\tilde{C}')$ is minus the form on $H_2(\tilde{C})$, we know from (5) that the form on $V_-^\perp$ is $-b_-$. Thus we have an isometric embedding $V_- \oplus V_-^\perp = b_- \oplus (-b_-)$ into $H(\mathbb{Z}^{10})$ and we are searching for a hamiltonian complement of V_- in $H(\mathbb{Z}^{10})$.

The different isometry classes of embeddings of a pair of direct summands V_- and $V_-^\perp$ (they are direct summands since V and $V^\perp$ are so) into $H(\mathbb{Z}^{10}) = H$ are equivalently classified by analyzing in how many different ways the hyperbolic form can be reconstructed from the sublattice $V_- \oplus V_-^\perp$. To do this we consider the adjoint $\mathrm{Ad}b_- : V_- \longrightarrow V_-^*$. Denote the cokernel of $\mathrm{Ad}b_-$ by L, a finite abelian group since $\mathrm{Det}\, b_- \neq 0$. On L we have an induced quadratic form $q: L \longrightarrow \mathbb{Q}/\mathbb{Z}$ given by $q([x]) = \frac{1}{2|L|} b_-((\mathrm{Ad}b_-)^{-1}(|L| \cdot x), (\mathrm{Ad}\, b_-)^{-1}(|L| \cdot x))$.

Similarly starting with $V_-^\perp$ we get a quadratic form denoted by $(L^\perp,q^\perp)$. Of course (L,q) and $(L^\perp,-q^\perp)$ are isometric and by means of this isometry identify them with (L,q). We can reconstruct H and the embeddings of V_- and $V_-^\perp$ as follows. $H = \operatorname{Ker}(V_-^* \times (V_-^\perp)^* \longrightarrow L)$, $V_- = \operatorname{Ker} p_2 : V_-^* \times (V_-^\perp)^* \longrightarrow (V_-^\perp)^*$, $V_-^\perp = \operatorname{Ker} p_1 : V_-^* x (V_-^\perp)^* \longrightarrow V_-^*$. Here the map $V_-^* \times (V_-^\perp)^* \longrightarrow L$ is the difference of the projections onto L. This reconstruction follows from a standard argument similar to ([W], p. 285 ff).

Thus we have to analyze the isometries between (L,q) and $(L^\perp,-q^\perp) = (L,q)$ modulo those which can be lifted to isometries of V_-^*. Indeed, (H,V_-) is zero bordant if and only if the corresponding isometry of (L,q) can be lifted to V_-^*. This follows since if V_- has a hamiltonian complement, (H,V_-) is isomorphic to an element which corresponds to Id on L. On the other hand the element corresponding to a liftable isometry of (L,q) has an obvious hamiltonian complement.

Unfortunately there exist isometries of (L,q) which cannot be lifted to V_-^*. We have to show that the corresponding elements of $\ell_5(\mathbb{Z}_2)$ don't occur in our geometric situation. The key for this is that we know that since C and C' are bordant rel. boundary in $\Omega_4(B,p)$ they are stably diffeomorphic [K], i.e. $C\#r(S^2\times S^2)$ is diffeomorphic to $C'\#r(S^2\times S^2)$ for some r and in particular there exists a bordism $\hat{W}$ between $C\#r(S^2\times S^2)$ and $C'\#r(S^2\times S^2)$ with $\Theta(\hat{W},C\#r(S^2\times S^2))$ zero bordant. Obviously $\hat{W}$ is bordant to $W\#r(S^2\times D^3)$ $\#r(S^2\times D^3)$ where the boundary connected sum takes place along C and C' respectively and W is appropriately chosen. If $(H(\Lambda^2),V)$ represents $\Theta(W,C)$ then $(H(\Lambda^{s+2r}),\ V\oplus H(\Lambda^r \times \{0\}))$ represents $\Theta(\hat{W},C\#r(S^2 \times S^2))$. Denote $\hat{V}_- := V_- \oplus H(\Lambda^r \times \{0\})_-$. Then $L = L \oplus H(\mathbb{Z}^r)/2$. We know that the isometry of (L,q) corresponding to $\Theta(W,C)_-$ can after adding Id on $H(\mathbb{Z}^r)/2$ be lifted to an isometry of $\hat{V}^*$. We call an isometry (L,q) with this property a <u>restricted</u> isometry.

<u>Lemma:</u> The group of restricted isometries of (L,q) modulo those induced by isometries of V_-^* is trivial.

Before we prove this Lemma we finish our argument that $\Theta(W,C)$ is zero bordant, i.e. V in $H(\Lambda^{10})$ has a hamiltonian complement T. We know that $V_{\pm}$ have hamiltonian complements $T_{\pm}$. We also know that V is a direct summand (over Λ) in $H(\Lambda^{10}) = H$. Choose $\mathbb{Z}$–bases a_i of V_+, b_i of V_-, c_i of T_+ and d_i of T_-, such that $(a_i + b_i)/2$ is a Λ–base of V and $a_i \circ c_j = b_i \circ d_j = 2\delta_{ij}$. Then we know that for each d_i there are elements $\alpha_i \in V_+$, $\beta_i \in V_-$ and $\gamma_i \in T_+$ such that $\alpha_i + \beta_i + \gamma_i + d_i = 0 \bmod 2$ in H and $\rho_i := (\alpha_i + \beta_i + \gamma_i + d_i)/2$ form a Λ–basis of H/V. We want to choose these elements so that they generate a hamiltonian, i.e. the form is trivial between those base elements.

Since $a_i + b_i = 0 \bmod 2$ we can assume $\beta_i = 0$. Write $\alpha_i = \Sigma \alpha_{ij} a_i$ and $\gamma_i = \Sigma \gamma_{ij} c_j$ with $\alpha_{ij} \in \{0, \pm 1\}$ and $\gamma_{ij} \in \{0,1\}$. A simple computation with evaluation of the form implies $\gamma_{ij} = \delta_{ij}$ and thus $\gamma_i = c_i$. Similarly one can show $\alpha_{ij} = \alpha_{ji} \bmod 2$ and $\alpha_{ii} = 0$. Since we are free to change the sign of α_{ij} we can assume $\alpha_{ij} = -\alpha_{ji}$ for $i \neq j$. With these assumptions it is easy to check that $\rho_i \circ \rho_j = 0$ for all i,j and we are finished.

4. Proof of the Lemma. In an equivalent formulation we have to study the following situation. Consider in $H(\mathbb{Z}) \oplus E_8$ the lattice $4 \cdot H(\mathbb{Z}) \oplus 2 \cdot E_8$ and consider $L = H(\mathbb{Z})/4H(\mathbb{Z}) \oplus E_8/2E_8 = L_1 \oplus L_2$ with the induced quadratic form q which is on L_1 given by $q[x] = \frac{1}{8} b(x,x)$ and on L_2 by $q[x] = \frac{1}{4} b(x,x)$ and $L_1 \perp L_2$. A simple calculation shows that the only isometries of $(L_1, q \mid L_1)$ are ± 1 and $\pm \begin{bmatrix} 0 & 1 \\ 1 & 0 \end{bmatrix}$, which obviously can be lifted to $L_1 = H(\mathbb{Z})$. The nontrivial analogous lifting statement holds for L_2 ([BS], p. 416). Thus we are finished if modulo isometries of $H(\mathbb{Z}) \oplus E_8$ each restricted isometry of L preserves L_1 and L_2.

We denote the standard symplectic basis of $H(\mathbb{Z})$ by e and f. Let $g:(L,q) \longrightarrow (L,q)$ be a restricted isometry. Write $g[e] = a[e] + b[f] + [x]$ with

$x \in E_8$. Since $g[e]$ has order 4, a or b must be odd. Since g is restricted, $g \oplus Id$ on $L \oplus H(\mathbb{Z}^r)/2$ can be lifted to an isometry of $H(\mathbb{Z}) \oplus E_8 \oplus H(\mathbb{Z}^r)$ under which e is mapped to $\tilde{a}e + \tilde{b}f + x + 2y + 2z$ where $a = \tilde{a} \bmod 4$, $b = \tilde{b} \bmod 4$, $y \in E_8$ and $z \in H(\mathbb{Z}^r)$. Computing the quadratic form of this element yields $2ab + (x + 2y) \circ (x + 2y) = 0 \bmod 8$.

Since a or b is odd we can after acting with an appropriate liftable isometry assume $a = 1$ or $g[e] = [e] + b[f] + [x]$. Now consider $\hat{g}(e) := e + (b - 4c)f + x + 2y$, where $2b + (x + 2y) \circ (x + 2y) = 8c$. Then $\hat{g}(e) \cdot \hat{g}(e) = 0$. We can extend $\hat{g}$ to an isometry of $H(\mathbb{Z}) \oplus E_8$ by setting $\hat{g}(f) = f$. Then $\hat{g}(e)$ and $\hat{g}(f)$ span a hyperbolic plane in $H(\mathbb{Z}) \oplus E_8$ whose orthogonal complement is isometric to E_8 and we use this isometry to extend $\hat{g}$.

After composing with $\hat{g}^{-1}$ we obtain h with $h[e] = [e]$. Since $h[e] \circ h[f] = \frac{1}{4}$ we must have $h[f] = a[e] + [f] + [y]$. By the same argument as above we obtain an isometry $\hat{h}$ of $H(\mathbb{Z}) \oplus E_8$ with $\hat{h}(e) = e$ and $\hat{h}[f] = a[e] + [f] + [y]$ and after composing again with $\hat{h}^{-1}$ we obtain an isometry which preserves $H(\mathbb{Z})/4H(\mathbb{Z})$ finishing our proof.

5. Some knottings in D^4. Let K be the punctured compact Klein bottle with boundary S^1. We consider smooth embeddings of $(K, \partial K)$ into (D^4, S^3) with fixed relative normal number, $\pi_1(D^4 - K) = \mathbb{Z}_2$, intersection form of the 2–fold ramified covering equal to $<1> \oplus <-1>$ and $(S^3, \partial K)$ a fixed knot. We claim that two such knottings (D^4, K) and (D^4, K') are homeomorphic rel. boundary. The proof is similar as for our Theorem and we indicate the necessary changes.

As in section 2 we choose linear identifications of open tubular neighborhoods of K and K′ and denote their complements by C and C′. We identify $\partial C = \partial C' = M$ and choose our identification such that the Spin structures on M agree and the diagram (1) commutes. A similar consideration as in section 2 shows that $H_2(\tilde{C}) = \mathbb{Z}_- \oplus \Lambda$ and the radical of the intersection form is $\mathbb{Z}_+ = H_2(\tilde{C})_+$ and the form on $H_2(\tilde{C})/\mathrm{rad}$ is $2\begin{bmatrix} 0 & 1 \\ 1 & 0 \end{bmatrix}$.

Then we proceed as in section 3. Most of the arguments there don't make any special assumptions which are not fulfilled in our situation. The only difference is in the analysis of $(H(\Lambda^2)_-, V_-)$. Again this is determined by an isometry of $(L = \text{coker}\, 4\begin{bmatrix} 0 & 1 \\ 1 & 0 \end{bmatrix}, q)$. The situation is easier than in section 4, since the lifting problem is simpler. The problem is here whether any isometry on (L,q) is induced from an isometry of $H(\mathbb{Z})$. But as mentioned in section 4 this holds, finishing the argument.

I would like to thank R. Gompf, O. Viro and C.T.C. Wall for useful conversation and M. Kneser for the information about a reference.

References

[BS] F. van der Blij and T. Springer: The arithmetic of octaves and of the group G_2, Indag. Math. 21 (1959), 406–418

[D] S. Donaldson: Irrationality and the h–cobordism conjecture, J. Diff. Geom. 26 (1987), 141–168.

[FKV] S.M. Finashin, M. Kreck, O.Ya. Viro: Non–diffeomorphic but homeomorphic knottings in the 4–sphere, in SLN 1346 (1988), 157–198

[F] M.H. Freedman: The disk theorem for 4–manifolds, Proc. Int. Congress Math., Warsaw 1983, 647–663

[FM] R. Friedman and J. Morgan: On the diffeomorphism type of certain algebraic surfaces I, J. Diff. Geom. 27 (1988), 297–398

[G] R. Gompf: Nuclei of elliptic surfaces, preprint 1989

[GM] L. Guillou and A. Marin: Une extension d'un theorem Rochlin sur la signature, C.R. Acad. Sci, 258 (1977), A 95–98

[K] M. Kreck: An extension of results of Browder, Novikov and Wall, preprint 1985 (to appear under the title surgery and duality in the Aspects series, Vieweg).

[OV] C. Okonek and A. van de Ven: Stable bundles and differentiable structures on certain elliptic surfaces, Inv. Math. 86 (1986), 357–370

[V] O. Viro: Compact 4–dimensional exotica with small homology, to appear in the Leningrad Math. J. vol 1: 4, 1989

[W] C.T.C. Wall: Quadratic forms on finite groups, and related topics, Topology 2 (1964), 281–298

Flat Algebraic Manifolds

F.E.A. JOHNSON

DEPARTMENT OF MATHEMATICS, UNIVERSITY COLLEGE LONDON. LONDON WC1E, 6BT

The relationship between the class $\mathcal{K}$, of fundamental groups of compact Kähler manifolds, and the class $\mathcal{P}$, of fundamental groups of smooth complex projective varieties, is not well understood; one clearly has $\mathcal{P} \subset \mathcal{K}$, but, although some compact Kähler manifolds are non-algebraic, there is, at present, no known example of a group in $\mathcal{K}$ which is definitely not in $\mathcal{P}$. It is known that membership of $\mathcal{K}$ is severely restricted [5].

In this paper, we consider the subclasses $\mathcal{K}_{flat}$, $\mathcal{P}_{flat}$ consisting of fundamental groups of compact Kähler (resp. complex projective) manifolds whose underlying Riemannian manifold is flat ; we show

<u>Theorem I:</u> The classes $\mathcal{K}_{flat}$ and $\mathcal{P}_{flat}$ are identical.

This follows easily from

<u>Theorem II:</u> A smooth compact flat Riemannian manifold X admits the structure of a flat Kähler manifold if and only if it also admits the structure of a smooth flat complex projective variety.

In a previous paper with E.G. Rees ([6]), we showed that $\mathfrak{X}_{flat}$ may be characterised as the class of extension groups G of the form

$$0 \to \mathbb{Z}^{2n} \longrightarrow G \longrightarrow \Phi \to 1$$

in which Φ is finite, G is torsion free, and the operator homomorphism $\rho : \Phi \longrightarrow GL_{2n}(\mathbb{Z})$ admits a complex structure; that is, the image of ρ may be conjugated by a real matrix so as to be contained within the subgroup $GL_n(\mathbb{C})$. By an amalgamation of Bertini's Theorem and the Lefschetz Hyperplane Theorem, we see that every group in $\mathcal{P}$ is the fundamental group of a smooth complex algebraic surface ; (see, for example, (1.4) of [7]). Thus we obtain ;

<u>Theorem III :</u> Let G be a torsion free group occurring in an extension of the form $0 \to \mathbb{Z}^{2n} \longrightarrow G \longrightarrow \Phi \to 1$ where Φ is finite and the operator homomorphism admits a complex structure ; then there exists a smooth complex projective surface X such that $G = \pi_1(X)$.

Our starting point is the formal similarity between the rational group ring of a finite group and the ring of rational endomorphisms of an abelian variety ; each is semisimple and admits a positive involution. The proof proceeds by an analysis of the *rational holonomy representation* $\rho : \Phi \longrightarrow GL_{2n}(\mathbb{Q})$, using Albert's classification of positively involuted semisimple algebras [1]. For restricted classes of holonomy group Φ, for example, the symmetric groups, the full complication of the proof may not emerge ; when Φ is nilpotent, a short proof using only classical representation theory has been given by the author's student N.C. Carr ([3]).

The paper is organised as follows ; positively involuted algebras are dealt with in §1, abelian varieties in §2, and rational representation theory in §3. Theorems I and II are proved in §4, as (4.3) and (4.2) respectively.

The problem investigated here arose out of joint work with Elmer Rees, to whom the author would like to express his gratitude for many interesting conversations and much help and good advice, not least his unsparing (and often unsuccessful) efforts to instill the virtues of brevity into the author's prose.

§ 1 : Positive division algebras :

Let $\mathbb{A}$ be a finite dimensional semisimple algebra over a field $\mathbb{K}$. An involution τ on $\mathbb{A}$ is an isomorphism of $\mathbb{A}$ with its opposite algebra such that $\tau^2 = 1_{\mathbb{A}}$. When $\mathbb{K}$ is a real field, the involution τ is said to be *positive* when $\mathrm{Tr}_{\mathbb{K}}(x\tau(x)) > 0$ for all nonzero $x \in \mathbb{A}$, where '$\mathrm{Tr}_{\mathbb{K}}$' denotes *'reduced trace'*. When $\mathbb{K} = \mathbb{Q}$, the class of such positively involuted algebras has been determined by Albert [1], [12]. We recall his results.

We may express $\mathbb{A}$ in the form $\mathbb{A} = \mathbb{A}_1 \oplus \ldots \oplus \mathbb{A}_m$, where each $\mathbb{A}_i$ is a simple two-sided ideal. If τ is a positive involution on $\mathbb{A}$, it follows easily that $\tau(\mathbb{A}_i) = \mathbb{A}_i$ for each i, so it suffices to consider the case where $\mathbb{A}$ is simple; that is, $\mathbb{A} = M_n(D)$, where D is a finite dimensional division algebra over $\mathbb{K}$. An involution σ on D extends to an involution $\hat{\sigma}$ on $\mathbb{A}$ thus ;

$$\hat{\sigma}((x_{ij})) = (\sigma(x_{ji}))$$

with transposed indices as indicated. By the Skolem-Noether Theorem, each involution on $M_n(D)$ has this form. Moreover, $\hat{\sigma}$ is positive if and only if σ is positive : that is ;

Proposition 1.1: Let $(\mathbf{A},\tau)$ be a positively involuted finite dimensional semisimple algebra over a real field $\mathbf{K}$. Then there is an isomorphism of involuted $\mathbf{K}$-algebras $(\mathbf{A},\tau) \cong (M_{n_1}(D_1),\hat{\tau}_1)\times\ldots\times(M_{n_m}(D_m),\hat{\tau}_m)$ where (D_i,τ_i) is a positively involuted division algebra .

Let $\mathbf{F}$ be a field, s a field automorphism of $\mathbf{F}$ of order n , and let a be a nonzero element of $\mathbf{F}$ such that s(a) = a. The *cyclic algebra* $(\mathbf{F},s,a)$ is constructed as follows ; $(\mathbf{F},s,a)$ is a two sided $\mathbf{F}$-vector space of dimension n, with basis $([X^r])_{0\leq r\leq n-1}$, subject to the relations

$$[X^r]\lambda \quad = \quad s^r(\lambda)[X^r] \qquad (\lambda \in \mathbf{F})$$

and is an algebra with centre $\mathbf{E} = \{ x \in \mathbf{F}: s(x) = x \}$ and multiplication

$$[X][X^r] = \begin{cases} [X^{r+1}] & 0 \leq r < n-1 \\ a[X^0] & r = n-1 \end{cases} .$$

In the case n = 2, we may take $\mathbf{F} = \mathbf{E}\sqrt{b}$ for some nonzero $b \in \mathbf{E}$. Then $(\mathbf{F},s,a)$ is isomorphic to the *quaternion algebra* $\left(\frac{a,b}{\mathbf{E}}\right)$ with basis $\{1,i,j,k\}$ over $\mathbf{E}$, subject to the relations $ij = -ji = k$; $i^2 = a.1$; $j^2 = b.1$. An involution τ of a simple algebra $\mathbf{A}$ is said to be *of the first kind* when it restricts to the identity on the centre $\mathfrak{Z}$ of $\mathbf{A}$; otherwise, τ is said to be *of the second kind.* A quaternion algebra $\left(\frac{a,b}{\mathbf{E}}\right)$ admits two essentially distinct involutions of the first kind, namely *conjugation*, c, and *reversion*, r, defined thus ;

$$c(x_0 + x_1 i + x_2 j + x_3 k) = x_0 - x_1 i - x_2 j - x_3 k$$

$$r(x_0 + x_1 i + x_2 j + x_3 k) = x_0 + x_1 i - x_2 j + x_3 k$$

Albert [1],[12], showed that a positively involuted division algebra (D,τ), of finite dimension over $\mathbf{Q}$, falls into one of four classes ; here $\mathbb{E}$ and $\mathbb{K}$ are algebraic number fields.

I: $D = \mathbb{E}$ is totally real and $\tau = 1_{\mathbb{E}}$;

II: $D = \left(\frac{a,b}{\mathbb{E}}\right)$, where $\mathbb{E}$ is totally real, a is totally positive, b is totally negative, and τ is reversion ;

III : $D = \left(\frac{a,b}{\mathbb{E}}\right)$, where $\mathbb{E}$ is a totally real, a and b are both totally negative, and τ is conjugation;

IV : $D = (\mathbb{K},s,a)$, where s is an automorphism of $\mathbb{K}$ whose fixed point field $\mathbb{E}$ is an imaginary quadratic extension, $\mathbb{E} = \mathbb{E}_0(\sqrt{b})$, of a totally real field $\mathbb{E}_0$, and $a \in \mathbb{E}$; moreover, if $\mathbb{L}$ is a maximal totally real subfield of $\mathbb{K}$, there exists a totally positive element $d \in \mathbb{L}$ such that $N_{\mathbb{E}/\mathbb{E}_0}(a) = N_{\mathbb{L}/\mathbb{E}_0}(d)$.

Albert's results [12] may be summarised thus ;

Theorem 1.2 : The finite dimensional rational division algebras which admit a positive involution of the *first kind* are precisely those of type I , II or III ; those which admit a positive involution of the *second kind* are precisely those of type IV .

§2 : Riemann matrices and abelian varieties :

Let $\mathbb{K}$ be a subring of the real number field $\mathbf{R}$; a *Riemann matrix over* $\mathbb{K}$ is a pair (A,t) where A is a free $\mathbb{K}$-module of finite rank, and t is a complex structure on the real vector space $A\otimes_E\mathbf{R}$; that is, $t : A\otimes_{\mathbb{K}}\mathbf{R} \longrightarrow A\otimes_{\mathbb{K}}\mathbf{R}$ is a real linear map such that $t^2 = -1$; $rk_{\mathbb{K}}(A)$ must then be even. There is a category $\mathcal{RM}_{\mathbb{K}}$ of Riemann matrices over $\mathbb{K}$ whose morphisms $\phi : (A_1,t_1) \longrightarrow (A_2,t_2)$ are $\mathbb{K}$-linear maps $\phi : A_1 \longrightarrow A_2$ such that $t_2(\phi\otimes 1) =$

$(\phi\otimes 1)t_1$. An object (A,t) in $\mathcal{RM}_{\mathbf{K}}$ is *simple* over $\mathbf{K}$ if and only if A contains no nontrivial $\mathbf{K}$-submodule B such that $t(B\otimes_{\mathbf{K}}\mathbf{R}) = B\otimes_{\mathbf{K}}\mathbf{R}$. $\mathcal{RM}_{\mathbf{K}}$ has finite products given by $(A_1,t_1) \times (A_2,t_2) = (A_1 \times A_2, t_1 \times t_2)$. By a *Riemann form* on $(A,t) \in \mathcal{RM}_{\mathbf{K}}$ we mean a nonsingular skew $\mathbf{K}$-bilinear form $\beta : A \times A \longrightarrow \mathbf{K}$ such that

(i) if $\beta_{\mathbf{R}}: A\otimes_{\mathbf{K}}\mathbf{R} \times A\otimes_{\mathbf{K}}\mathbf{R} \longrightarrow \mathbf{R}$ denotes the $\mathbf{R}$-bilinear form obtained from β, then for each $x,y \in A\otimes_{\mathbf{K}}\mathbf{R}$,

$$\beta_{\mathbf{R}}(tx,ty) = \beta_{\mathbf{R}}(x,y) \ ;$$

(ii) the associated form $\tilde{\beta} : A\otimes_{\mathbf{K}}\mathbf{R} \times A\otimes_{\mathbf{K}}\mathbf{R} \longrightarrow \mathbf{R}$, $\tilde{\beta}(x,y) = \beta_{\mathbf{R}}(tx,y)$ is positive definite symmetric.

A Riemann matrix is *algebraic* when it admits a Riemann form. If $(A,t) \in \mathcal{RM}_{\mathbf{K}}$, we write $(A,t)^* = (A^*,t^*)$ where $A^* = \mathrm{Hom}_{\mathbf{K}}(A,\mathbf{K})$ is the $\mathbf{K}$-*dual* of A, t^* is the $\mathbf{R}$-*dual* of t, making the identifications $A^*\otimes_{\mathbf{K}}\mathbf{K} \cong \mathrm{Hom}_{\mathbf{K}}(A,\mathbf{R}) \cong \mathrm{Hom}_{\mathbf{R}}(A\otimes_{\mathbf{K}}\mathbf{R},\mathbf{R})$. The correspondence $(A,t) \mapsto (A,t)^*$ gives a contravariant functor $* : \mathcal{RM}_{\mathbf{K}} \longrightarrow \mathcal{RM}_{\mathbf{K}}$. A Riemann form β on (A,t) gives rise to an injective $\mathbf{K}$-linear map $\hat{\beta} : A \longrightarrow A^*$ such that $\hat{\beta}(x)(y) = \beta(x,y)$ for all $x, y \in A$. When $\mathbf{K}$ is a field, $\hat{\beta}$ is bijective, and, identifying A with A^{**}, we obtain a Riemann form β^* on $(A,t)^*$ thus ;

$$\beta^*(\xi,\eta) = (\hat{\beta})^{-1}(\xi)(\eta) \ .$$

(A,t) is algebraic precisely when $(A,t)^*$ is algebraic ; moreover, a Riemann form on (A,t) induces an isomorphism $\hat{\beta} : (A,t) \longrightarrow (A,t)^*$ in $\mathcal{RM}_{\mathbf{K}}$. In this case, if $\mathbf{K}$ is a subfield of $\mathbf{R}$, the $\mathbf{K}$-algebra $\mathrm{End}_{\mathbf{K}}(A,t)$ admits a positive involution, denoted by '$\sim$', thus : $\tilde{\phi} = (\hat{\beta})^{-1}\phi^*(\hat{\beta})$. This is the classical theorem of Rosati [10]. Taken together with the complete reducibility theorem of Poincaré [8], [9], we obtain :

Theorem 2.1 : Let $\mathbf{K}$ be a subfield of $\mathbf{R}$, and let (A,t) be an algebraic Riemann matrix over $\mathbf{K}$. Then (A,t) is isomorphic in $\mathcal{RM}_{\mathbf{K}}$ to a product

$$(A,t) \cong (A_1,t_1)^{(e_1)} \times \ . \ . \ . \times (A_m,t_m)^{(e_m)}$$

where (A_i,t_i) $(1\leq i\leq m)$ are simple Riemann matrices over $\mathbf{K}$, and the isomorphism types (A_i,t_i) and multiplicities e_i are unique up to order. Moreover, $\mathrm{End}_{\mathbf{K}}(A,t)$, the algebra of $\mathbf{K}$-endomorphisms of (A,t), is a product

$$\mathrm{End}_{\mathbf{K}}(A,t) \cong \prod_{i=1}^{m} M_{e_i}(D_i)$$

where $D_i = \mathrm{End}_{\mathbf{K}}(A_i,t_i)$ is a positively involuted division algebra over $\mathbf{K}$. In particular, $\mathrm{End}_{\mathbf{K}}(A,t)$ is a positively involuted semisimple $\mathbf{K}$-algebra.

We construct algebraic Riemann matrices in two ways :

I : CM-algebras :

By a *CM-algebra* we mean a quadruple $(\mathbb{A},\mathbb{E},\tau,a)$ where

(i) $\mathbb{E}$ is a totally real algebraic number field of finite degree over $\mathbf{Q}$;

(ii) $(\mathbb{A},\tau)$ is a finite dimensional positively involuted $\mathbb{E}$-algebra such that $\tau_{\mathbb{E}} = 1_{\mathbb{E}}$;

(iii) $a\in\mathbb{A}$ has the property that a^2 is a totally negative element of $\mathbb{E}$; in particular,

(iv) $\mathbb{F} = \mathbb{E}(a)$ is a purely imaginary extension of $\mathbb{E}$;

(v) $\tau(\mathbb{F}) = \mathbb{F}$ and τ restricts to the nontrivial element of $\mathrm{Gal}(\mathbb{F}/\mathbb{E})$.

In a CM-algebra $(\mathbb{A},\mathbb{E},\tau,a)$, we do not assume that $\mathbb{E}$ is the centre of $\mathbb{A}$. With a CM-algebra $(\mathbb{A},\mathbb{E},\tau,a)$, we associate the *canonical complex structure* $t : A\otimes_{\mathbb{E}}\mathbf{R} \longrightarrow A\otimes_{\mathbb{E}}\mathbf{R}$ given by $t(x\otimes 1) = ax \otimes \frac{1}{\sqrt{-a^2}}$.

Proposition 2.2: Let $(\mathbf{A},E,\tau,a)$ be a CM-algebra with canonical complex structure t. Then $(\mathbf{A},t)$ is an algebraic Riemann matrix over $\mathbf{E}$.

Proof : It is easy to check that $\beta : \mathbf{A} \times \mathbf{A} \longrightarrow \mathbf{E}$ defined by

$$\beta(x,y) = Tr_E(ay\tau(x) - ax\tau(y))$$

is a Riemann form for $(\mathbf{A},t) \in \mathcal{RM}_E$. □

For algebras of type III or IV, the following statement is tautologous; for algebras of type II, it is a restatement of an observation of Shimura ([12], p.153, Proposition 2).

Proposition 2.3: Let $(\mathbf{A},\tau)$ be a positively involuted finite dimensional division algebra of type II, III or IV over $\mathbf{Q}$; then there exists a subfield $\mathbf{E} \subset \mathbf{A}$, and $\xi \in \mathbf{A}$ such that $(\mathbf{A},E,\tau,\xi)$ is a CM-algebra.

II : The doubling construction :

Let $\mathbf{K}$ be a subfield of $\mathbf{R}$. To each finite dimensional $\mathbf{K}$-vector space V , we associate a Riemann matrix D(V), the *double* of V, over $\mathbf{K}$ thus ; $D(V) = (V\oplus V,\tilde{t}\otimes 1)$ where $\tilde{t}$: $V\oplus V \longrightarrow V\oplus V$ is the $\mathbf{K}$-linear map $\tilde{t}(x_1,x_2) = (-x_2,x_1)$. Observe that D(V) is algebraic since

$$\beta((x_1,x_2), (y_1,y_2)) = \sum_{j=1}^{n} \left(f_j(x_2)f_j(y_1) - f_j(x_1)f_j(y_2)\right)$$

is a Riemann form for D(V), where $(f_i)_{1\leq i\leq n}$ is a basis for the $\mathbf{K}$-dual of V. The doubling construction can be obtained from the CM-algebra construction as follows; $D(\mathbf{K})$ is the Riemann matrix obtained from the CM-algebra $\mathbf{K}(\sqrt{-1})$. However, $V \cong \mathbf{K}^n$ so that $D(V) \cong D(\mathbf{K})^n$. Suppose that E, F are rings such that $E \subset F \subset \mathbf{R}$. There is an *extension of scalars* functor $\mathcal{E}_E^F : \mathcal{RM}_E \longrightarrow \mathcal{RM}_F$,

$\mathcal{S}_{\mathbf{E}}^{\mathbf{F}}(A,t) = (A\otimes_{\mathbf{E}}\mathbf{F},t) \ \left(= (A,t)\otimes_{\mathbf{E}}\mathbf{F}\right)$ on making the identification $\mathbf{F}\otimes_{\mathbf{E}}\mathbf{R} \approx \mathbf{R}$. When $\mathbf{E}$ is also a field, and $\mathbf{F}$ is a finite algebraic extension of $\mathbf{E}$, we write

$$\mathcal{F}_{\mathbf{E}} = \left\{ \begin{array}{l} \sigma : \sigma \text{ is a field imbedding} \\ \qquad \sigma : \mathbf{F} \longrightarrow \overline{\mathbf{E}} \ : \ \sigma_{\mathbf{E}} = 1_{\mathbf{E}} \end{array} \right\}$$

where $\overline{\mathbf{E}}$ is the algebraic closure of $\mathbf{E}$. Suppose that, in addition, $\mathbf{F}$ is also *real over* $\mathbf{E}$, in the sense that for each $\sigma \in \mathcal{F}_{\mathbf{E}}$, $\sigma(\mathbf{F}) \subset \mathbf{R}$. We may identify

$$\mathbf{F}\otimes_{\mathbf{E}}\mathbf{R} \approx \prod_{\sigma \in \mathcal{F}_{\mathbf{E}}} {}^{\sigma}\mathbf{R} \quad ; \quad (A\otimes_{\mathbf{E}}\mathbf{F})\otimes_{\mathbf{F}}\mathbf{R} \approx A\otimes_{\mathbf{E}}\mathbf{R} \approx \prod_{\sigma \in \mathcal{F}_{\mathbf{E}}} {}^{\sigma}A\otimes_{\sigma(\mathbf{F})}\mathbf{R}$$

where, for each $\sigma \in \mathcal{F}_{\mathbf{E}}$, ${}^{\sigma}\mathbf{R} \approx \mathbf{R}$. Thus, if t is a complex structure on A, there exists a natural complex structure

$$\prod_{\sigma \in \mathcal{F}_{\mathbf{E}}} {}^{\sigma}t \ : \ A\otimes_{\mathbf{E}}\mathbf{R} \longrightarrow A\otimes_{\mathbf{E}}\mathbf{R}$$

so that we obtain a *restriction of scalars* functor

$$\mathcal{R}_{\mathbf{F}/\mathbf{E}} : \mathcal{RM}_{\mathbf{F}} \longrightarrow \mathcal{RM}_{\mathbf{E}} \ ; \ \mathcal{R}_{\mathbf{F}/\mathbf{E}}(A,t) = (A, \prod_{\sigma \in \mathcal{F}_{\mathbf{E}}} {}^{\sigma}t)$$

The constructions $\mathcal{S}_{\mathbf{E}}^{\mathbf{F}}$, $\mathcal{R}_{\mathbf{F}/\mathbf{E}}$ preserve algebraicity. Let $\mathbf{K}$ be a finite algebraic extension of $\mathbf{Q}$. Applying the construction $\mathcal{R}_{\mathbf{K}/\mathbf{Q}}$ to a Riemann matrix $(A,t) \in \mathcal{RM}_{\mathbf{K}}$, enables us to construct a Riemann matrix over $\mathbf{Q}$. We denote by $A(t)$ the complex vector space obtained from $A\otimes_{\mathbf{K}}\mathbf{R}$ by means of t, and regard A as being imbedded in $A(t)$ by means of $x \mapsto x\otimes 1$. A Riemann matrix (A,t) over $\mathbf{Z}$ determines a *complex torus* $A(t)/A$; clearly every complex torus may be so described. An *abelian variety* is a complex torus which is *algebraic*, that is, which admits a holomorphic imbedding into some $\mathbf{P}_n(\mathbf{C})$. The following result from the classical theory of theta-functions [8] justifies the usage *"algebraic Riemann matrix"*.

Proposition 2.4: Let (A,t) be a Riemann matrix over $\mathbb{Z}$. Then the following conditions are equivalent ;

(i) $A(t)/A$ is an abelian variety ;

(ii) (A,t) is an algebraic Riemann matrix over $\mathbb{Z}$;

(iii) $(A,t)\otimes_{\mathbb{Z}}\mathbb{Q}$ is an algebraic Riemann matrix over $\mathbb{Q}$.

Conversely, if (A,t) is an algebraic Riemann matrix over $\mathbb{Q}$, and Λ_1 , $\Lambda_2 \subset A$ are free abelian subgroups of maximal rank, then $A(t)/\Lambda_1$ is an abelian variety isogenous to $A(t)/\Lambda_2$, so that extension of scalars from $\mathbb{Z}$ to $\mathbb{Q}$ yields a bijection

$$\left\{ \begin{array}{c}\text{isogeny classes of} \\ \text{abelian varieties}\end{array} \right\} \xrightarrow{-\otimes\mathbb{Q}} \left\{ \begin{array}{c}\mathbb{Q}\text{-isomorphism classes of} \\ \text{algebraic Riemann matrices over } \mathbb{Q}\end{array} \right\}.$$

§3 : Representations of finite groups :

Throughout this section we fix a finite group Φ; let $\mathbf{K}$ be a subfield of $\mathbb{R}$. The group algebra $\mathbf{K}[\Phi]$ is semisimple, by Maschke's Theorem, and has a positive involution τ given by ;

$$\tau(a) = \sum_{g\in\Phi} a_g g^{-1} \qquad (a = \sum_{g\in\Phi} a_g g)$$

By (1.1), there is an isomorphism of involuted $\mathbf{K}$-algebras

$$(\mathbf{K}[\Phi],\tau) \cong (M_{n_1}(D_1),\hat{\tau}_1) \times \ldots \times (M_{n_m}(D_m),\hat{\tau}_m)$$

where each D_i is a finite dimensional division algebra over $\mathbf{K}$, admitting the positive involution τ_i. Let $\mathbf{V}$ be a simple (left) $\mathbf{K}[\Phi]$ module; for some unique i $(1\leq i\leq m)$, $\mathbf{V}$ is isomorphic to a simple left ideal of $M_{n_i}(D_i)$, and $\mathrm{End}_{\mathbf{K}[\Phi]}(\mathbf{V}) \cong D_i$. We are principally interested in the two cases $\mathbf{K} = \mathbb{Q}$; $\mathbf{K} = \mathbb{R}$.

When $\mathbf{K} = \mathbf{R}$, each finite dimensional division algebra is isomorphic to $\mathbf{R}$, $\mathbf{H}$ or $\mathbf{C}$; an isotypic module $\mathbf{V} \cong \mathbf{W}^{(e)}$ is said to be of type $\mathbf{R}$, $\mathbf{H}$ or $\mathbf{C}$ according to the type of the division algebra $End_{\mathbf{R}[\Phi]}(\mathbf{W})$. Similarly when $\mathbf{K} = \mathbf{Q}$, an isotypic module $\mathbf{V} \cong \mathbf{W}^{(e)}$ is ascribed the type (I, II, III or IV) of the division algebra $End_{\mathbf{R}[\Phi]}(\mathbf{W})$. A *complex structure* for the $\mathbf{K}[\Phi]$-module $\mathbf{V}$, is an element $t \in End_{\mathbf{R}[\Phi]}(\mathbf{V}\otimes_{\mathbf{K}}\mathbf{R})$ such that $t^2 = -1$.

<u>Proposition 3.1 :</u> Let $\mathbf{V}$ be a finitely generated $\mathbf{R}[\Phi]$-module: the following conditions on $\mathbf{V}$ are equivalent ;

(i) $\mathbf{V}$ admits a complex structure ;

(ii) each isotypic component of $\mathbf{V}$ admits a complex structure ;

(iii) each simple summand of type $\mathbf{R}$ occurs with even multiplicity in $\mathbf{V}$.

<u>Proof : (i) ⇒ (ii) :</u> Write $\mathbf{V}$ in its isotypic decomposition $\mathbf{V} = \mathbf{W}_1\oplus\ldots\oplus\mathbf{W}_m$ where each $\mathbf{W}_i$ is isotypic, with $Hom_{\mathbf{R}[\Phi]}(\mathbf{W}_i,\mathbf{W}_j) = \{0\}$ for $i\neq j$. Corresponding to the Wedderburn decomposition of $\mathbf{R}[\Phi]$ as a product of simple two sided ideals, we may write the identity element 1 of $\mathbf{R}[\Phi]$ as a sum of central idempotents $e_1 + e_2 + . . + e_n$ with $m \leq n$, such that $e_ie_j = 0$ for $i \neq j$, and , for $1 \leq i \leq m$,

$$(*) \qquad \mathbf{W}_i = \bigcap_{i\neq j} A(j)$$

where $A(j) = \left\{v \in \mathbf{V} : e_j.v = 0 \right\}$. Let t be a complex structure on $\mathbf{V}$. Since t is $\mathbf{R}[\Phi]$-linear, it commutes with each e_j. Hence $t(A(j)) \subset A(j)$ for each j, and so, by (*) above, $t(\mathbf{W}_i) \subset \mathbf{W}_i$ for each i. That is, each isotypic component of $\mathbf{V}$ admits a complex structure. <u>Q.E.D. (i) ⇒ (ii)</u> .

(ii) ⇒ (iii): Let $\mathbf{W} \cong \mathbf{V}^{(n)}$ be an isotypic $\mathbf{R}[\Phi]$-module with $\mathrm{End}_{\mathbf{R}[\Phi]}(\mathbf{V}) \cong \mathbf{R}$, so that $\mathrm{End}_{\mathbf{R}[\Phi]}(\mathbf{W}) \cong M_n(\mathbf{R})$. A complex structure t on $\mathbf{W}$ induces a complex structure t_* on the real vector space $\mathrm{End}_{\mathbf{R}[\Phi]}(\mathbf{W})$ thus; $t_*(f) = t \circ f$. It follows that $n^2 = \dim_{\mathbf{R}}(\mathrm{End}_{\mathbf{R}[\Phi]}(\mathbf{W}))$ is even, and hence n is even. Q.E.D. (ii) ⇒ (iii) .

(iii) ⇒ (i) : Let $\mathbf{V} \cong \mathbf{V_R} \oplus \mathbf{V_C} \oplus \mathbf{V_H}$ where

$$\mathbf{V_R} \cong \mathbf{V}_1^{a_1} \oplus \mathbf{V}_2^{a_2} \oplus \ldots\ldots \oplus \mathbf{V}_r^{a_r}$$

$$\mathbf{V_C} \cong \mathbf{U}_1^{b_1} \oplus \mathbf{U}_2^{b_2} \oplus \ldots\ldots \oplus \mathbf{U}_s^{b_s}$$

$$\mathbf{V_H} \cong \mathbf{W}_1^{c_1} \oplus \mathbf{W}_2^{c_2} \oplus \ldots\ldots \oplus \mathbf{W}_t^{c_t}$$

and where $\mathbf{V}_i$, $\mathbf{U}_j$, $\mathbf{W}_k$ are all simple $\mathbf{R}[\Phi]$-modules, with $\mathrm{End}_{\mathbf{R}[\Phi]}(\mathbf{V}_i) \cong \mathbf{R}$, $\mathrm{End}_{\mathbf{R}[\Phi]}(\mathbf{U}_j) \cong \mathbf{C}$, and $\mathrm{End}_{\mathbf{R}[\Phi]}(\mathbf{W}_k) \cong \mathbf{H}$. Clearly, each $\mathbf{U}_j$ admits a complex structure. Moreover, since $\mathbf{C} \subset \mathbf{H}$, each $\mathbf{W}_k$ admits a complex structure. However, since each a_i is even, by taking complex structures on doubles, each $\mathbf{V}^{a_i}$ admits a complex structure. $\mathbf{V}$ is now a direct sum of submodules each of which admits a complex structure, and so admits a complex structure. Q.E.D. (ii) ⇒ (iii): □

Recall that if $\mathbf{W}$ is a finitely generated $\mathbf{Q}[\Phi]$-module, $\mathbf{W}$ contains a Φ-invariant $\mathbb{Z}$-lattice $\mathbf{L}$, and that any two such are commensurable. We say that a complex structure t on $\mathbf{W}$ is a *𝒫-structure* for $\mathbf{W}$ when for some (and hence for any) Φ-invariant $\mathbb{Z}$-lattice, $\mathbf{W}(t)/\mathbf{L}$ is an abelian variety. We wish to give an analogue of (3.1) for the existence of 𝒫-structures ; first we note :

Proposition 3.2: Let $\mathbf{V}$ be a simple $\mathbf{Q}[\Phi]$-module; if $\mathbf{V}$ is of type II, III or IV then $\mathbf{V}$ admits a $\mathcal{P}$-structure; if $\mathbf{V}$ is of type I then $\mathbf{V}\oplus\mathbf{V}$ admits a $\mathcal{P}$-structure.

Proof: The second statement follows easily from the doubling construction of §2. To prove the first, put D = $\mathrm{End}_{\mathbf{Q}[\Phi]}(\mathbf{V})$. By (2.2), there exists a subfield $\mathbf{E}$ of the division algebra D, and an element $\xi \in D$ such that $(D,\mathbf{E},\tau,\xi)$ is a CM-algebra, where τ is the canonical involution on D inherited from $\mathbf{Q}[\Phi]$. Then $(D,\mathbf{E},\tau,\xi)$ has the *canonical complex structure* $t: D\otimes_{\mathbf{E}}\mathbf{R} \longrightarrow D\otimes_{\mathbf{E}}\mathbf{R}$ defined thus; $t(x\otimes 1) = \xi x \otimes \frac{1}{\sqrt{-\xi^2}}$. (D,t) is an algebraic Riemann

algebraic over $\mathbf{E}$. However, $\mathbf{V}$ is a vector space over D, of dimension m, say, and also has a canonical complex structure $T:\mathbf{V}\otimes_{\mathbf{E}}\mathbf{R} \longrightarrow \mathbf{V}\otimes_{\mathbf{E}}\mathbf{R}$ defined by the same formula as t, namely $T(\underset{\sim}{v}\otimes 1) = \xi\underset{\sim}{v} \otimes \frac{1}{\sqrt{-\xi^2}}$. Clearly there is

an isomorphism of Riemann matrices $(\mathbf{V},T) \cong (D,t)^m$, so that $(\mathbf{V},T)$ is also algebraic. Since $\xi \in D$, and the D-action on $\mathbf{V}$ commutes with that of Φ, it follows easily that T commutes with the Φ-action on $\mathbf{V}$; that is, T is a $\mathcal{P}$-structure for $\mathbf{V}$. □

Theorem 3.3: Let $\mathbf{W}$ be a finitely generated $\mathbf{Q}[\Phi]$-module. Then the following conditions on $\mathbf{W}$ are equivalent;

(i) each $\mathbf{Q}[\Phi]$-simple summand of type I has even multiplicity in $\mathbf{W}$;

(ii) $\mathbf{W}$ admits a $\mathcal{P}$-structure ;.

(iii) $\mathbf{W}$ admits a complex structure.

Proof: Write $\mathbf{W} = \mathbf{V}_{(I)} \oplus \mathbf{V}_{(II)} \oplus \mathbf{V}_{(III)} \oplus \mathbf{V}_{(IV)}$ where, for T = I, II, III, IV, $\mathbf{V}_{(T)}$ denotes a direct sum of simple $\mathbf{Q}[\Phi]$-

modules of type T.

(i) ⇒ (ii) : By (3.2), $V_{(II)} \oplus V_{(III)} \oplus V_{(IV)}$ admits a $\mathcal{P}$-structure; by hypothesis, each simple summand of $V_{(I)}$ has even multiplicity, so that, again by (3.2), $V_{(I)}$ admits a $\mathcal{P}$-structure. Hence V admits a $\mathcal{P}$-structure.

(ii) ⇒ (iii): Obvious .

(iii) ⇒ (i): Let $W \cong V_1^{(e_1)} \oplus .. \oplus V_m^{(e_m)}$ be the isotypic decomposition of W, with V_i a simple left ideal of $M_{n_i}(D_i)$, where $Q[\Phi] \cong M_{n_1}(D_1) \times .. \times M_{n_m}(D_m)$ is the Wedderburn decomposition of $Q[\Phi]$. On writing the identity element 1 of $Q[\Phi]$ as a sum of primitive central idempotents $1 = \epsilon_1 + \epsilon_2 + \ldots + \epsilon_k$, we see that

$$\epsilon_j \,|\, V_i = \begin{cases} 0 & \text{if } i \neq j \\ 1_{V_j} & \text{if } i = j \end{cases} .$$

Hence $\mathrm{Hom}_{R[\Phi]}(V_i \otimes_Q R, V_j \otimes_Q R) = \{0\}$ when $i \neq j$. On extending scalars, we obtain

$$\epsilon_j \otimes 1 \,|\, V_i \otimes R = \begin{cases} 0 & \text{if } i \neq j \\ 1_{V_j \otimes R} & \text{if } i = j \end{cases} .$$

Since $\epsilon_j \otimes 1$ is a sum of primitive central idempotents in $R[\Phi]$, no $R[\Phi]$-simple summand of V_j is isomorphic to any $R[\Phi]$-simple summand of V_i, and hence $\mathrm{Hom}_{R[\Phi]}(V_i \otimes_Q R, V_j \otimes_Q R) = 0$ for $i \neq j$. Now suppose that V_i is of type I; that is, D_i is a totally real algebraic number field with $d_i = \dim_Q(D_i)$. Then

$$V_i \otimes_Q R \cong U_1 \oplus U_2 \oplus \ldots \oplus U_{d_i} \text{ where } (U_r)_{1 \le r \le d_i}$$

are isomorphically distinct simple $R[\Phi]$-modules : for we may write

$$D_i \otimes_Q R \cong R \underbrace{\times \ldots \times}_{d_i} R .$$

and identify $V_i\otimes_{\mathbf{Q}}\mathbf{R}$ with the subspace $U_1\oplus U_2\oplus\ .\ .\oplus U_{d_i}$ of $\underbrace{M_{n_i}(\mathbf{R})\ \times\ .\ .\ .\ \times\ M_{n_i}(\mathbf{R})}_{d_i}$ where U_r is the subspace of the rth copy of $M_n(\mathbf{R})$ consisting of matrices concentrated in the first column. Clearly each U_r is $\mathbf{R}[\Phi]$-simple, and for $r \neq s$, U_r and U_s correspond to distinct simple factors of $\mathbf{R}[\Phi]$, and so are not $\mathbf{R}[\Phi]$-isomorphic.

We assume that $\mathbf{W}$ admits a complex structure; that is, $\mathbf{W}\otimes_{\mathbf{Q}}\mathbf{R}$ admits a complex structure. Since $\mathrm{Hom}_{\mathbf{R}[\Phi]}(V_i\otimes_{\mathbf{Q}}\mathbf{R}, V_j\otimes_{\mathbf{Q}}\mathbf{R}) = \{0\}$ for $i \neq j$, it follows that the multiplicity of each U_r in $\mathbf{W}\otimes_{\mathbf{Q}}\mathbf{R}$ is the same as the multiplicity of V_i in $\mathbf{W}$, namely e_i. Now each U_r is a type $\mathbf{R}$-summand of $\mathbf{W}\otimes_{\mathbf{Q}}\mathbf{R}$, so that, by (3.1), e_i is even when V_i is a type I summand. □

§4 : Flat Kähler manifolds and flat algebraic manifolds :

Recall that a closed flat Riemannian manifold X is isometric to one of the form $X = {}_G\backslash E(n)/_{O(n)}$ where $E(n)$ is the group of Euclidean motions of $\mathbf{R}^n$, $O(n)$ is the isotropy group of the origin, and $G = \pi_1(X)$ is a torsion free discrete cocompact subgroup of $E(n)$. Moreover, there is a natural exact sequence

$$0 \longrightarrow \Lambda \longrightarrow G \longrightarrow \Phi \longrightarrow 1$$

in which Φ, the holonomy group of X, is finite, and $\Lambda \cong \mathbf{Z}^n$ is the translation subgroup of G. Conversely, given any such *torsion free* extension, G imbeds as a discrete cocompact subgroup of $E(n)$, and $X_G = {}_G\backslash E(n)/_{O(n)}$ is a compact flat Riemannian manifold. For these details, we refer the reader to [2], [4], [13].

Since the Kähler condition is purely local, a closed flat Riemannian manifold of real dimension 2n which admits a compatible complex structure is automatically Kähler. We denote by $\mathcal{K}_{flat}$ the class of groups which occur as the fundamental group of some compact flat

Kähler manifold, and by $\mathcal{P}_{flat}$ the subclass of $\mathcal{K}_{flat}$ consisting fundamental groups of smooth flat complex projective varieties. We may represent a compact flat Kähler manifold X in the form $X = {}_{G}\backslash \mathbb{H}(n)/_{U(n)}$ where $\mathbb{H}(n) = \mathbf{C}^n \rtimes U(n)$ is the group of "Hermitian" isometries of $\mathbf{C}^n$, and G is a torsion free discrete cocompact subgroup of $\mathbb{H}(n)$. The classes $\mathcal{P}_{flat}$, $\mathcal{K}_{flat}$ may be characterised in the following way, which for $\mathcal{K}_{flat}$ is already known (see [6]).

Theorem 4.1 : $\mathcal{P}_{flat}$ (resp. $\mathcal{K}_{flat}$) consists precisely of those torsion free groups G which occur in an extension

$$\mathcal{E} : \qquad 0 \longrightarrow \mathbb{Z}^{2n} \longrightarrow G \longrightarrow \Phi \longrightarrow 1$$

in which the operator homomorphism $\rho : \Phi \longrightarrow GL_{2n}(\mathbb{Z})$ admits a $\mathcal{P}$-structure (resp. complex structure), and in which Φ is finite. Given any such pair $(\mathcal{E},\rho)$, there is a smooth flat complex projective variety (resp. compact, complex flat Kähler manifold) X whose fundamental group is G, and whose holonomy representation is ρ.

Proof : We give the proof for smooth projective varieties: the proof for Kähler manifolds is slightly simpler [6].

Let X be a smooth flat projective variety of complex dimension n; the universal covering $\tilde{X}$ of X is holomorphically equivalent to $\mathbf{C}^n$. Considering X as a Hermitian manifold, $\pi_1(X) = G$ occurs in an extension

$$0 \longrightarrow \Lambda \longrightarrow G \longrightarrow \Phi \longrightarrow 1$$

in which Φ is finite; Λ, the kernel of the holonomy representation of $\pi_1(X)$, is isomorphic to $\mathbb{Z}^{2n}$ and acts on $\mathbf{C}^n$ as a group of translations. Let $\hat{X}$ be the finite covering of X with $\pi_1(\hat{X}) = \Lambda$. Then $\hat{X} = \mathbf{C}^n/\Lambda$ is a complex torus which, being a finite holomorphic covering of a smooth projective variety, is algebraic. The operator homomorphism $\rho : \Phi \longrightarrow GL_{2n}(\mathbb{Z})$ extends to the holonomy representation $\rho : \Phi \longrightarrow GL_{2n}(\mathbf{R})$ of the Riemannian manifold X. However, the Hermitian metric on X is preserved by the holonomy representation so that $\mathrm{Im}(\rho)$ is contained in $U(n)$. Thus ρ admits a complex structure, which, since $\mathbf{C}^n/\Lambda$ is algebraic, is also a $\mathcal{P}$-structure.

Conversely, if $0 \to \mathbb{Z}^{2n} \to G \to \Phi \to 1$ is a torsion free extension in which the operator homomorphism ρ admits a $\mathcal{P}$-structure $t : \mathbb{Z}^{2n}\otimes_{\mathbb{Z}}\mathbf{R} \longrightarrow \mathbb{Z}^{2n}\otimes_{\mathbb{Z}}\mathbf{R}$, and where Φ is finite, let i denote the inclusion, $i : \mathbb{Z}^{2n} \subset V = \mathbb{Z}^{2n}\otimes_{\mathbb{Z}}\mathbf{R}$. We have a corresponding inclusion $i : GL_{2n}(\mathbb{Z}) \subset GL_{\mathbf{R}}(V)$, and induced maps $i_*: H^*(\Phi,\underset{\sim}{\mathbb{Z}}^{2n}) \longrightarrow H^*(\Phi,\underset{\sim}{V})$ where $\underset{\sim}{\mathbb{Z}}^{2n}$ (resp. $\underset{\sim}{V}$) denotes the $\mathbb{Z}[\Phi]$ (resp. $\mathbf{R}[\Phi]$) module in which Φ acts by ρ (resp. $i\rho$). Up to congruence, G is classified by the pair (i,c), where $c \in H^2(\Phi,\underset{\sim}{\mathbb{Z}}^{2n})$ is the characteristic class of the extension defining G . Let $L(G)$ be the extension

$$0 \longrightarrow \underset{\sim}{V} \longrightarrow L(G) \longrightarrow \Phi \longrightarrow 1$$

classified by the pair $(i\rho,\ i_*(c))$. Then there is a morphism of exact sequences

$$\begin{array}{ccccccccc} 0 & \longrightarrow & \mathbb{Z}^{2n} & \longrightarrow & G & \longrightarrow & \Phi & \longrightarrow & 1 \\ & & \cap & & \cap & & \| & & \\ 0 & \longrightarrow & \underset{\sim}{V} & \longrightarrow & L(G) & \longrightarrow & \Phi & \longrightarrow & 1 \end{array}$$

in which G is imbedded as a discrete cocompact subgroup of the Lie group $L(G)$. The identity component of $L(G)$ is V, and $L(G)$ has finitely many connected components, indexed by the elements of Φ. Since Φ is finite and V is divisible, $i_*(c) = 0$, so that $L(G)$ splits as a semidirect product $L(G) = V \rtimes \Phi$. Write $X = {}_G\backslash L(G)/_\Phi$. Then X is a compact flat Riemannian manifold having a finite covering $\hat{X} = {}_{\mathbb{Z}^{2n}}\backslash V(t)$ which, since ρ admits a $\mathcal{P}$-structure, is a complex algebraic torus, and on which Φ acts freely by complex analytic diffeomorphisms to give X. Thus X is also a smooth flat complex projective variety ([11] pp. 395-398). By construction, $\pi_1(X) = G$, and the holonomy representation of X is ρ. □

Thus we obtain

Theorem 4.2 : Let X be a smooth compact flat Riemannian manifold; if X admits the structure of a flat Kähler manifold, then X also admits the structure of a flat smooth complex projective variety.

Proof : Let $\mathcal{E} = (0 \to \mathbb{Z}^{2n} \to \pi_1(X) \to \Phi \to 1)$ be the exact sequence defining the fundamental group of X, with holonomy representation $\rho : \Phi \longrightarrow GL_{2n}(\mathbb{Z})$. Since X admits a flat Kähler structure, ρ admits a complex structure. By (3.3), ρ admits a $\mathcal{P}$-structure, and, by (4.1), X admits the structure of a smooth flat complex projective variety. □

Corollary 4.3 : The classes $\mathcal{K}_{flat}$ and $\mathcal{P}_{flat}$ are identical .

By the Bertini-Lefschetz Theorem , we obtain ;

Corollary 4.4 : Every element of $\mathcal{K}_{flat}$ is the fundamental group of a smooth compact complex algebraic surface.

REFERENCES

[1]: A.A.Albert ; Involutorial simple algebras and real Riemann matrices: Ann. of Math. 36 (1935) 886 - 964 .

[2]: L.Bieberbach ; Uber die Bewegungsgruppen der Euklidischen Raume I : Math. Ann. 70 (1911) 297 - 336.

[3]: N.C. Carr ; Ph.D. Thesis, Univ.of London, (in preparation).

[4]: L.S. Charlap ; Compact flat Riemannian manifolds I : Ann. of Math. 81 (1965) 15-30.

[5]: F.E.A. Johnson and E.G. Rees ; On the fundamental group of a complex algebraic manifold: Bull. L.M.S. 19 (1987) 463 - 466.

[6]: F.E.A. Johnson and E.G. Rees ; Kähler groups and rigidity phenomena : (to appear) .

[7]: F.E.A. Johnson and E.G. Rees ; The fundamental groups of algebraic varieties : (to appear in Proceedings of the International Conference on Algebraic Topology, Poznan, 1989).

[8]: S. Lang ; Introduction to algebraic and abelian functions : Graduate Texts in Mathematics , vol.89 : Springer-Verlag, 1982.

[9]: D.Mumford ; Abelian varieties : Oxford University Press, 1985.

[10]: C.Rosati ; Sulle corrispondenze fra i punti di una curva algebrica : Annali di Matematica, 25 (1916) 1 - 32 .

[11]: I.R. Shafarevitch ; Basic algebraic geometry : Springer-Verlag .

[12]: G.Shimura ; On analytic families of polarised abelian varieties and automorphic functions : Ann. of Math. 78 (1963) 149 - 192.

[13]: J.A. Wolf ; Spaces of constant curvature : McGraw-Hill, 1967.

PART 2

FLOER'S INSTANTON HOMOLOGY GROUPS

The seminal ideas of Andreas Floer have attracted a great deal of interest over the past few years. In his work Floer has developed a number of important new insights, notably through the novel use of ideas from Morse theory. His work also fits perfectly into an overall theme of these Proceedings by illustating the parallels between symplectic geometry and gauge theory in 3 and 4 dimensions. On the symplectic side a decisive achievement of Floer's programme was his proof of Arnol'd's conjecture on fixed points, a conjecture which had its origins in Hamiltonian mechanics and the work of Poincaré. See also the introduction to the section on symplectic geometry in Volume 2. Floer's lecture on this work is described in the notes by Kotschick in this section.

The other papers in this section are related to Floer's work in gauge theory. The ideas here are closely related to those involved in the section above, on the use of Yang-Mills instantons in 4-manifold theory. The Floer instanton homology groups of a 3-manifold Y are defined by instantons on the cylinder $Y \times \mathbf{R}$, interpolating between flat connections at the two ends. The space of flat connections over Y, or representations of the fundamental group, occupies a central place in the theory, and Floer's groups can be regarded as a refinement of this space of representations. Roughly speaking, if one tries to extend an argument for instantons over closed 4-manifolds to instantons over a 4-manifold-with-boundary, one finds that the new phenomena that arise, which have to do with connections which are flat but not trivial over the 3-dimensional boundary, are captured by the Floer homology groups of the boundary. Similarly, if a 4-manifold X is split into two pieces by a 3-dimensional submanifold Y, then the instantons on X can be analysed in terms of those on the two pieces and the Floer homology of Y.

For some time after Floer's work first appeared there was a dearth of explicit examples on which to test the theory, due to the difficulty of performing calculations. This picture is now beginning to change, with progress on a number of fronts which is well-illustrated by the papers in this section. On the one hand, in the new developments which he described in his lectures in Durham, Floer has found exact sequences for his homology groups which open up the possibility of making systematic calculations from a Dehn surgery description of a 3-manifold. (The development of a similar programme in one dimension higher, for the Yang-Mills invariants of 4-manifolds, also forms an important goal of current research.) On the other hand, for a particular class of 3-manifolds - the Seifert fibred manifolds- special geometric features have been used to calculate the Floer homology groups. Two approaches to this are described in the contributions of Fintushel and Stern and of Okonek below. The simplifying feature here is that, as shown by Fintushel and Stern, one can obtain the Floer groups directly from the representations, without considering instantons. The technique of Okonek brings the problem of describing the representations into the realm of algebraic geometry, and holomorphic bundles on a complex algebraic surface: a natural problem is to see if algebro-geometric techniques can be brought to bear on other calculations of the Floer groups (and of the "cup products" which can be defined in Floer homology). The paper of Fintushel and Stern also describes an application of the calculations for Seifert manifolds to obtain a result on 4-manifolds (a result which has in turn been used by Akbulut to detect exotic structures on open 4-manifolds). Here one should refer also to the contribution of Gompf in the preceding section.

Before Floer introduced his more general theory, ideas which can now be seen as going in a similar direction were developed by Fintushel and Stern, Furuta and others. Here the 3-manifolds concerned are lens spaces and, instead of 4-manifolds with boundaries, one can consider compact 4-dimensional orbifolds. These developments, and their striking topological applications, are described in the paper of Fintushel and Stern. Yang-Mills theory over orbifolds has had a number of other applications: for example Furuta and Steer have developed a theory for 2-dimensional orbifolds, generalising the work of Atiyah and Bott, which gives another, rather complete, description of the Floer homology of Seifert manifolds.

The contribution of Furuta deals with some more geometrical aspects. As we have mentioned, Floer's homology groups depend in general on knowledge of the instantons over cylinders, and these are normally quite inaccessible. The work of Atiyah, Drinfeld, Hitchin and Manin (ADHM) gives a complete description of the instantons over the 4-sphere in terms of matrix data. This can then be used to describe instantons on quotients of S^4, and in particular (by conformal invariance) on the cylinder with a lens space as cross-section. This is at present the only kind of example where one can obtain such explicit information. There are many different reductions of the ADHM description which can be made in this fashion and the investigation of these is at present an active area of research, with work by Furuta, Braam, Austin, Kronheimer and others. These descriptions involve an attractive blend of geometry, representation theory and matrix algebra. In his paper below Furuta gives a number of applications of these ideas, including an analogue of Floer's homology groups for lens spaces in which the groups can in principle be computed completely in terms of matrix algebra, via the ADHM description.

Instanton Homology, Surgery, and Knots

Andreas Floer

Department of Mathematics
University of California
Berkeley, CA 94720

We describe a long exact sequence relating the instanton homology of two homology 3-spheres which are obtained from each other by ± 1-surgery. The third term is a $\mathbf{Z}_4$-graded homology associated to knots in homology 3-spheres.

1. Instanton homology.

Let M be a homology 3-sphere, i.e. an oriented closed 3-dimensional smooth or topological manifold whose first homology group $H_1(M, \mathbf{Z})$ vanishes. Poincaré, who first conjectured that M would have to be the standard 3-sphere, the first nontrivial example, now known as the Poincaré sphere. Its fundamental group is of order 120. Since then, many other examples were found, all of which have an infinite fundamental group. For example,

$$M(p,q,r) = \{x \in \mathbf{C}^3 \mid |x| = 1 \text{ and } x_1^p + x_2^q + x_3^r = 0\}$$

is a homology 3-sphere if p, q, and r are relative prime. In this case, $M(p,q,r)$ is called a Brieskorn sphere. Properties of Brieskorn spheres were studied e.g. in [**M**]. Recently, Donaldson's theory of instantons on 4-manifolds applied successfully to the study of 3-manifolds. First, Fintushel and Stern [**FS**] proved that the Poincaré-sphere has infinite order in "integral cobordism". Pursuing the same approach Furata [**Fu**] proved that all manifolds $M(2,3,6k-1)$ are linearly independent for any $k \in \mathbf{N}$ (see also [**FS2**]). There is a strong feeling that instantons have more to say about 3-manifolds, even though the above results rely very much on special properties of the Brieskorn spheres and of their fundamental groups. In [**F1**] and in the present paper, we therefore approach the problem from the other side, by constructing instanton-invariants on 3-manifolds which can be defined rather generally, leaving computations and applications (some luck provided) to the future. The invariant, as defined in [**F1**], takes on the form of a graded abelian group $I_*(A)$ graded by $\mathbf{Z}_8$. This as well as the definition of I_* suggests that one should consider it as a homology theory, and we will in fact refer to it as instanton homology. It is the purpose of the present paper to expose further properties of I_* to justify this terminology.

Since we will need a slight extension of instanton homology, we briefly review the construction. Let A be a general SO_3-bundle over an oriented closed 3-manifold M. Instanton homology is a result of applying methods of Morse theory to the following (infinite dimensional) variational problem: Consider the space $\mathcal{A}(A)$ of L_1^4-Sobolev connections on A. Choosing a reference connection (which we will always assume to be the product connection θ if A is the product bundle), we can identify $\mathcal{A}(A)$ with the space $L_1^4(\Omega^1(A))$ of Sobolev 1-forms with values in the adjoint bundle $\mathrm{ad}(A) = (A \times so_3)/SO_3$. The Sobolev coefficients are fixed rather arbitrarily to ensure that each connection is actually continuous on M. $\mathcal{A}(M)$ is acted upon by the gauge group

$$\mathcal{G} = L_2^4(A \times_{\mathrm{Ad}} SO_3).$$

We rather want to restrict ourselves to the subgroup

$$\mathcal{G}_S = L_2^4(A \times_{\mathrm{Ad}} SU_2)/\mathbf{Z}_2.$$

(Note that SO_3 is the group of inner automorphisms of SU_2 as well as of SO_3.) The double covering $SO_3 = SU_2/\mathbf{Z}_2$ defines an extension

$$1 \to \mathcal{G}_S(A) \to \mathcal{G}(A) \xrightarrow{\eta} H^1(M, \mathbf{Z}_2) \to 0, \tag{1.1}$$

where $H^1(M, \mathbf{Z}_2)$ has the usual additive group structure. The homomorphism η can be described topologically as the obstruction to deforming g to the identity over the 1-skeleton of M. In fact, we can define $\mathcal{G}_S$ as the set of all gauge transformations in M which are homotopic to one of the "local" transformations which map the exterior of some 3-ball B^3 in M to the identity. We therefore have a natural isomorphism $\pi_0(\mathcal{G}_S) \simeq \mathbf{Z}$, through the degree $\deg(g)$ of the map $M/(M - B^3) \to SU_2 \simeq S^3$. The quotient space $\mathcal{B}(A) = \mathcal{A}(A)/\mathcal{G}_S$ is then a finite covering of the space of gauge equivalence classes of connections on M, with covering group $H_1(M, \mathbf{Z}_2)$.

To define the Chern Simons function (see [**CS**]) note that $T_\alpha \mathcal{A}(A) = L_1^4(\Omega^1(A))$, so that the integral of $\mathrm{tr}(F_a \wedge \alpha)$ over M, for $(a, \alpha) \in T\mathcal{A}(M)$ defines a canonical one-form on $\mathcal{A}(M)$. It turns out to be closed; in fact, there exists a function $\mathfrak{s}$ on $\mathcal{A}(A)$ such that

$$d\mathfrak{s}(a)\alpha = \int \mathrm{tr}(F_a \wedge \alpha). \tag{1.2}$$

It is almost gauge invariant in the sense that

$$\mathfrak{s}(g(a)) = \mathfrak{s}(a) + \deg(g). \tag{1.3}$$

Hence it defines, up to an additive constant, a function $\mathbf{a} : \mathcal{B}(A) \to \mathbb{R}/\mathbb{Z}$. It can also be described as the "secondary Pontrjagin class". Recall that on a principal bundle X over a closed 4-manifold, the first Pontrjagin class is represented by the 4-form $p_1(a) = \operatorname{tr}(F_a \wedge F_a)$ for any connection a. Here, $F_a = da + a \wedge a$ is the curvature 2-form. (For forms with values in $\operatorname{ad}(X)$, the exterior product is here extended by matrix composition in the adjoint representation of so_3.) It follows that the integral $\int p_1(a)$ is an integer and is independent of A. If the boundary $\partial X = A$ is not empty, then this is generally not true any more, but $\int p_1(a)$ modulo the integers depends only on the restriction of a to M and is given by $\mathfrak{s}$. This can actually be used to define $\mathfrak{s}$, since every SO_3-bundle M can be extended to some SO_3-bundle X as above.

By definition, the critical set of $\mathfrak{s}$ is the set of flat connections on M, which we will denote by $\mathcal{R}(A)$. It is well known that the holonomy yields an injective map

$$\mathcal{R}(A) \to \operatorname{Hom}(\pi_1(M), SO_3)/\operatorname{ad}(SO_3).$$

Flat connections are therefore sometimes referred to as representations (of the fundamental group). Conversely, for each representation one can construct an SO_3-bundle with a flat connection whose holonomy is prescribed by the representation.

It is $\mathcal{R}(A)$ which will become the set of simplices in instanton homology. To understand this, recall the following statement of (finite dimensional) Morse theory.

THEOREM (Thom–Smale–Witten). *Let f be a function on a closed manifold B with nondegenerate critical set $C(f)$. Let $C_p(f)$ denote the free abelian group over all $x \in C(f)$ with Morse index p. Then there exist homomorphisms*

$$\partial_p : C_p(f) \to C_{p-1}(f)$$

such that $\partial_p \partial_{p+1} = 0$ and

$$\ker \partial_p / \operatorname{im} \partial_{p+1} = H_p(B).$$

In fact, if g is a metric on B such that the gradient field on f induces a Morse–Smale flow on B, then the matrix elements $\langle \partial a, b \rangle$ with respect to the natural basis can *be defined as the intersection number of the unstable manifold of x and the stable manifold of y in arbitrarily level sets between $f(a)$ and $f(b)$.*

In particular, it follows that $|C_p(f)| \geq \dim H_p$. It is the defining property of Morse–Smale flows that the intersections above are transverse. The manifolds involved can all be given natural orientations, so that the matrix elements are integers. They would define homomorphisms with any $\mathbb{Z}$-module (i.e. any abelian group) as coefficients, and the same is true for instanton homology. We will, however, restrict ourselves to coefficients in $\mathbb{Z}$ for the sake of brevity.

We want to apply a similar procedure to the Chern Simons function. To define the gradient flow, note that the set $\mathcal{B}^*(A)$ of irreducible (i.e. nonabelian) connections is a smooth Banach manifold with tangent spaces

$$T_a\mathcal{B}^* = \{\alpha \in L_1^4(\Omega^1(A)) \mid d_a^*\alpha = 0\}, \tag{1.3}$$

see e.g. **[FU]** or **[F1]**. Note moreover that for any metric σ on the base manifold, $\mathbf{f}_\sigma(a) := *F_a \in L^p(\Omega^1 \otimes su_2)$ satisfies $d_a^*\mathbf{f}_\sigma(a) = 0$ due to the "Bianchi identity" $d_aF_a = 0$. Finally, the gauge equivariance of F_a implies that $f_\sigma(g(a)) = g\mathbf{f}_\sigma(a)g^{-1}$, so that f_σ is in fact a section of the bundle $\mathcal{L}$ whose fibres L_a are obtained by replacing L_1^4 by L^4 in (1.4). Even though it is not a tangent field over $\mathcal{B}^*$ in the sense of Banach manifolds, it has properties similar to vector fields on finite dimensional manifolds. The reason is that the flow trajectories of $\mathbf{f}_\sigma$, i.e. the solutions of the "flow equation"

$$\frac{\partial a(\tau)}{\partial \tau} + \mathbf{f}_\sigma(a(\tau)) = 0 \tag{1.4}$$

are in 1–1 correspondence to self dual connections a on the infinite cylinder $M \times \mathbb{R}$ with vanishing τ-component and with $a|_{M\times\{\tau\}} = a(\tau)$. One can also show (see **[F1]**) that they "connect" two critical points if and only if their Yang Mills action $\|F_A\|_2^2$ is finite. Three problems arise if we try to fit this gradient flow into the framework of the above theorem. First, we have mentioned above that the Chern Simons function is well defined only locally. Surprisingly, it turns out that we can simply ignore this point, since not only the function, but also the Morse index is ill defined along nontrivial loops in $\mathcal{B}(A)$. Second, since flat connections are not necessarily nondegenerate as critical points of $\mathfrak{s}$, we perturb $\mathfrak{s}$ by a function of the form

$$h_\kappa : \mathcal{B}(A) \to \mathbb{R};\ h_\kappa(a) \int_D h(\kappa_\theta(a))d\mu(\theta) \tag{1.5}$$

where h is a character of G, and $\kappa_\theta(a)$ the parallel transport along a thickened knot

$$\kappa : T \times D \times SO_3 \to A.$$

Here, D is the two disc and $T = \partial D$ the 1-sphere. The measure $d\mu(\theta)$ can be assumed to be smooth and supported in the interior of D. Let us denote by s a triple $s = (\sigma, \lambda, h)$, where σ is a metric on A and (λ, h) a disjoint collection of knots labeled by characters of G. It defines a perturbation $\mathfrak{s}$ of the Chern–Simons function, with L^2-gradients

$$f_s(a) = *F_a + s'(a)$$

where s' is a smooth section of $T\mathcal{B}^*(A)$ and in this sense a compact perturbation of $*F_a$. The critical points and trajectories of f_s in $\mathcal{B}^*(A)$ are now given by

(1.6)

$$\mathcal{R}_s(A) = \{\alpha \in \mathcal{B}^*(A) \mid f_s(a) = 0\}$$

$$\mathcal{M}_s(A) = \{a : \mathbb{R} \to \mathcal{A}(A) \mid \frac{\partial a(\tau)}{\partial \tau} + f_s(a(\tau)) = 0 \text{ and } \lim_{\tau \to \pm\infty} a(\tau) \in \mathcal{R}_s(A)\}.$$

Analytically, we consider $\mathcal{M}_s(A)$ as a perturbation of the space of self-dual connections on $\mathbb{R} \times A$. In fact, the temporal gauge (see [**F1**]) defines a bijection

$$\mathcal{M}_s(A) \subset \{a \in \mathcal{A}(\mathbb{R} \times A) \mid F_a + *F_a = s'(a) \text{ and } \|F_a + s'(a)\|_2 < \infty\}/\mathcal{G}(\mathbb{R} \times A)$$

where

$$\mathcal{A}(\mathbb{R} \times A) = \cup_{\alpha,\beta \in \mathcal{R}_s(A)} \alpha_+ + \beta_- + L_1^p(\Omega'_{\mathrm{ad}}(A))$$

and for each $\alpha \in \mathcal{R}_s(A)$, $\alpha_\pm$ are chosen such that for $\pi : \mathbb{R} \times A \to A$,

$$\alpha_\pm = \pi^* \hat{\alpha} \text{ on } \mathbb{R}_\pm \times A$$

for some representative $\hat{\alpha}$ of α. The flow equation is then given by a non-linear $\mathcal{G}(\mathbb{R} \times A)$-equivariant map

$$f_s : \mathcal{A}(\mathbb{R} \times A) \to L^p(\Omega^-(\mathbb{R} \times A))$$

$$f_s(a) = \frac{1}{2}(F_a + *F_a) + s'(a).$$

We call s stable if the operators

$$f'_s(a) = d^-_{a,s} + s''(a) : L_1^2(\Omega^1_{\mathrm{ad}}(\mathbb{R} \times A) \to L^2(\Omega^2_+(\mathbb{R} \times A))$$

are surjective for all $a \in \mathcal{M}_s(A)$. This implies (see [**F1**]) that $\mathcal{R}_s(A)$ is non-degenerate as the critical set and contains no non-trivial reducible representations. By compactness, it is then also finite. The set of stable parameters plays the role of the Morse–Smale gradient flows, and is denoted by $\mathcal{S}(A)$. Then we proved in [**F1**]:

THEOREM 1. *$\mathcal{S}(A)$ is not empty and contains for any metric σ on A elements (σ, h) such that*

$$\|h\| = \sum_{\kappa\in\lambda} |h_\kappa|^2$$

is arbitrarily small. Then $\mathcal{R}_s(A)$ is finite and $\mathcal{M}_s(A)$ decomposes into smooth manifolds of nonconstant dimensions satisfying

$$\dim \mathcal{M}_s(\alpha,\beta) = \mu(\alpha) - \mu(\beta)(\operatorname{mod} 8)$$

for some function $\mu = \mathcal{R}_s(A) \to \mathbf{Z}_8$. There exists a natural orientation on $\mathbf{M}_s$ which is well defined up to a change of orientation of $\alpha \in \mathcal{R}_s(A)$ (meaning a simultaneous change of the orientations on $\mathcal{M}(\alpha,\beta)$ and $\mathcal{M}(\beta,\alpha)$ for all β in $\mathcal{R}_s(A)$) and which has the following property: Denote by R_p, $p \in \mathbf{Z}_8$, the free abelian group over the elements a of $\mathcal{R}_s$ with $\mu(a) = p$. Define the homomorphism

$$\partial_{s,p} : R_p \to R_{p-1}$$

$$\partial_{s,p}(\alpha) = \sum_{A\in\hat{M}(\alpha,\beta)} \mathbf{o}(a)\beta,$$

where $\mathbf{o}(a) = 0$ for $\dim_a \mathcal{M}_s(\alpha,\beta) \neq 0$. Then if A allows no non-trivial abelian connections, we have $\partial_p\partial_{p+1} = 0$, and the homology groups

$$I_p(A,s) := \ker\partial_p / \operatorname{im}\partial_{p+1}$$

are canonically isomorphic for any $s \in \mathcal{S}(A)$.

The only point in the proof of Theorem 1 that differs from the case of trivial SU_2-bundles is the question of orientations. This is not a property of the moduli space alone, but follows from a global property of the operator family D_A over $\mathbf{B}(M)$. To be more precise, define the determinant line bundle Λ of D as the real line bundle with fibers

$$\Lambda_a = \det D_a = \det(\ker D_a) \otimes_{\mathbf{R}} \det(\operatorname{cok} D_a).$$

Local trivializations can be defined as follows. For any finite dimensional subspace $E \subset L_a$ such that $\operatorname{cok} D_a$ maps injectively into $\operatorname{Hom}(E,\mathbf{R})$, define the projection $\pi_E^\perp : L_a \to L_a/E$. Then $\det\ker \pi_E^\perp D_a$ can be identified naturally with $\det(D_A)$, and can clearly be extended smoothly on a neighborhood of A in $\mathbf{B}$. If now in addition $\mathbf{f}_s(A) \in E$, then the set

$$\mathcal{M}_E = \{a + \xi \mid \xi \in T_a\mathcal{B} \text{ and } \mathbf{f}_s(a+\xi) \in E\}$$

is a smooth manifold locally at A, and its orientation is determined by an orientation of Λ and E. In particular, any orientation on Λ defines an orientation on $\mathcal{M}$.

LEMMA 1.1. *Let X be an SO_3-bundle over a compact 4-manifold X, and let $\mathbf{G}_S(X)$ denote the group of associated SU_2-gauge transformations defined as in* (1.1). *Then Λ is orientable on $\mathcal{A}(X)/\mathcal{G}_S(X)$.*

PROOF: Since $\mathcal{G}_S$ contains only the SU_2-gauge transformations, orientability can be proved in the same way as Proposition 3.20 and Corollary 3.22 of [**D**]. □

The canonical isomorphism of the instanton homology groups can be understood best in the following functional framework.

DEFINITION 1.1: $SO_3[3]$ is the category whose objects are principal SO_3-bundles A, B over closed oriented 3-manifolds and whose morphisms $X : A \to B$ are principal SO_3 bundles over oriented *smooth* 4-manifolds together with an oriented bundle isomorphism $g_X : A \cup (-B) \to \partial X$. In this case we also write $A = X_+$ and $B = X_-$.

Note that every equivariant diffeomorphism g of M defines an "endomorphism" $Z_g = (M \times [0,1], \bar{g})$ of M, where $g : M \cup \bar{M} \to M \times \{0,1\}$ is the union of $g_0(x) = (g(x), 0)$ and $g_1(x) = (x, 1)$. We will restrict ourselves to the (full) subcategory $SO_3^*[3]$ consisting of bundles M which do not admit a nontrivial abelian connection. That is, the objects in $SO_3^*[3]$ are either trivial bundles over homology 3-spheres, or nontrivial bundles with no reducible flat connections. We define a functor I_* from $SO_3^*[3]$ into the $\mathbf{Z}_8$-graded abelian groups by means of the following auxiliary structures on M. Consider conformal structures σ on

$$X_\infty = \mathbb{R}_- \times A_- \cup X \cup \mathbb{R}_+ \times B_+$$

which are equivalent to product metrics on the ends, and on a collection $\lambda \times D \times \mathbb{R}$ of thickened cylinders in X_∞. Let h be a time dependent character of G associated to the components of λ, and define

$$\mathcal{M}_s(X) = \{a \in \mathcal{A}(X) \mid F_a + *_\sigma F_a = s'(a) \text{ and } \|F_a + s'(a)\|_2 < \infty\}/\mathcal{G}(X)$$

as in (1.5). Denote by $\mathcal{S}(X)$ the set of all such (σ, λ, h) such that $\mathcal{M}_s(X)$ is stable, and which define stable limits $s_\pm \in \mathcal{S}(X_\pm)$.

THEOREM 2. *$\mathcal{S}(X)$ is nonempty and for $s \in \mathcal{S}(X)$,*

$$v_{X,s} : R_*(A, s_A) \to R_*(B, s_B)$$
$$v_{X,s}(\alpha) = \sum_{a \in \mathbf{M}(\alpha,\beta)} o(a)\beta$$

has the following properties

(1) $v_{X,s}\partial_s = \partial_{s_+} v_{X,s}$.

(2) *Let* $(X, s) : (A, s_A) \to (B, s_B)$ *and* $(Y, s_Y) : (B, s_B) \to (C, s_C)$ *be two stable cobordisms. For* $\rho \in \mathbf{R}_+$ *large enough define on the bundle*

$$X\sharp_\rho Y = (\mathbf{R}_- \times A) \cup X \cup ([-\rho, \rho] \times B) \cup Y \cup (\mathbf{R}_+ \times C) : A \to C.$$

Then for ρ *large enough, the obvious perturbed parameter* $\pi_X \sharp_\rho \pi_Y$ *on* $X\sharp_\rho Y$ *is regular, and induces the composite*

$$v_\rho = v_{Y,s_Y} \circ v_{X,s}.$$

(3) *The homomorphism*

$$(X, s)_* : I_*(A, s_A) \to I_*(B, s_B)$$

does not depend on the choice of $s_X \in \mathcal{S}(X)$.

These are three properties that allow us to consider I_* as a functor on the category $SO_3^*[3]$ rather than a functor on the category of pairs (A, s). (The same argument is used e.g. in the definition of algebraic homologic theories through projective resolutions.) I_* has a cyclic $\mathbf{Z}_8$-grading, meaning that $\mathbf{Z}_8$ acts freely on I_* by increasing the grading. If M is a nontrivial bundle, then there usually does not exist a canonical identification of the grading label p with an element of $\mathbf{Z}_8$. However, if $H_1(M) = 0$ then the gauge equivalence class of the product connection θ_M is a nondegenerate (though reducible) element of $\mathbf{R}$. In this case, we can therefore define a canonical $\mathbf{Z}_8$-grading of $I_*(M)$ through the convention that

$$\mu(a) = \dim \mathcal{M}(a, \theta_M).$$

2. Dehn surgery.

For a knot κ in a homology 3-sphere M, denote by κM the homology 3-sphere obtained by (+1)-surgery on κ. We want to show that $I_*(M)$ and $I_*(\kappa M)$ are related by an exact triangle (a long exact sequence) whose third term is an invariant of κ. In fact, it will turn out that it depends only on the knot complement K of κ, so that it is the same for all exact sequences relating $\kappa^q M$ and $\kappa^{q+1} M$. We may say that "integral closures" $K(\alpha)$ with $H_1(K(\alpha)) = 0$ is an affine $\mathbf{Z}$-family of manifolds. The only closure of K which is canonical is $\bar{K} := K(\kappa)$, which is homology-equivalent to $S^2 \times S^1$. Although there exist, consequently, reducible representations of $\pi_1(\bar{K})$, none of them can be represented by flat connections in the SO_3-bundle P over $\bar{K}$ with nonzero $w_2(P) \in H^2(\bar{K}, \mathbf{Z}_2)$. In fact, restriction yields a 1–1 correspondence

$$\mathcal{R}(\bar{K}, P) \simeq \{\rho \in \mathrm{Hom}(\pi_1(K), SU_2) \mid \rho(\kappa) = -1\}.$$

In fact, the elements of this set do not extend to SU_2-representations on $\bar{\kappa}$, but their induced SO_3-representations do. Since κ is trivial in $H_1(K)$, all abelian representations of $\pi_1(K)$ would assign it the identity $1 \in SU_2$. It is easy to see that if $\eta(g) \neq 0$, then g^2 generates $\pi_0(\mathcal{G}_S(P))$. Our knot invariant now takes on the following form:

DEFINITION 2.1. *The instanton homology of a knot κ in a homology 3-sphere with complement κ is defined as the $\mathbf{Z}_4$-graded abelian group*

$$I_*(K) = I_*(K \cup_\alpha T \times D \times SO_3),$$

where $\alpha(\{1\} \times \partial D \times 1)$ is non-trivial in the fibre but trivial in $H_1(K)$. The surgery triangle of κ is the triple

$$I_*(K) \to I_*(M) \to I_*(\kappa M) \to I_*(K) \tag{2.1}$$

of surgery cobordisms, where w is trivial relative to the boundaries and with the ends identified in such a way that its total degree is -1.

The extensions of the framed surgery cobordisms to SO_3-bundles is unique determined by the prescription in Definition 2.1. That the total degree of (2.1) is equal to -1 modulo 4 follows from an index calculation. The construction becomes perfectly symmetric in the three maps of (2.1) if one formulates the surgery problem for general knots (κ, A), which are principal SO_3-bundle A over a closed oriented 3-manifold $M = A/SO_3$, together with an equivariant embedding

$$\kappa \cong T \times D \times SO_3 \to A.$$

Viewed from the knot complement K, (κ, A) defines sections κ_0, κ_1 of the bundle ∂K over two simple closed curves in $|\partial K| = T^2$, through

$$\kappa_0 = \{1\} \times \partial D \times \{1\}$$

$$\kappa_1 = T \times \{1\} \times \{1\}.$$

This class of knots is acted upon by the group

$$\begin{aligned}\Gamma &= \mathbf{Z}_2^2 \ltimes Sl_2\mathbf{Z} \\ &= (\mathrm{Diff}(T^2)/\,\mathrm{Diff}_1(T^2)) \rtimes H^1(T^2,\mathbf{Z}_2)\end{aligned} \tag{2.2}$$

of isotopy classes of equivariant diffeomorphisms of the bundle $T^2 \times SO_3$. In the complementary picture,

$$\Theta(\iota, K) = (\iota \circ \Theta^{-1}, K).$$

For $\Theta = \begin{pmatrix} n & p \\ m & q \end{pmatrix} \rtimes (a,b) \in Sl_2\mathbf{Z} \rtimes \mathbf{Z}_2^2$ with $\mathbf{Z}_2 = \{0,1\}$, we have

$$(\Theta_K)_0 = \{(t^n, t^p, \rho(at) \mid t \in T\} \simeq a(n\kappa_\nu + p\kappa_\pi)$$

$$(\Theta_K)_1 = \{(t^m, t^q, \rho(bt) \mid t \in T\} \simeq b(m\kappa_\nu + q\kappa_\pi).$$

EXAMPLE 1: The restriction

$$\mathrm{Aut}(T \times D \times SO_3) = \mathbf{Z}_2 \ltimes \mathbf{Z} \to \mathrm{Aut}(T^2 \times SO_3)$$

defines the subgroup Γ_0 of surgeries leaving $\kappa_\nu = \iota_\partial(\{1\} \times T \times \{1\})$ invariant. Hence it describes the subgroup of all surgeries on κ which do not change the topology of the bundle $A = K(\kappa_\nu)$. In fact, Γ_0 is generated by the operator ε changing the orientation of κ, the operator Φ changing the framing of κ and the operator Ψ of order 2 changing the trivialization of A over the core of κ:

$$\Gamma_0 = \{\varepsilon \begin{pmatrix} 1 & 0 \\ m & 1 \end{pmatrix} \rtimes (0,b) \mid b, \varepsilon \in \mathbf{Z}_2,\, m \in \mathbf{Z}\}.$$

EXAMPLE 2: The element

$$\Sigma = \begin{pmatrix} 0 & 1 \\ -1 & 0 \end{pmatrix}$$

of order 2 in $PSl(2,\mathbf{Z})$ corresponds to framed ("honest") surgery on the framed (and lifted) knot κ.

Σ together with Γ_0 spans Γ. However, there is a more canonical generating system

$$PSl_2\mathbf{Z} = \mathbf{Z}_2 * \mathbf{Z}_3$$

where $\mathbf{Z}_2$ is generated by Σ and $\mathbf{Z}_3$ by

$$\Xi = \Sigma\varphi = \begin{pmatrix} 0 & -1 \\ 1 & 0 \end{pmatrix} \begin{pmatrix} 1 & 0 \\ 1 & 1 \end{pmatrix} = \begin{pmatrix} -1 & -1 \\ 1 & 0 \end{pmatrix} \tag{2.3}$$

satisfying

$$\Xi^2 = \begin{pmatrix} -1 & -1 \\ 1 & 0 \end{pmatrix}\begin{pmatrix} -1 & -1 \\ 1 & 0 \end{pmatrix} = \begin{pmatrix} 0 & 1 \\ -1 & -1 \end{pmatrix} = \Xi^{-1}$$

$$1 + \Xi + \Xi^2 = 1 + \Xi + \Xi^{-1} = 0.$$

Hence $(\psi\Xi)^3 = \mathrm{id}$ for any $\psi \in H^1(T^2, \mathbb{Z}_2)$. If $\psi = (0,1)$ changes the lifting along the longitudinal, i.e. $\psi \in \Gamma_0$, then $\psi\Xi$ corresponds to a framed surgery on κ. Denote by X_κ the corresponding surgery cobordism. Replacing κ by $\psi\Xi\kappa$ and repeating the process, we obtain a "surgery triangle" associated to a framed knot κ. We abbreviate it as

$$(2.4) \qquad \begin{array}{ccc} & A_\kappa & \\ Y_\kappa \swarrow & & \nwarrow Z_\kappa \\ A & \xrightarrow[X_\kappa]{} & \kappa A, \end{array}$$

where

$$A_\kappa = K \cup_{\tilde\alpha} (D^2 \times T \times SO_3)$$

$$\kappa A = K \cup_\alpha (D \times T \times SO_3).$$

The attaching maps α and $\tilde\alpha$ are determined by

$$\tilde\alpha(\partial D \times \{1\}) = \psi\kappa_1$$

$$\alpha(\partial D \times \{1\}) = \kappa_0 + \kappa_1.$$

The surgery triangle of K coincides with (2.4) if $H_1(K) = \mathbb{Z}$ and κ_1 is canonical, i.e. $\kappa_1 = \partial S_K$ for some subbundle $(S_K, \partial S_K) \subset (K, \partial K)$ with two-dimensional base. Therefore exactness of (2.1) follows from the following result

THEOREM 3. *If all bundles M, κM, and M_κ are in $SO_3^*[3]$, then $(X, Y, Z) := (X_K, Y_K, Z_K)$ induces an exact triangle.*

Because of the symmetry of (1.7) with respect to Ξ, we only need to prove exactness of the two homomorphisms

$$I_*A_\kappa \xrightarrow{X_*} I_*A \xrightarrow{Y_*} I_*(\kappa A).$$

$$\operatorname{im} Y_* = \ker X_* \subset I_*(A).$$

The sequence property $X_*Y_* = 0$ follows from a connected sum decomposition. (This is simlar to the vanishing theorem for Donaldson's invariants.)

LEMMA 2.1.

$$XY = \bar Z \sharp P$$

where P is the SO_3-bundle over $\mathbb{C}P^2$ with $w_1(P) = 1$ and $p_1(P) = 0$. There exists some $s \in \rho(X)$ such that the discrete part of $\mathcal{M}_s(X)$ is empty.

3. Instanton homology for knots.

All 3-manifolds are interrelated through surgery on knots, and are interrelated in many ways. What is needed is a method to "decrease the complexity" of 3-manifolds by surgery, such that the third term which arises is simpler, too. A way of doing this is to consider the instanton homology of certain 3-manifolds as "relative" homologies $I_*(A, \lambda)$ for a link (a disjoint collection of knots) in A. Then by surgery we can reduce the problem to $I_*(\lambda)$ for a link in S^3. At least in S^3 there is an organized scheme of simplification of links, by the skein move, where two parts of λ pass through each other. In knot theory, such a move often involves a third link λ/γ, which is described as follows.

DEFINITION 3.1. *A crossing γ of a link λ in A is an embedding $\gamma : D \times I \to A$ such that $\gamma^{-1}[\lambda]$ is given by the graphs of the functions*

$$e_\pm : [0,1] \to D;\ e_\pm(t) = e^{\pm i\pi t}.$$

Then define $\gamma\lambda$ by replacing the graph of $e_\pm$ by the graph of

$$\bar{e}_\pm(t) = e_\pm(1-t).$$

To define λ/γ, replace $e_\pm$ by the sets $[-1/2, 1/2] \times \{0,1\}$.

Topologically, $\gamma\lambda$ is obtained by a ± 1-surgery on the circle γ defined by the crossing, and which does not change κ. We will therefore use the operative notation

$$\gamma(A, \lambda) = (\gamma \cdot A, \lambda) \approx (A, \gamma \cdot \lambda),$$

where A is omitted if it is the three sphere. Of course, A_λ differs from A by the connected sum with the non-trivial bundle over $T \times S^2$. One still obtains skein relations between links in A, if one extends the concept of instanton homology for links: For any 3-manifold Q_λ with $\partial Q_\lambda = \lambda \times T$, one can define invariants for framed knots by "closing up" $A \backslash \lambda \times D$ by Q_λ. It turns out that the following two types of closures are needed.

DEFINITION 3.2. *For a knot $\kappa : T \times D \times SO_3 \to A$, we define $A(\kappa)$ to be the bundle obtained by (framed) surgery on κ. Moreover, let $S_{g,b}$ be the oriented surface of genus g with b boundary components, and let U_0, U_1 denote the non-trivial SO_3-bundles over $S_{0,2} \times T$ and $S_{1,1} \times T$, respectively. (U_0, U_1 are unique up to automorphism). Let $A[\kappa]$ be obtained by gluing U_1 to the complement of κ, where*

T is identified with the normal fibre. These definitions extend componentwise to links λ. If ξ is a link of two components in M and A is a bundle over its complement which does not extend to M, then we define $A\langle\xi\rangle$ to be obtained by gluing U_0 to A.

All gluings are performed according to the given framings, or more precisely by using the parametrizations of the normal neighborhoods that we assume to be given for every knot. We sometimes refer to a link of the type ξ in Definition 2 as a "charged link in A", and consider A as a "discontinuous SO_3-bundle over M". If κ, λ, and ξ are disjoint links, then we define

$$I_*(A\backslash\xi) = I_*(A\langle\xi\rangle)$$
$$I_*(A,\lambda) = I_*(A[\lambda]).$$

A general combination of these three constructions leads to homology groups

$$I_*(A(\kappa)\backslash\xi,\lambda) = I_*(A(\kappa)\langle\xi\rangle[\lambda]).$$

The reason for this definition is the fact that there exists a self-contained skein theory for link homologies defined in this way.

Actually, I_* does not depend on the framing of λ, and it is entirely independent of the parametrization of A near ξ. This will follow from the "excision" property of Theorem 6. Note that the construction is natural with respect to "strict" link cobordisms, which are cobordism of pairs of the form

$$(X,\lambda\times I):(A,\lambda)\to(A',\lambda').$$

In fact, we can define for any closure U_λ with $\partial U_\lambda = \partial\hat{\lambda}$,

$$X_* = X\cup_{\lambda\times I}(U_\lambda\times I): A\cup_\lambda U_\lambda\times\{-1\}\to A'\cup_{\lambda'}U_\lambda\times\{\pm 1\}.$$

This applies in particular to all cobordisms obtained by framed surgery on a link disjoint from λ, e.g. for the surgery triangle

$$\begin{array}{ccc} & A_\gamma\cup_\lambda U_\lambda & \\ \swarrow & & \nwarrow \\ A\cup_\lambda U_\lambda & \longrightarrow & \gamma A\cup_\lambda U_\lambda = A\cup_{\gamma\lambda}U_\lambda \end{array} \tag{3.1}$$

of surgery on a skein loop $\gamma = \partial D_\gamma$. Now a redefinition of the top term of (3.1) yields the following skein relations for I_*:

THEOREM 4. *Assume that γ crosses κ and a component of a link λ with boundary conditions given by U_λ, and let $\lambda + \kappa$ be the connected sum along γ and $(\kappa, \lambda') = \gamma(\kappa, \lambda)$. Then we have an exact triangle*

$$\begin{array}{ccc} & I_*(A \cup_{\lambda+\kappa} U_\lambda) & \\ \swarrow & & \nwarrow \\ I_*(A(\kappa) \cup_\lambda U_\lambda) & \longrightarrow & I_*(A(\kappa') \cup_{\lambda'} U_\lambda). \end{array} \tag{1}$$

In particular,

$$\begin{array}{ccc} & I_*(A, \lambda + \kappa) & \\ \swarrow & & \nwarrow \\ I_*(A(\kappa), \lambda) & \longrightarrow & I_*(A(\kappa'), \lambda') \end{array} \tag{1a}$$

$$\begin{array}{ccc} & I_*(A \backslash (\lambda + \kappa)) & \\ \swarrow & & \nwarrow \\ I_*(A(\kappa) \backslash \lambda) & \longrightarrow & I_*(A(\kappa') \backslash \lambda') \end{array} \tag{1b}$$

If γ crosses two components λ_0, λ_1 of λ, then we have an exact triangle

$$\begin{array}{ccc} & I_*(A(\lambda/\gamma)) & \\ \swarrow & & \nwarrow \\ I_*(A(\lambda)) & \longrightarrow & I_*(A(\gamma \cdot \lambda)) \end{array} \tag{1c}$$

If γ is a crossing of a knot, and if the lifting of γ extends to D_γ, then $(A(\kappa))_\gamma = A[\kappa/\gamma]$ and we have an exact sequence

$$\begin{array}{ccc} & I_*(A \backslash (\kappa/\gamma)) & \\ \gamma_- \swarrow & & \nwarrow \gamma_0 \\ I_*(A(\kappa)) & \underset{\gamma_*}{\longrightarrow} & I_*(A(\gamma\kappa)) \end{array} \tag{2}$$

with $\gamma_* = (X(\gamma))_*$. Finally, for a crossing between two components of a link λ we have an exact triangle

$$\begin{array}{ccc} & I_*(A, \lambda/\gamma) & \\ \swarrow & & \nwarrow \\ I_*(A \backslash \lambda) & \longrightarrow & I_*(A \backslash \gamma \cdot \lambda) \end{array} \tag{3}$$

There is no skein relation for a crossing of a single charged component. In fact, it is unnecessary to unknot a charged component since whenever λ_+ and λ_- are separated by a two-sphere in A, $I_*(A \backslash \lambda) = 0$. (There are no flat connections on the non-trivial bundle over S^2.) It therefore suffices to consider crossings of λ_+ with λ_- or normal or framed components.

THEOREM 5. *Set* $\Lambda = H^*(T^2) \cong \mathbb{Z}^4$. *Then in the situations of Theorem 1, we have exact triangles*

$$(1') \qquad \begin{array}{ccc} & I_*(A\backslash\kappa/\gamma)\otimes\Lambda & \\ \gamma_- \swarrow & & \nwarrow \gamma_0 \\ I_*(A,\kappa) & \xrightarrow[\gamma_*]{} & I_*(A,\gamma\cdot\kappa) \end{array}$$

$$(2') \qquad \begin{array}{ccc} & I_*(A,\lambda/\gamma)\otimes\Lambda & \\ \tilde\gamma_- \swarrow & & \nwarrow \tilde\gamma_0 \\ I_*(A,\lambda) & \xrightarrow[\tilde\gamma_*]{} & I_*(A,\gamma\cdot\lambda) \end{array}$$

$$(3') \qquad \begin{array}{ccc} & I_*(A\backslash(\lambda/\gamma))\otimes\Lambda & \\ \swarrow & & \nwarrow \\ I_*(A\backslash\lambda_1,\lambda_0) & \longrightarrow & I_*(A\backslash\lambda_1',\lambda_0');\ \lambda'=\gamma\cdot\lambda \end{array}$$

The proof of Theorem 4 proceeds by elementary topology, showing that the top term of the surgery triangle (4) of γ is diffeomorphic to the top terms in the triangles of Theorem 4. The proof of Theorem 5 involves the Künneth formula of I_*. Let U denote the non-trivial SO_2-bundle over the two-torus. Then I_* is multiplicative with respect to connected sums along U-bundles in the following sense:

THEOREM 6. *A, B be closed connected bundles containing two-sided homologically non-trivial thickened U-bundles U_A and U_B. Define*

$$C = A +_U B = A\backslash U_A \cup_U B\backslash U_B.$$

Here, the identification is uniquely determined by requiring that the boundary of

$$\tilde W = ((-\infty,0]\times(A \amalg B))\cup([0,\infty)\times C)$$

has boundary

$$\partial\tilde W = U\times T = \partial(U\times D).$$

Then the cobordism

$$W := \tilde W \cup (U\times D) : A \amalg B \to C$$

induces an isomorphism

$$W_* : I^*A \otimes_{g_U} I^*B \xrightarrow{\cong} I^*(A +_U B),$$

where $g_U(a\otimes b) = g_{U_A} a\otimes b + a\otimes g_{U_B} a$.

Naturality with respect to surgery cobordisms and the canonical excision isomorphisms q_0, q_1 determine the functor I_* in a similar way as ordinary homology is determined by the Eilenberg axioms: Let $SO_3^*[3]$ denote the category of bundles over A over closed oriented 3-manifolds such that either $H_1(A/G)=0$ or $w(A)[S]=0$ on some oriented surface in A/G. Let the morphisms in $SO_3^*[3]$ be given by arbitrary SO_3-bundles over 4-dimensional cobordisms.

THEOREM 7. *Let J_* be a functor on $SO_3^*[3]$ such that the surgery triangle induces an exact triangle and such that W induces an isomorphism. Then any functor transformation $I_* \to J_*$ which is an isomorphism on $I_*(S^3) = 0$ and $I_*(U \times T)/g_U = \mathbf{Z}$ is a functor equivalence.*

PROOF: Note that α is an isomorphism $I_*(A(\kappa)\backslash\xi, \lambda)$ if $\kappa \cup \xi \cup \lambda$ have no crossing. Assume that α is an isomorphism whenever $\kappa \cup \xi \cup \lambda$ have less than k crossings, and consider a link μ with k-crossings. Then for any $\mu = \lambda \cup \xi$, the 5-lemma applied to the exact triangle

$$\begin{array}{ccc} I_*(S^3\backslash\xi \cup_\lambda U_\lambda) & \longrightarrow & I_*(S^3\backslash\xi^1 \cup_\lambda U_\lambda) \\ \nwarrow & & \swarrow \\ & I_*(S^3\backslash\xi_0 \cup_\lambda \tilde{U}_\lambda) & \end{array}$$

corresponding to a skein between ξ_+ and λ implies that if α is an equivalence on $I_*(S^3\backslash\xi \cup_\lambda U_\lambda)$, then it is an equivalence on $I_*(S^3\backslash\xi^1 \cup_\lambda U_\lambda)$ where ξ^1 is obtained from ξ by ξ_+ passing through an arbitrary number of components of λ and ξ_-. Since I_* as well as J_* are trivial if ξ_+ is separated from $\lambda\backslash\xi_+$ by a sphere, we conclude that α is an isomorphism on $I_*(S^3(\kappa)\backslash\xi, \lambda)$ for all ξ.

Similarly, α is an isomorphism $I_*(S^3(\kappa), \lambda)$ if it is an isomorphism on $I_*(S^3(\kappa^1), \lambda^1)$ for (κ^1, λ^1) obtained from κ, λ by skein. Since (κ, λ) is skein related to the trivial link, this proves by induction that α is a functor equivalence. Since all 3-manifolds can be obtained as $A = S^3(\kappa)$ for some framed link κ in S^3, this also proves that α is an equivalence on $I_*(A\backslash\xi, \lambda)$ for any link $\xi \cup \lambda$ with $\xi = \emptyset$ or $\xi = (\xi_+, \xi_-)$ in 3-manifold A. □

Let us apply Theorem 4 to some simple examples. First, note that if τ is the trivial knot, λ the standard link, and 3_1 the trefoil knot, then

$$S^3 \cup_\tau U_1,\ S^3 \cup_{\lambda_1} U_0 = S^3(3_1)$$

are U-bundle over T. Hence

$$I_*(S^3, \tau) \cong I_*(S^3\backslash\lambda_1) = I_*(S^3(3_1)) \cong \mathbf{Z}^2.$$

This isomorphism is also represented in the skein triangles

$$\begin{array}{ccc} & I_*(\tau) & \\ \swarrow & & \nwarrow \\ 0 = I_*(S^3\backslash 2\tau) & \longrightarrow & I_*(S^3\backslash\lambda_1) \end{array}$$

$$\begin{array}{ccc} & I_*(S^3\backslash\lambda_1) & \\ \swarrow & & \nwarrow \\ I_*(3_1) & \longrightarrow & I_*(\tau) \end{array}.$$

More generally, we obtain

$$\begin{array}{ccc} & I_*(\tau) & \\ 0\swarrow & & \nwarrow\cong \\ I_*(S^3\backslash\lambda_k) & \longrightarrow & I_*(S^3\backslash\lambda_{k+1}) \end{array}$$

and hence $I_*(S^3\backslash\lambda_{k+1}) = \mathbb{Z}^k(q)$. Note that

$$\chi(I_*(S^3\backslash\xi)/g) = lk(\xi).$$

This is essentially the relation that was used by Casson to determine the knot invariant

$$\begin{aligned} \lambda^1(\kappa) &= \lambda(\kappa PS^3) - \lambda(\kappa P^{-1}S^3) \\ &= \chi((I_*(S^3(\kappa))/g) \\ &= \Delta''_\kappa(1), \end{aligned}$$

where $\Delta_\kappa(t)$ is the Alexander polynomial.

References

[**A**] Atiyah, M. F., *New invariants of* 3 *and* 4 *dimensional manifolds*, in "Symp. on the Mathematical Heritage of Hermann Weyl," University of North Carolina, May 1987 (eds. R. Wells et al.).

[**AM**] Akbulut, S. and McCarthy, J., *Casson's invariant for oriented homology* 3*-spheres – an exposition*, Mathematical Notes, Princeton University Press.

[**C**] Cassen, A., *An invariant for homology* 3*-spheres*, Lectures at MSRI, Berkeley (1985).

[**CS**] Chern, S. S. and Simons, J., *Characteristic forms and geometric invariants*, Ann. Math. **99** (1974), 48–69.

[**D**] Donaldson, S. K., *The orientation of Yang–Mills moduli spaces and* 4*-manifold topology*, J. Diff. Geom. **26** (1987), 397–428.

[**F1**] Floer, A., *An instanton-invariant for* 3*-manifolds*, Commun. Math. Phys. **118** (1988), 215–240.

[**F2**] ———, *Instanton homology and Dehn surgery*, Preprint, Berkeley (1989).

[**FS1**] Fintushel, R. and Stern, R. J., *Pseudofree orbifolds*, Ann. Math. **122** (1985), 335–346.

[**FS2**] ———, *Instanton homology of Seifert fibered homology 3-spheres*, Preprint.

[**FS3**] ———, *Homotopy $K3$ surfaces containing $\Sigma(2,3,7)$*, Preprint.

[**FU**] Freed, D. and Uhlenbeck, K. K., "Instantons and Four-Manifolds," Springer, 1984.

[**G1**] Goldman, W. M., *The symplectic nature of the fundamental group of surfaces*, Adv. in Math. **54** (1984), 200–225.

[**G2**] ———, *Invariant functions on Lie groups and Hamiltonian flows of surface group representations*, MSRI Preprint (1985).

[**H**] Hempel, J., *3-manifolds*, Ann. of Math. Studies **86**, Princeton University Press, Princeton (1967).

[**K**] Kirby, R., *A calculus for framed links in $S3$*, Invent. Math. **45** (1978), 35–56.

[**Ko**] Kondrat'ev, V. A., *Boundary value problems for elliptic equations in domains with conical or angular points*, Transact. Moscow Math. Soc. **16** (1967).

[**M**] Milnor, J., *On the 3-dimensional Brieskorn manifolds $M(p,q,r)$, knots, groups, and 3-manifolds*, Ann. of Math. Studies **84**, Princeton University Press (1975), 175–225.

[**S**] Smale, S., *An infinite dimensional version of Sard's theorem*, Amer. J. Math. **87** (1973), 213–221.

[**T1**] Taubes, C. H., *Self-dual Yang–Mills connections on non-self-dual 4-manifolds* J. Diff. Geom. **17** (1982), 139–170.

[**T2**] ———, *Gauge theory on asymptotically periodic 4-manifolds*, J. Diff. Geom. **25** (1987), 363–430.

[**W**] Witten, E., *Supersymmetry and Morse theory*, J. Diff. Geom. **17** (1982), 661–692.

Instanton Homology

Lectures by **ANDREAS FLOER**

Department of Mathematics, University of California, Berkeley

Notes by **DIETER KOTSCHICK**[1)]

The Institute for Advanced Study, Princeton, and
Queens' College, Cambridge

(Editors Note. *Floer gave three lectures at Durham; two on his work in Yang-Mills theory and one on his work in symplectic geometry. As an addition to Floer's own contribution to these Proceedings, immediately above, Dieter Kotschick kindly agreed to write up these notes, which give an overview of Floer's three talks.)*

These are notes of the lectures delivered by Andreas Floer at the LMS Symposium in Durham. After a general introduction, the three sections correspond precisely to his three lectures. The reference for the first lecture is [F1], and for the third [F2] and [F3]. For details of the second lecture see Floer's article in this volume.

The unifying theme behind the topics discussed here is Morse theory on infinite dimensional manifolds. Recall that classical Morse theory on finite-dimensional manifolds, as developed by M. Morse, R. Thom, S. Smale and others, can be viewed as deriving the homology of a manifold from a chain complex spanned by the critical points of a Morse function, with boundary operator defined by the flow lines between critical points. This was the approach taken by J. Milnor in his exposition of Smale's work on the structure of high-dimensional smooth manifolds [M].

This point of view was described in the language of quantum field theory by E. Witten [W]. To him, critical points are the groundstates of a theory, and flow lines

1) Supported by NSF Grant No. DMS–8610730.

between them represent tunneling by "instantons". These ideas form the background against which Floer developed the theories described in these lectures, in which the manifold considered is infinite-dimensional. In the first of these theories the manifold under consideration is the space of gauge equivalence classes of (irreducible) connections on a bundle over a 3-manifold. The gradient lines are given, literally, by the instantons of 4-dimensional Yang-Mills theory. In the second theory the manifold considered is the loopspace of a symplectic manifold, and the gradient lines or instantons are Gromov's pseudoholomorphic curves [G].

A common feature of these "Morse theories" is that there is no Palais-Smale condition satisfied by the Morse function. In each case the Hessian at a critical point has infinitely many positive and negative eigenvalues, and a "Morse index" is defined by a suitable renormalisation. This is quite different from the classical infinite-dimensional variational problems, such as the energy function on loops, for example. In fact these instanton homology theories on infinite dimensional manifolds should be considered as "middle dimensional" homology theories, as suggested by M. F. Atiyah [A] and others.

Instanton Homology for 3-manifolds

Let M be an oriented 3-manifold. Considering finite-action instantons over $M \times \mathbf{R}$, we want to construct invariants of M from moduli spaces. Let P be an SO(3)-bundle over M, extended trivially to $M \times \mathbf{R}$, and denote by $\mathcal{M}_g(M \times \mathbf{R})$ the space of gauge equivalence classes of connections on $P \to M \times \mathbf{R}$ satisfying

(1) $$\int_{M\times\mathbf{R}} \|F\|^2 < \infty$$

(2) $$F + *F = 0,$$

where (2) is the anti-self-duality equation with respect to the product metric ($g\times$ canonical) on $M \times \mathbf{R}$.

The condition of finite energy (1) forces elements of $\mathcal{M}_g(M \times \mathbf{R})$ to converge to well-defined flat connections on the ends of the cylinder $M \times \mathbf{R}$. Denote by $\mathcal{R}(M)$ the space of gauge equivalence classes of flat connections on $P \to M$. Then

$$\mathcal{M}_g(M \times \mathbf{R}) = \coprod_{(A,B)\in\mathcal{R}(M)^2} \mathcal{M}(A,B),$$

where $\mathcal{M}(A, B)$ denotes the moduli space of finite-action instantons interpolating between A and B.

We always assume that there are no nontrivial reducible flat connections on $P \to M$. This is the case in particular if M is a homology 3-sphere, and also if M has the homology of $S^1 \times S^2$ and $w_2(P) \neq 0$. Moreover, just for the sake of exposition, we assume that $\mathcal{R}(M)$ is discrete. (If it is not we can still carry out all the constructions after replacing $\mathcal{R}(M)$ by the solution space of a suitable perturbation of the flatness equation, cf. [F1].)

Now all the moduli spaces $\mathcal{M}(A, B)$ can be oriented in a coherent fashion by considering extensions to a 4-manifold X with boundary $\partial X = M$, and using Donaldson's theory [D].

The action of $\mathbf{R}$ by translation on $M \times \mathbf{R}$ induces an action on each $\mathcal{M}(A, B)$, and we denote by $\hat{\mathcal{M}}(A, B)$ the reduced moduli space $\mathcal{M}(A, B)/\mathbf{R}$. Uhlenbeck's compactness principle [U] has the following consequences:

(a) 0-dimensional components of $\hat{\mathcal{M}}(A, B)$ are compact, and

(b) if a sequence in a 1-dimensional component of $\hat{\mathcal{M}}(A, B)$ has no convergent subsequence, then some subsequence decomposes asymptotically into instantons in $\hat{\mathcal{M}}(A, C)$ and $\hat{\mathcal{M}}(C, B)$, where C is a nontrivial flat connection.

With all these technicalities in place we can define instanton homology. Let $\mathcal{R}^*(M)$ be the set of gauge equivalence classes of nontrivial flat connections on $P \to M$. Then counting, with signs given by the orientations, the number of 0-dimensional components of $\hat{\mathcal{M}}(A, B)$ gives a linear map ∂_g on the free abelian group generated by the elements of $\mathcal{R}^*(M)$ (via (a) above). Moreover, (b) can be used to prove $\partial_g^2 = 0$. Thus

Definition [F1]: The instanton homology of M is

$$I_*(M, g) = \ker \partial_g / \operatorname{im} \partial_g .$$

The two cases we are interested in are when M is a homology 3-sphere, in which case there is only one SO(3)-bundle over M (the trivial one), and when $H_*(M, \mathbf{Z}) =$

$H_*(S^1 \times S^2, \mathbf{Z})$ and $P \to M$ is the unique SO(3)-bundle with $w_2 \neq 0$. In each case there is a grading on the instanton homology, which we have indicated by writing I_*. In the first case $* \in \mathbf{Z}_8$, and in the second $* \in \mathbf{Z}_4$.

The whole construction underlying the definition of instanton homology has certain functorial properties. Most importantly, if X^4 is a cobordism between M^3 and N^3, then counting points in zero-dimensional moduli spaces over X gives a chain homomorphism $\mathcal{R}^*(M) \cdot \mathbf{Z} \to \mathcal{R}^*(N) \cdot \mathbf{Z}$. Applying this to $M \times \mathbf{R}$ with a "twisted" metric gives homomorphisms

$$I_*(M, g_+) \rightleftarrows I_*(M, g_-)$$

which can be seen to be inverses of each other. This proves that $I_*(M)$ is, in fact, independent of the metric g.

The definition of $I_*(M)$ given above does not bring out completely the variational origin of the whole theory. There is a function (the "Chern-Simons functional") on the space of connections on $P \to M^3$, such that $\mathcal{R}(M)$ is the critical set of the function, and the $\hat{\mathcal{M}}(A, B)$ parametrize gradient lines between the critical points A and B. We will come back to this point of view in the third lecture, to explain the analogy with constructions in symplectic geometry.

Instanton Homology and Dehn Surgery

The definition of instanton homology given in the previous lecture is enough to allow one to calculate it completely in simple cases. Indeed, such calculations have been carried out by several people for the Seifert fibered homology 3-spheres, with the most complete results in [FS].

Leaving aside the study of examples, we want to describe a more systematic approach to such calculations by producing an "exact triangle" relating the instanton homology of a homology 3-sphere M with that of M', obtained from M by Dehn surgery on a knot. This is in the spirit of the usual exact sequences for standard homology theories. It gives, in principle, a universal computational tool, because every homology 3-sphere, or, more generally, every orientable 3-manifold, can be obtained from S^3 by a sequence of Dehn surgeries [L].

Let $\mathcal{K}$ be a knot in M. We think of this as an embedded solid torus $\mathcal{K}\colon S^1 \times D^2 \hookrightarrow M$. The knot complement is $K = M \setminus \mathcal{K}(S^1 \times D^2)$ with boundary $\partial K = T^2$, a 2-torus.

We denote by $\bar{K}$ the "closure" of K obtained by gluing in $S^1 \times D^2$ interchanging parallels and meridians on T^2. Then $H_*(\bar{K}, \mathbf{Z}) = H_*(S^1 \times S^2, \mathbf{Z})$.

We want to do (+1)-Dehn surgery on $\mathcal{K}$. To this end we take out $\mathcal{K}(S^1 \times D^2)$ and glue it back in mapping the meridian to the diagonal in T^2. This gives another homology 3-sphere M'.

Now for any such surgery we have a surgery cobordism obtained as follows: take $M \times [0,1]$ and attach a 2-handle $D^2 \times D^2$ to $M \times \{1\}$ using $\mathcal{K}$. The resulting 4-manifold X has two boundary components; one is M, and the other is the surgered manifold. As explained in the previous lecture, each cobordism gives a homomorphism between the instanton homologies of the ends. We apply this to the triple M, M', $\bar{K}$ related by surgery cobordisms. For $\bar{K}$ we use a $\mathbf{Z}_8$-graded double cover $\bar{I}_*(\bar{K})$ of the $\mathbf{Z}_4$-graded $I_*(\bar{K})$. The result is:

Theorem (Floer): *The triangle*

$$\begin{array}{ccc} I_*(M) & \longrightarrow & I_*(M') \\ \nwarrow & & \swarrow \\ & \bar{I}_*(\bar{K}) & \end{array}$$

is exact.

It turns out that this can be proved by analyzing the effect of Dehn surgery on the representation space of the fundamental group. No detailed understanding of instantons on tubes is needed.

First, to see that for two consecutive maps α, β in the triangle $\ker \beta \subset \operatorname{im} \alpha$ holds, one looks at the representation spaces $\mathcal{R}(\cdot) = \operatorname{Hom}(\pi_1(\cdot), \mathrm{SO}(3))/\mathrm{SO}(3)$ for all the manifolds involved. (For $\bar{K}$ the condition $w_2 \neq 0$ has to be imposed.) Now each of the manifolds M, M', $\bar{K}$ contains $K \supset \partial K = T^2$. Thus for each of these manifolds $\mathcal{R}(\cdot) \subset \mathcal{R}(K) \subset \mathcal{R}(\partial K) = T^2/\mathbf{Z}_2$. It can be shown that a suitable perturbation of $\mathcal{R}(M')$ splits into $\mathcal{R}(M)$ and $\mathcal{R}(\bar{K})$ and that this implies $\ker \beta \subset \operatorname{im} \alpha$.

To prove $\operatorname{im} \alpha \subset \ker \beta$, i.e. $\beta \circ \alpha = 0$, one uses a connected sum decomposition of the 4-manifold defining $\beta \circ \alpha$, in the spirit of Donaldson's vanishing theorem. For example, one could use one of the arguments in [K], §6. The point is that this 4-manifold splits off $\overline{\mathbf{C}P^2}$ as a connected summand, and that the relevant SO(3)-bundle has $w_2 \neq 0$ on $\overline{\mathbf{C}P^2}$. This means that any 0-dimensional moduli space on the 4-manifold giving $\beta \circ \alpha$ must be empty, which implies $\beta \circ \alpha = 0$. For a more detailed account of Floer's proof of this theorem, we refer to his article in these Proceedings.

Finally, we note that the above theorem ties in nicely with Casson's work, described in [AM]. Casson defined a numerical invariant $\lambda(M)$ for homology 3-spheres M by counting, with suitable signs, the number of nontrivial representations of $\pi_1(M)$ in SU(2). In fact, $\lambda(M)$ is one half this number, and Casson showed that $\lambda(M)$ is always an integer. Casson gave the following formula for the change of his invariant under Dehn surgery:

$$\lambda(M') = \lambda(M) + \lambda'(\mathcal{K}),$$

where $\lambda'(\mathcal{K})$ is a knot invariant extracted from the Alexander polynomial.

The relation of Casson's invariant with instanton homology is simply that the Euler characteristic,$\chi(I_*(M)) = \sum(-1)^i \dim I_i(M)$, is $2\lambda(M)$, as proved by Taubes [T], [F1]. The mystery of the integrality of λ now becomes the mystery of the evenness of $\chi(I_*)$. However, the above theorem, and the fact that going around the triangle once gives a shift in the grading by -1, show

$$\begin{aligned}\chi(I_*(M')) &= \chi(I_*(M)) + \chi(\bar{I}_*(\bar{K})) \\ &= \chi(I_*(M)) + 2\chi(I_*(\bar{K})).\end{aligned}$$

Now working inductively from S^3 to any given M using [L], we find that $\chi(I_*(M))$ is always even, as $\chi(I_*(S^3)) = 0$, and Dehn surgery changes the Euler characteristic by $2\chi(I_*(\bar{K}))$. Note that this proves $\lambda'(\mathcal{K}) = \chi(I_*(\bar{K}))$.

Symplectic Instanton Homologies

There are deep analogies between gauge theory on 3- and 4-dimensional manifolds as used in the previous two lectures, and the theory of J-holomorphic curves in symplectic manifolds, as initiated by Gromov [G] and developed in [F2,3]. We summarize some of these analogies in the following table, in which X is a smooth oriented 4-manifold, and V is a symplectic $2n$-manifold with symplectic form ω.

	$\mathrm{SO}(3) \to P \to X$	(V, ω)
1)	A choice of conformal structure gives the ASD equation	A choice of almost complex structure gives a $\bar{\partial}$-operator

		for maps
	$$F + *F = 0,$$	$$u: C \to V,$$
	for connections on P.	C a Riemann surface.
2)	If $X = M \times \mathbf{R}$, the ASD equation is $$\frac{\partial A}{\partial t} + *_M F_A^{(M)} = 0.$$	If $C = S^1 \times \mathbf{R}$, the $\bar{\partial}$-equation is $$\frac{\partial u}{\partial s} + J\frac{\partial u}{\partial t} = 0.$$
3)	In fact, $*_M F_A^{(M)}$ is the gradient of the Chern-Simons function.	Similarly, $J\frac{\partial u}{\partial t}$ is the gradient of the symplectic action defined below.
4)	This setup for $M^3 \times \mathbf{R}$ leads to instanton homology.	Here we obtain an "index cohomology" which turns out to be $H^*(V, \mathbf{Z}_2)$.

Roughly speaking, the space of connections on M^3 on the left hand side is replaced on the right by the loop space LV of V. Point loops correspond to flat connections, and holomorphic maps $S^1 \times \mathbf{R} \to V$ to instantons on $M^3 \times \mathbf{R}$. The variational problem underlying the symplectic theory comes from the symplectic action function a defined as follows. Fix a loop $z_0 \in LP$, and set

$$a(z) = \int_{S^1 \times [0,1]} u^*\omega,$$

where $u: S^1 \times [0,1] \to V$ interpolates between z and z_0.

In both of these theories the PDEs considered are elliptic, we have Fredholm equations and finite-dimensional trajectories. Moreover, the spaces of trajectories have similar compactness properties.

Now let us look at some applications. Let $H: S^1 \times V \to \mathbf{R}$ be a time-dependent Hamiltonian function on P. Then the critical points of

$$a(z) + \int_{S^1} H(t, z(t))dt$$

are solutions of

$$(*) \qquad \frac{\partial z(t)}{\partial t} = J\nabla H(t, z(t)),$$

i.e. they are the fixed points of the exact symplectic diffeomorphism defined by H.

Assume $\pi_2(V) = 0$. Then it turns out that the cohomology of the complex generated by the solutions of $(*)$, with boundary operator defined by the flow lines of the modified function, is isomorphic to $H^*(V, \mathbf{Z}_2)$. This gives the following

Theorem [F2]: *Let V be a closed symplectic manifold with $\pi_2(V) = 0$, and ϕ an exact diffeomorphism of V with nondegenerate fixed points. Then the number of fixed points is at least the sum of the $\mathbf{Z}_2$-Betti numbers of V.*

This result was previously conjectured by V. I. Arnold, and special cases were proved by Conley and Zehnder, Hofer and others. A more general theorem, covering the case of degenerate fixed points, was obtained in [F3].

In fact, the above theorem can be recovered from a more general relative version for Lagrangian submanifolds $L \subset V$. This says that if $\pi_2(V, L) = 0$, then

$$|L \cap \phi(L)| \geq \sum_i \dim H^i(L, \mathbf{Z}_2).$$

To deduce the fixed-point theorem above from this Lagrangian intersection result, one considers V and the graph of a symplectomorphism as two Lagrangian submanifolds of the product $V \times V$. To prove the Lagrange intersection result Floer used a homology constructed with a boundary operator defined by "holomorphic strips"; i.e.pseudo- holomorphic maps from $[0,1] \times \mathbf{R}$ to V which map one boundary, $\{0\} \times \mathbf{R}$, to L and the other, $\{1\} \times \mathbf{R}$, to $\phi(L)$.These can be regarded as the gradient lines of a function on the space of paths in V beginning on L and ending on $\phi(L)$, and the critical points of the function are the constant paths , mapping to the intersection points of L and $\phi(L)$,

Among the various interesting questions raised by these results, we mention the following :

(1) Is it possible to remove the assumption on π_2 ? Evidence that this can be done in some cases comes from Fortune's proof [Fo] of the above theorem for $V = \mathbf{C}P^n$.

On the other hand the "relative theorem" is false for a nullhomologous small circle on a surface (the Lagrangian condition is vacuous here). The methods described in this lecture have been pushed through in the case that $[\omega]$ is zero in $\pi_2(V)$, or more generally if, on $\pi_2(V)$, ω is a multiple of the *first Chern class* of V (defined by the almost-complex structure).

(2) What can be said for symplectic diffeomorphisms which are not exact, i.e. not deformations of the identity? Similarly, can one estimate $|L \cap L'|$ by something better than the intersection number of the homology classes of L, L' if $L' \neq \phi(L)$?

(3) Let S be a surface and V the representation space of $\pi_1(S)$ in a compact semisimple Lie group. The representations coming from a handlebody H_1 with boundary S make up a Lagrangian submanifold $L_1 \subset V$. Thus the representation space of a 3-manifold M is identified via a Heegard splitting with $L_1 \cap L_2$. This is the approach that Casson took to define his invariant. It was suggested by Atiyah [A] that this should extend to give a link between instanton homology in the symplectic and 3-manifold cases. Can this be made rigorous?

(4) If, in answer to (2), instanton homology for Lagrangian intersections in the non-exact case is a new invariant, are there exact sequences etc. which can be used to calculate the homology groups?

References

[AM] S. Akbulut and J. McCarthy, *Casson's Invariant for Oriented Homology 3-spheres – an exposition.* Princeton University Press (to appear).

[A] M. F. Atiyah, *New Invariants for 3- and 4-Dimensional Manifolds.* Proc. Symp. Pure Math. 48 (1988) 285–299.

[D] S. K. Donaldson, *The orientation of Yang-Mills moduli spaces and 4-manifold topology.* J. Differential Geometry 26 (1987) 397–428.

[F1] A. Floer, *An Instanton-Invariant for 3-Manifolds.* Commun. Math. Phys. 118, 215–240 (1988).

[F2] ———, *Morse theory for Lagrangian intersections.* J. Differential Geometry 28 (1988) 513–547.

[F3] ———, *Cuplength Estimates on Lagrangian Intersections.* Commun. Pure Applied Math. XLII 335–356 (1989).

[F4] ———, *Witten's complex and infinite dimensional Morse theory.* J. Differential Geometry 30 (1989) 207–221.

[Fo] B. Fortune, *A symplectic fixed point theorem for* CP^n. Invent. Math. 81 (1985) 29–46.

[FS] M. Furuta and B. Steer, *Seifert Fibred Homology 3-spheres and the Yang-Mills Equations on Riemann Surfaces with Marked Points.* (preprint)

[G] M. Gromov, *Pseudoholomorphic curves in symplectic manifolds.* Invent. Math. 82 (1985) 307–347.

[K] D. Kotschick, *SO(3)-invariants for 4-manifolds with* $b_2^+ = 1$. (preprint)

[L] W. B. R. Lickorish, *A representation of orientable combinatorial 3-manifolds.* Ann. of Math. 76 (1962) 531–538.

[M] J. W. Milnor, *Lectures on the h-Cobordism Theorem.* Princeton University Press, 1965.

[T] C. H. Taubes, *Casson's invariant and gauge theory.* (preprint)

[U] K. K. Uhlenbeck, *Connections with* L^p *Bounds on Curvature.* Commun. Math. Phys. 83, 31–42 (1982).

[W] E. Witten, *Supersymmetry and Morse theory.* J. Differential Geometry 17 (1982) 661–692.

Invariants for Homology 3-Spheres

RONALD FINTUSHEL[1] AND RONALD J. STERN[2]

1. Introduction

In this paper we survey our work of the last few years concerning invariants for homology 3-spheres. We shall pay special attention to the role of gauge theory, and we shall try to place this work in a proper context. Homology 3-spheres exist in abundance. Let us begin with a list of some examples and constructions.

i. The binary icosohedral group I is a subgroup of $SU(2)$ of order 120. The quotient of its action on S^3 is the Poincaré homology 3-sphere P^3, and so $\pi_1(P^3) = I$. In fact, it is known that the only finite non-trivial group that can occur as the fundamental group of a homology 3-sphere is I, and it is still unknown if P^3 is the only homology 3-sphere Σ with fundamental group I. (But of course this depends on the 3-dimensional Poincaré conjecture.)

ii. The Brieskorn homology 3-sphere $\Sigma(p,q,r)$ is defined as $\{z_1^p + z_2^q + z_3^r = 0\} \cap S^5$ where p, q, r are pairwise relatively prime. It is the link of the Brieskorn singularity of type (p,q,r) [**B**]. In fact $\Sigma(2,3,5) = P^3$, and if $1/p + 1/q + 1/r < 1$ then $\pi_1(\Sigma(p,q,r))$ is infinite.

iii. The Seifert fibered homology 3-spheres $\Sigma = \Sigma(a_1, \ldots, a_n)$ (see [**NR**]) are a rather ubiquitous collection of homology 3-spheres. These 3-manifolds possess an S^1-action with orbit space S^2. If $\Sigma \neq S^3$, then the S^1-action has no fixed points and has finitely many exceptional orbits (multiple fibers) with pairwise relatively prime orders $a_1, \ldots, a_n$. If $n \geq 3$, then $\Sigma \neq S^3$ and the orders classify $\Sigma = \Sigma(a_1, \ldots, a_n)$ up to diffeomorphism. If $n \leq 2$, then $\Sigma = S^3$. As our notation predicts, the Brieskorn sphere $\Sigma(p,q,r)$ is Seifert fibered with 3 exceptional orbits of orders p, q, and r. In fact the Seifert fibered homology 3-spheres are the links of singularities in Brieskorn complete intersections [**NR**]. Also one can show that for $\Sigma = \Sigma(a_1, \ldots, a_n)$ we have

$$(1.1)\quad \pi_1(\Sigma) = < x_1, ..., x_n, h | h \text{ central}; x_i^{a_i} = h^{-b_i}, i = 1, ..., n; x_1 \cdots x_n = h^{-b_0} > .$$

Here $b_0, b_1, ..., b_n$ are chosen so that

$$a(-b_0 + \sum_{i=1}^{n} \frac{b_i}{a_i}) = 1$$

where $a = a_1 \cdots a_n$. We say that Σ has Seifert invariants $\{b_0; (a_1, b_1), \ldots, (a_n, b_n)\}$. (These are, of course, not unique.)

---

[1]Partially supported by NSF Grant DMS8802412
[2]Partially supported by NSF Grant DMS8703413

iv. Given a knot K in S^3 one can perform a $1/n$, $n \in \mathbf{Z}$, Dehn surgery on K to obtain a homology 3-sphere. The homology 3-spheres $\Sigma(p,q,pqn \pm 1)$ are obtained by $\pm 1/n$ surgery on the (p,q) torus knot; the homology 3-spheres $\Sigma(p,q,r,s)$ with $qr - ps = \pm 1$ are obtained by a ± 1 surgery on the connected sum of the (q,r) and $(-p,s)$ torus knots. It is not known if every irreducible homology 3-sphere can be obtained by a Dehn surgery on some knot in S^3. However, every homology 3-sphere can be obtained as an integral surgery on a link in S^3. Recently, Gordon and Luecke [**GL**] have shown that nontrivial Dehn surgery on a nontrivial knot never yields S^3. Furthermore, any homology 3-sphere obtained by Dehn surgery on a knot is irreducible.

v. Given a knot K in S^3 one can take the n-fold cyclic cover of S^3 branched over K, denoted K_n. This is a homology 3-sphere when $|\prod_{i=1}^{n} \Delta(\omega^i)| = 1$, where $\Delta(t)$ is the Alexander polynomial of K normalized so that there are no negative powers of t and has non-zero constant coefficients and $\omega = e^{\frac{2\pi i}{n}}$. The homology 3-sphere $\Sigma(p,q,r)$ is the r-fold cover of the (p,q) torus knot and every $\Sigma(a_1,\ldots,a_n)$ is the 2-fold cover of S^3 branched over a rational knot (see [**BZ**]). Not every homology 3-sphere is a cyclic branched cover [**My**]. However, every 3-manifold is an irregular branched cover of the figure eight knot [**HLM**].

Oriented homology 3-spheres Σ_1 and Σ_2 are said to be *homology cobordant* if there is an oriented 4-manifold W with $\partial W = \Sigma_1 \coprod -\Sigma_2$ and such that the inclusions $\Sigma_i \to W$ induce isomorphisms in integral homology. Equivalently, Σ_1 and Σ_2 are homology cobordant provided $(-\Sigma_1)\sharp\Sigma_2$ bounds an acyclic 4-manifold. The set of equivalence classes of oriented homology 3-spheres under this relation is denoted Θ_3^H. With the operation of connected sum "#", Θ_3^H is an abelian group, the *homology cobordism group of oriented homology 3-spheres.* The additive inverse in Θ_3^H is obtained by a reverse of orientation.

Until recently the only fact known concerning the group Θ_3^H was the existence of the Kervaire-Milnor-Rochlin homomorphism $\mu : \Theta_3^H \to \mathbf{Z}_2$ defined as $\mu(\Sigma) = \mathrm{sign}(W^4)/8 \pmod 2$ where W is a parallelizable 4-manifold with boundary Σ. The proof that $\mu(\Sigma)$ is independent of the choice of W^4 and depends only on the class of Σ in Θ_3^H utilizes Rochlin's theorem, which states that the signature of a closed spin (*i.e.* almost-parallelizable) 4-manifold is divisible by 16. Since $\Sigma(2,3,5)$ is the E_8-singularity, $\mu(\Sigma(2,3,5)) = 1$ and μ is a surjection.

The group Θ_3^H has a distinguished history in the study of manifolds. Its structure is closely related to the question of whether a topological n-manifold M^n, $n \geq 5$, is a polyhedron. In [**GS**] and [**Mat**] it is shown that M^n is a polyhedron if and only if an obstruction $\tau_M \in H^5(M^n; \ker(\mu : \Theta_3^H \to \mathbf{Z}_2))$ vanishes, and that if $\tau_M = 0$ there are $|H^5(M^n; \ker\mu)|$ triangulations up to concordance. Furthermore, $\tau_M = 0$ for all M if and only if there is a homology 3-sphere Σ with $\mu(\Sigma) = 1$ and such that $\Sigma\sharp\Sigma$ bounds a smooth acyclic 4-manifold. At the time that these papers were written (*circa* 1978) a reasonable conjecture was that $\Theta_3^H = \mathbf{Z}_2$, so that $\ker\mu = 0$. To date, the existence of a homology sphere with the above properties is unknown.

However, in §5 we shall utilize techniques from gauge theory to show that the group Θ_3^H is infinite and, in fact, infinitely generated.

Another importance of Θ_3^H arises in 4-manifold theory. One can study 4-manifolds by splitting them along embedded homology 3-spheres. In §9 we shall give an example of this approach. In the other direction, one can attempt to construct 4-manifolds by studying the bounding properties of homology 3-spheres, for example their image in Θ_3^H. If a homology 3-sphere Σ bounds the 4-manifold U with intersection form I_U, and if $-\Sigma$ bounds V with intersection form I_V, then $X = U \cup V$ has intersection form $I_U \oplus I_V$. Conversely, if the intersection form of a closed 4-manifold X decomposes as $I_1 \oplus I_2$, then there is a homology 3-sphere Σ in X splitting it into two 4-manifolds W_1 and W_2 with intersection forms I_1 and I_2 respectively [**FT**]. This has been useful in constructing exotic 4-manifolds and group actions. For example in [**FS1**] it is shown that $\Sigma(3,5,19)$ bounds a contractible manifold W^4 and that the double of $W^4 \cup_t W^4$ along the free involution $t : \Sigma \to \Sigma$ contained in the S^1-action on Σ is S^4 with a free involution τ (obtained by interchanging the copies of W^4) that is not in any sense smoothly equivalent to the antipodal map. Thus S^4/τ is a smooth homotopy RP^4 that is not s-cobordant to RP^4. Other related constructions are given in [**FS2**].

Prior to the 1980's there were few invariants for homology 3-spheres. One had the Kervaire-Milnor-Rochlin invariant discussed above and the η and ρ invariants introduced by Atiyah-Patodi-Singer [**APS1-3**] and discussed in §6. Another, not so well-used, invariant is the Chern-Simons invariant discussed in §3. As we shall see, it is the Chern-Simons invariant that motivates many of the exciting new insights in 3-manifold topology. Of course the fundamental group also plays an important role in 3-manifold topology. However, its role was not underscored until the introduction of Casson's invariant in 1983.

2. Casson's Invariant

It is natural to study $\pi_1(\Sigma)$ to obtain invariants of a homology 3-sphere Σ. One way to do this is via representation spaces. Consider the compact space $\mathcal{R}(\Sigma)$ of conjugacy classes of representations of $\pi_1(\Sigma)$ into $SU(2)$. "Generically" this is a finite set of points which can be assigned orientations. Casson's invariant, $\lambda(\Sigma)$, is half the count (with signs) of those points corresponding to nontrivial representations. (Of course the construction of this invariant when $\mathcal{R}(\Sigma)$ is not finite is considerably more difficult. See [**AM**] for an exposition.) Casson showed that $\lambda(\Sigma) \equiv \mu(\Sigma) \pmod 2$ and used this new invariant to settle an outstanding problem in 3-manifold topology; namely, showing that if Σ is a homotopy 3-sphere, then $\mu(\Sigma) = 0$.

The natural correspondence

$$\mathcal{R}(\Sigma) = \frac{\text{representations of } \pi_1(\Sigma) \text{ into } SU(2)}{\text{conjugation}} \leftrightarrow \frac{\text{flat SU(2) connections over } \Sigma}{\text{gauge equivalence}}$$

indicates that there should be a differential-geometric approach to the definition of $\lambda(\Sigma)$, which was discovered by Taubes [**T2**]. ("Gauge equivalence" means equivalence under the action of the automorphism group of the (trivial) $SU(2)$ bundle supporting these connections.) First fix a Riemannian metric on Σ. Then from this point of view, one computes $\lambda(\Sigma)$ by counting equivalence classes of nontrivial flat connections with sign given by the parity of the spectral flow of the elliptic operator:

$$D_b : (\Omega^0_\Sigma \oplus \Omega^1_\Sigma) \otimes \mathfrak{su}(2) \to (\Omega^0_\Sigma \oplus \Omega^1_\Sigma) \otimes \mathfrak{su}(2)$$

given by $(\alpha, \beta) \mapsto (d_b^*\beta, d_b\alpha + \star d_b\beta)$ as the connection b varies along a path from the trivial connection θ to the given flat connection a, and d_b denotes the covariant derivative corresponding to the connection b and d_b^* is its formal adjoint. The *spectral flow* is the net number of negative eigenvalues of D_b which become positive as b varies along the path. (Since D_θ has three zero eigenvalues, one must fix a convention for dealing with them.) At a flat connection, a, the kernel of the operator D_a measures the dimension of the Zariski tangent space of $\mathcal{R}(\Sigma)$.

Let $\mathcal{A}_\Sigma$ be the space of all $SU(2)$ connections over Σ, and let $\mathcal{B}_\Sigma$ be the quotient of $\mathcal{A}_\Sigma$ modulo gauge equivalence. An appropriate Sobolev norm on $\mathcal{A}_\Sigma$ turns $\mathcal{B}_\Sigma$ into a Hilbert manifold (with a positive codimensional singular set which meets $\mathcal{R}(\Sigma)$ only in the trivial connection). The tangent space to $\mathcal{A}_\Sigma$ at a point a is $\Omega^1_\Sigma \otimes \mathfrak{su}(2)$, and the normal space to the orbit of a under gauge equivalence may be identified with the solutions of the equation $d_a^*\beta = 0$. Thus, loosely speaking, we may view the map sending a connection a to the Hodge star of its curvature $\star F_a$ as a vector field on $\mathcal{A}_\Sigma$ whose critical set consists of the flat connections. In the next section we shall see that this is actually the gradient vector field of a function on $\mathcal{A}_\Sigma$. The Hessian of this function at a is thus $\star d_a$, and at a critical point it preserves the equation $d_a^*\beta = 0$. It is easy to see that for a nontrivial flat connection a, the kernel of $\star d_a$ on $\{d_a^*\beta = 0\}$ may be identified with the kernel of D_a. Since D_a is self-adjoint, we can add a compact perturbation term so that the corresponding vector field has a zero-dimensional critical set. This explains the statement above that $\mathcal{R}(\Sigma)$ is generically a finite set of points.

3. Chern-Simons Invariants

Let Σ be a homology 3-sphere. Each principal $SU(2)$-bundle P over Σ is trivial, *i.e.* is isomorphic to $\Sigma \times SU(2)$. As we have alluded in the last section, given a trivialization, one can identify the space of connections $\mathcal{A}_\Sigma$ of Sobolev type L^p_k with the space $L^p_k(\Omega^1(\Sigma) \otimes \mathfrak{su}(2))$ of 1-forms on Σ with values in the Lie algebra $\mathfrak{su}(2)$ in such a way that the zero element of $\mathcal{A}_\Sigma$ corresponds to the product connection θ on $\Sigma \times SU(2)$. The gauge group of bundle automorphisms of P can be identified with $\mathcal{G} = L^p_{k+1}(\Sigma, SU(2))$ acting on $\mathcal{A}_\Sigma$ by the nonlinear transformation law

$$g(a) = gag^{-1} + (dg)g^{-1}.$$

We shall assume that $k+1 > 3/p$ so that $\mathcal{G}$ consists of continuous maps. The group $\mathcal{G}$ is not connected; in fact $\pi_0(\mathcal{G}) = \mathbf{Z}$ given by the degree of $g : \Sigma \to SU(2)$. The quotient $\mathcal{B}_\Sigma = \mathcal{A}_\Sigma/\mathcal{G}$ can be considered as an infinite dimensional manifold except near those connections a for which the isotropy group

$$\mathcal{G}_a = \{g \in \mathcal{G} | g(a) = a\}$$

is larger than $\{\pm\text{id}\}$. Such connections are called *reducible*. For example, the trivial connection θ is reducible since its isotropy group consists of all constant maps $g : \Sigma \to SU(2)$. *Irreducible* connections form an open dense set $\mathcal{B}^*_\Sigma$ in $\mathcal{B}_\Sigma$. The set of flat connections is invariant under $\mathcal{G}$.

Given any connection a, we can take a path $\gamma : I = [0,1] \to \mathcal{A}_\Sigma$ from the trivial connection θ to a. This path determines a connection A_γ in the trivial $SU(2)$ bundle over $\Sigma \times I$. Let

$$\text{CS}(\theta, a) = \frac{1}{8\pi^2} \int_{\Sigma\times I} \text{Tr}(F_{A_\gamma} \wedge F_{A_\gamma}).$$

This definition is independent of the choice of path γ because $\mathcal{A}_\Sigma$ is contractible. However, the function $\text{CS}(\theta, \cdot) : \mathcal{A}_\Sigma \to \mathbf{R}$ does depend on the trivialization of P. If θ' is the trivial connection with respect to another trivialization, and γ' is a path in $\mathcal{A}_\Sigma$ from θ' to a, we can glue the connections A_γ and $A_{\gamma'}$ together over $\Sigma \times \{0\}$ and along $\Sigma \times \{1\}$ via a gauge transformation to obtain a connection A in an $SU(2)$-bundle E over $\Sigma \times S^1$ and

$$\text{CS}(\theta, a) - \text{CS}(\theta', a) = \frac{1}{8\pi^2} \int_{\Sigma\times S^1} \text{Tr}(F_A \wedge F_A) = c_2(E)$$

an integer. A similar argument shows that if $g \in \mathcal{G}$ then $\text{CS}(\theta, g(a)) = \deg(g) + \text{CS}(\theta, a)$, so that $\text{CS}(\theta, \cdot)$ descends to a function $\text{CS} : \mathcal{B}_\Sigma \to \mathbf{R}/\mathbf{Z}$, independent of the choice of trivialization. It has an L^2-gradient given by $a \mapsto \star F_a$; hence it is a function on $\mathcal{B}(\Sigma)$ whose critical set is $\mathcal{R}(\Sigma)$. At an $a \in \mathcal{R}(\Sigma)$, the Hessian is $\star d_a$. This $\mathbf{R}/\mathbf{Z}$ invariant can be regarded as a (mod $\mathbf{Z}$) charge of the connection A_γ, for $\text{Tr}(F_{A_\gamma} \wedge F_{A_\gamma})$ is the Chern-Weil integrand.

Chern-Simons invariants were overlooked by low dimensional topologists since it was shown in [**APS1-3**] that the ρ_a invariants discussed in §6 (well-defined as real numbers) were congruent to $\text{CS}(a)$ mod $\mathbf{Z}$. The only utility of $\text{CS}(a)$ appeared to be that it determined the nonintegral part of ρ_a. As it turns out, it is the Chern-Simons functional that plays a central role in the modern understanding of homology 3-spheres. As a simple starting point, noting that $\mathcal{R}(\Sigma)$ is compact, we define

$$\bar{\tau}(\Sigma) = \min\{\text{CS}(\alpha) | \alpha \in \mathcal{R}(\Sigma)\} \in [0, 1).$$

We shall see in §5 that coupled with the techniques of [**FS3**] these invariants are extremely useful.

4. Representations of $\Sigma(a_1,\ldots,a_n)$

A Seifert fibered homology sphere $\Sigma = \Sigma(a_1,\ldots,a_n)$ admits a natural S^1-action whose orbit space is S^2. Orient Σ as the link of an algebraic singularity or equivalently as a Seifert fibration with Seifert invariants $\{b_0;(a_i,b_i), i=1,\ldots,n\}$ as in §1. With this orientation Σ bounds the canonical resolution, a negative definite simply connected smooth 4-manifold. Let $W = W(a_1,\ldots,a_n)$ denote the mapping cylinder of the orbit map. It is a 4-dimensional orbifold with boundary $\partial W = \Sigma$ and W has n singularities whose neighborhoods are cones on the lens spaces $L(a_i,b_i)$ (see [**FS3**]). If we orient W so that its boundary is $-\Sigma$ its intersection form will be positive definite. Let W_0 denote W with open cones around the singularities removed. Then

$$\pi_1(W_0) = \pi_1(\Sigma)/ < h > = T(a_1,...,a_n)$$
$$=< x_1,...,x_n | x_i^{a_i} = 1, i = 1,...,n;\ x_1\cdots x_n = 1 > .$$

When $n = 3$ this is the usual triangle group and in general it is a genus zero Fuchsian group. The element $h \in \pi_1(\Sigma)$ is represented by a principal orbit of the S^1-action. It is central, and for any representation α of $\pi_1(\Sigma)$ into $SU(2)$ we have $\alpha(h) = \pm 1$. Thus α gives rise to a representation of $\pi_1(W_0)$ into $SO(3)$. Conversely, any flat $SO(3)$ bundle over W_0 restricts to one over Σ, and there it lifts to a flat $SU(2)$ bundle since Σ is a homology sphere. Thus $SU(2)$-representations of $\pi_1(\Sigma)$ are in one-to-one correspondence with $SO(3)$-representations of $\pi_1(W_0)$.

Given $\alpha \in \mathcal{R}(\Sigma)$, let $\mathbf{V}_\alpha$ denote the flat real 3-plane bundle over W_0 determined by α. When $\mathbf{V}_\alpha$ is restricted over $L(a_j,b_j) \subset \partial W_0$ it splits as $\mathbb{L}_{\alpha,j} \oplus \mathbb{R}$ where $\mathbb{R}$ is a trivial real line bundle and $\mathbb{L}_{\alpha,j}$ is the flat 2-plane bundle corresponding to the representation $\pi_1(L(a_j,b_j)) \to \mathbb{Z}_{a_j}$ of weight l_j, where $\alpha(x_j)$ is conjugate in $SU(2)$ to $e^{\pi i l_j/a_j}$. (The presentation (1.1) shows that $\alpha(x_j)$ is an a_jth or $2a_j$th root of unity.) The preferred generator of $\pi_1(L(a_j,b_j))$ corresponds to the deck transformation

$$(z,w) \longmapsto (\zeta z, \zeta^{b_j} w)$$

of S^3 where $\zeta = e^{2\pi j/a_j}$. Thus $\mathbb{L}_{\alpha,j}$ is the quotient of $S^3 \times \mathbb{R}^2$ by this $\mathbb{Z}_{a_j}$-action. The bundle $\mathbb{L}_{\alpha,j}$ extends over the cones $cL(a_j,b_j)$ as $(\mathbb{C}^2 \times_{\mathbb{Z}_{a_j}} \mathbb{R}^2) \oplus \mathbb{R}$, an $SO(3)$-V-bundle whose rotation number over the cone point is l_j (with respect to the preferred generator). So $\mathbf{V}_\alpha$ extends to an $SO(3)$ V-bundle over W. In [**FS6**] we determine which (l_1,l_2,l_3) can arise for representations of $\pi_1(W_0)$, $n=3$, thus determining $\mathcal{R}(\Sigma)$ for all Brieskorn spheres Σ.

Here is another way to think about representations $\alpha : \pi_1(\Sigma) \to SU(2)$. After conjugating in $SU(2)$, we may assume that $\alpha(x_1) = e^{\pi i l_1/a_1} \in S^1 \subset SU(2)$. For $j = 1,\ldots,n$ let S_j be the conjugacy class of $e^{\pi i l_j/a_j}$. This is a 2-sphere in $SU(2)$ which contains $\alpha(x_j)$. So $\alpha(x_1x_2)$ lies on the 2-sphere $\alpha(x_1)\cdot S_2$, and generally, $\alpha(x_1\ldots x_{j+1})$ lies on the 2-sphere $\alpha(x_1\ldots x_j)\cdot S_{j+1}$. Finally, since $\alpha(h) = \pm 1$, the presentation (1.1) implies that $\alpha(x_1 \dots x_n) = \pm 1$. Thus α corresponds to a mechanical linkage in $SU(2)$ with ends at $e^{\pi i l_1/a_1}$ and $\alpha(x_1 \dots x_n) = \pm 1$ and with arms corresponding to radii of the spheres $\alpha(x_1, \dots, x_j) \cdot S_{j+1}$. (See Figure 1, where $n = 5$.)

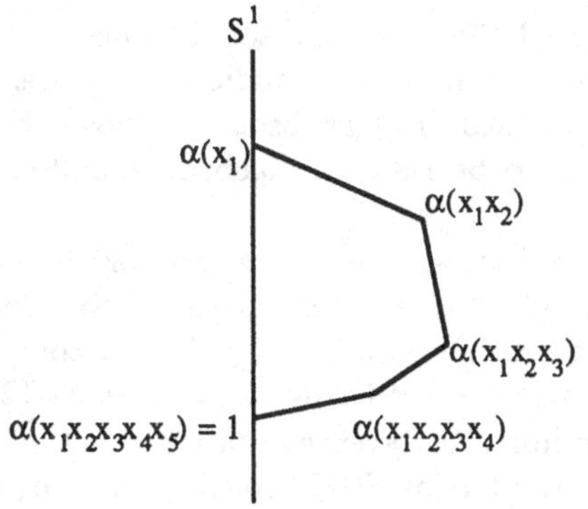

Figure 1

Representations $\pi_1(\Sigma) \to SU(2)$ thus correspond to choices of $(l_1, \dots, l_n)$ such that a linkage from $e^{\pi i l_1/a_1}$ to $\alpha(x_1 \dots x_n) = \pm 1$ exists; numerical criteria for this are given in [**FS6**]. The connected component of any $\alpha \in \mathcal{R}(\Sigma)$ is the corresponding component in the configuration space of mechanical linkages modulo rotations leaving S^1 invariant.

For example, consider a Brieskorn sphere $\Sigma(a_1, a_2, a_3)$. A representation corresponds to a linkage as in Figure 2.

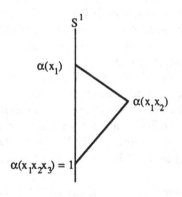

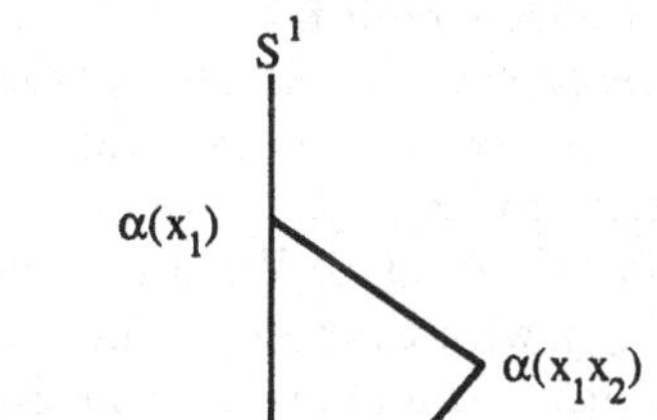

Figure 2

This linkage is rigid modulo rotations leaving S^1 fixed. Thus $\mathcal{R}(\Sigma(a_1, a_2, a_3))$ consists of a finite number of isolated representations. More generally we have:

PROPOSITION 4.1 [**FS6**]. *Let $\Sigma = \Sigma(a_1, ..., a_n)$. If $\alpha : \pi_1(\Sigma) \to SU(2)$ is a representation with $\alpha(a_i) \neq \pm 1$ for $i = 1, ..., m$, and $\alpha(a_i) = \pm 1$ for $i = m+1, ..., n$, then the connected component $\mathcal{R}_\alpha$ of α in the space $\mathcal{R}(\Sigma)$ is a closed manifold of dimension $2m - 6$.*

In [**FS6**] we conjectured that any connected component of $\mathcal{R}(\Sigma)$ has a Morse function with critical points only of even indices. This was proved to be the case in [**KK1**] using the mechanical linkages discussed above. Furthermore, these components were also shown to be rational algebraic manifolds in [**BO**] and Kähler manifolds in [**FuS**].

One can show further that, as the critical set for the Chern-Simons function, $\mathcal{R}(\Sigma)$ is nondegenerate. That is, the Hessian $\star d_a$ of the Chern-Simons function is nondegenerate normal to $\mathcal{R}(\Sigma)$. Our next goal is to compute the Chern-Simons invariant of a representation $\alpha : \pi_1(\Sigma(a_1, \ldots, a_n) \to SU(2)$. Our technique will work with the corresponding $SO(3)$-representation α'. If $\tilde{A}$ is an $SU(2)$-connection which interpolates from the trivial $SU(2)$-connection to α, and if A' is an $SO(3)$-connection interpolating from the trivial $SO(3)$-connection to α', then the Chern-Simons invariant of α is obtained by integrating the Chern form of $\tilde{A}$, *i.e.* $\mathrm{CS}(\alpha) = \int_{\Sigma \times I} c_2(\tilde{A}) \in \mathbf{R}/\mathbf{Z}$; whereas $\mathrm{CS}(\alpha') = \int_{\Sigma\times I} p_1(A')$ is obtained by integrating the ($SO(3)$) Pontryagin form of A'. We have $\mathrm{CS}(\alpha') \in \mathbf{R}/4\mathbf{Z}$ since $p_1(A') = 4c_2(\tilde{A})$ if $\tilde{A}$ is a lift of A' (*c.f.* [**HH**]).

An $SO(2)$ V-vector bundle L over W is classified by the Euler class $e \in H^2(W_0) \cong \mathbf{Z}$ of its restriction over $W_0 = W -$(neighborhood of singular points). Let L_e denote the V-bundle corresponding to the class e times a generator in $H^2(W_0, \mathbf{Z})$, and let B be any connection on L_e which is trivial near ∂W. Then the relative Pontryagin number of L_e is $\frac{e^2}{a} = \int_W p_1(B)$ where $a = a_1 \cdots a_n$. Let A be the $SO(3)$-connection on $\mathbf{V}_\alpha$ over $W \cup (\Sigma \times \mathbf{R}^+) \cong W$ which is built from the flat (V-) connection α' over W and from a connection A' over $\Sigma \times \mathbf{R}^+$ which interpolates from α' to the trivial connection. The rotation numbers l_i of the representation α' depend on choices of generators for the fundamental groups of the lens space links of W. These are then determined by the Seifert invariants of Σ. We shall suppose that Σ has Seifert invariants $\{b_0; (a_1, b_1), \ldots, (a_n, b_n)\}$ with b_0 even. (This can always be arranged.) If one of the a_i's is even, assume it is a_1, and arrange the Seifert invariants so that the b_i, $i \neq 1$, are even. This specifies rotation numbers l_i for α'. Let $e \equiv \sum_{i=1}^n l_i \frac{a}{a_i}$ (mod $2a$), and let L_e be the corresponding $SO(2)$ V-vector bundle. Then stabilize to get an $SO(3)$ V-vector bundle $L_e \oplus \mathbf{R}$ with connection A_e which is trivial over the end $\Sigma \times \mathbf{R}^+$.

Truncate W by removing neighborhoods of the singular points, leaving W_0, and let $\delta W_0 = \partial W_0 \setminus \Sigma$ (a disjoint union of lens spaces). Let E be the $SO(3)$ vector bundle over W_0 carrying the connection A, *i.e.* E is the restriction of $\mathbf{V}_\alpha$. Also let $E_{e,0}$ be the restriction of E_e over W_0, and let $Y_0 = W_0 \cup_{\delta W_0} W_0$.

Over Y_0 we can construct the $SO(3)$ vector bundle $E \cup E$ by gluing by the identity over δW_0, and we obtain the connection "$A \cup A$". The relative Pontryagin number of $E \cup E$ is $\int_{Y_0} p_1(A \cup A) = 0$, since orientations get reversed. The number e is chosen so that it is possible to form the bundle $E \cup E_{e,0}$ over Y_0, gluing by a connection-preserving bundle isomorphism over δW_0. In **[FS6]** it is shown that $w_2(E \cup E) = w_2(E \cup E_{e,0})$. Then, since the connections $A \cup A_{e,0}$ and $A \cup A$ are asymptotically trivial, we have (just as we would for a closed 4-manifold) $\int_{Y_0} p_1(A \cup A) \equiv \int_{Y_0} p_1(A \cup A_{e,0})$ (mod $4\mathbf{Z}$). Thus

$$0 = \int_{W_0} p_1(A) - \int_{W_0} p_1(A_e) \equiv \int_{W_0} p_1(A) - \frac{e^2}{a} \quad (\text{mod } 4\mathbf{Z}).$$

But

$$\int_{W_0} p_1(A) = \int_{W_0} p_1(\alpha') + \int_{\Sigma \times \mathbb{R}^+} p_1(A') = 0 - \mathrm{CS}(\alpha').$$

Hence we have

THEOREM 4.2. *Let $\Sigma = \Sigma(a_1, ..., a_n)$ and let α be a representation of $\pi_1(\Sigma)$ into $SU(2)$ with associated representation α' into $SO(3)$. Then $CS(\alpha') = -\frac{e^2}{a}$ mod $4\mathbf{Z}$ and $CS(\alpha) = -\frac{e^2}{4a}$ mod $\mathbf{Z}$ where $e \equiv \sum_{i=1}^n l_i \frac{a}{a_i}$ (mod $2a$).*

Another very interesting technique for computing Chern-Simons invariants is discussed in **[KK2]**. It applies to homology spheres which are obtained from surgery on a knot.

5. Gauge Theory for $\Sigma(a_1, \ldots, a_n)$

The success of Donaldson's approach to 4-manifolds **[D1-4]** has motivated the use of similar techniques in the study of homology cobordism properties of homology 3-spheres. The approach taken in **[FS3]** is to study the mapping cylinder, W, of the orbit map $\Sigma(a_1, \ldots, a_n) \to S^2$, which appeared prominently in the last section. Suppose, for example, that $\Sigma(a_1, \ldots, a_n)$ is the boundary of a simply connected, positive definite 4-manifold U. If we use the orientation on W which was described in the last section we can form the union $W \cup U$, obtaining a simply connected orbifold X. Even though X is not a manifold, it is a rational homology manifold, and thus has a well-defined rational intersection form, with respect to which it is positive definite. Let $X_0 = W_0 \cup U$. There is a preferred class $\omega \in H_2(X; \mathbf{Z}) \cong H_2(X_0, \partial X_0; \mathbf{Z})$ which is represented by the 2-sphere orbit space in W, and $\omega^2 = \frac{1}{a}$. As in §4, we form the $SO(3)$ V-bundle $L_\omega \oplus \mathbb{R}$ over X, where L_ω is the $SO(2)$ V-bundle whose Euler class in $H^2(X_0)$ is Poincaré dual to ω. In **[FS3]** we studied the moduli space $\mathcal{M}$ of self-dual connections on E_ω. For a fixed metric on X this is the solution space of the equation $F_A = \star F_A$ in $\mathcal{B}_X$. In **[FS3]** we showed that (perhaps after a compact perturbation) $\mathcal{M}$ is a compact manifold except at the single reducible self-dual connection, which has a neighborhood homeomorphic to a cone on a complex projective space. Its formal dimension as computed by the index theorem is

$$R(a_1,\dots,a_n;\omega)=\frac{2}{a}-3+n+\sum_{i=1}^{n}\frac{2}{a_i}\sum_{k=1}^{a_i-1}\cot(\frac{\pi ak}{a_i^2})\cot(\frac{\pi k}{a_i})\sin^2(\frac{\pi k}{a_i}).$$

If this integer is positive, it will actually give the dimension of $\mathcal{M}$. (Note that since $\dim\mathcal{M}=R(a_1,\dots,a_n;\omega)$, the existence of the singular point in $\mathcal{M}$ implies that the dimension $R(a_1,\dots,a_n;\omega)$ is odd. This argument applied to W, considered as an orbifold with a noncompact end, rather than to X works independently of the bounding properties of $\Sigma(a_1,\dots,a_n)$. In fact, it is shown in [**FS3**] that $R(a_1,\dots,a_n;\omega)$ is odd even when it is negative.) If $R(a_1,\dots,a_n;\omega)>0$, this means that the complex projective space which is the link of the unique singularity of $\mathcal{M}$ is null-cobordant. In the case that we have a $\mathbf{C}P^{2m}$ this is an immediate contradiction, for $\mathbf{C}P^{2m}$ has even Euler characteristic and thus cannot bound a compact manifold. In the case of $\mathbf{C}P^{2m+1}$ one can get a similar contradiction using the so-called "basepoint fibration". Actually one has a slightly stronger statement:

THEOREM 5.1 ([**FS3**]). *If $R(a_1,\dots,a_n;\omega)>0$ then $\Sigma(a_1,\dots,a_n)$ cannot bound a smooth positive definite 4-manifold whose first homology has no 2-torsion.*

Easy calculations show that $R(2,3,6k-1)=1$ for all $k>0$.

COROLLARY 5.2 ([**FS3**]). *If $R(a_1,\dots,a_n;\omega)>0$ then $\Sigma(a_1,\dots,a_n)$ has infinite order in the homology cobordism group Θ_3^H.*

PROOF: Suppose a multiple $m\Sigma(a_1,\dots,a_n)$ (connected sum) is trivial in Θ_3^H, then it bounds an acyclic 4-manifold U. Since $-U$ is also acyclic we may assume that $m>0$. By attaching 3-handles to U we obtain a 4-manifold V whose boundary is the disjoint union of m copies of $\Sigma(a_1,\dots,a_n)$, and with $H_1(V)=0=H_2(V)$. Recall that $\Sigma(a_1,\dots,a_n)$ bounds a negative definite simply connected 4-manifold — its canonical resolution, N. Now attach $m-1$ copies of $-N$ to the boundary components of V to obtain a positive definite 4-manifold W with $H_1(W)=0$ and with $\partial W=\Sigma(a_1,\dots,a_n)$. This contradicts Theorem 5.1.

Thus we see that the Brieskorn spheres $\Sigma(2,3,6k-1)$, $k>0$ all have infinite order in Θ_3^H. In particular this is true for the Poincaré homology sphere $\Sigma(2,3,5)$. For other interesting results concerning the moduli spaces of self-dual connections over orbifolds one should read the work of Furuta [**Fu1**],[**Fu2**].

Let us now return to the invariant $\bar{\tau}(\Sigma)$ defined in §3. Here we shall change its definition in order to allow us to use $SO(3)$-representations and the orientation conventions which we have adopted. Hence define

$$\tau(\Sigma)=\min\{-\mathrm{CS}(\alpha')|\alpha':\pi_1(\Sigma)\to SO(3)\}\in[0,4).$$

This may be interpreted as follows. Fix a trivial connection θ so that we may view the Chern-Simons function as integer-valued on $\mathcal{A}_\Sigma$. The application of a gauge transformation g to an $SO(3)$ connection changes its Chern-Simons invariant by $4 \cdot \deg(g)$. So gauge equivalent to a given flat connection there is a unique connection a whose Chern-Simons invariant $\mathrm{CS}(\theta, a)$ calculated with respect to θ lies in the interval $(-4, 0]$. We are minimizing $-\mathrm{CS}(\theta, a)$ over the compact space $\mathcal{R}(\Sigma)$. For Brieskorn spheres either of the two techniques for finding representations described in §4 will determine all of the (finitely many) numbers e that occur in Theorem 4.2, and thus give us a finite procedure for the calculation of $\tau(\Sigma(p,q,r))$. In fact Theorem 4.2 implies that $\tau(\Sigma(p,q,r))$ is the minimum of all $\frac{e^2}{pqr} \in [0,4)$ where e is chosen as in that theorem. Note that $pqr\tau(\Sigma(p,q,r)) \in \mathbf{Z}$.

For example, for p and q relatively prime, $\tau(\Sigma(p,q,pqk-1)) = 1/(pq(pqk-1))$, for $k \geq 1$. One way to see this is to use one of the algorithms presented in §4 to find a representation with associated Euler number $e = 1$. Another way to see this is to consider the orbifold W which is the mapping cylinder of the orbit map $\Sigma(p,q,pqk-1) \to S^2$ and study the moduli space $\mathcal{M}$ of asymptotically trivial self-dual $SO(3)$-connections with Pontryagin charge $\frac{1}{a}$ in the V-bundle $E = L_1 \oplus \mathbb{R}$ over W. (Here, and throughout the rest of the paper, we shall write W to mean $W \cup \Sigma \times \mathbb{R}^+$.) Then $\dim\mathcal{M} = R(p,q,r;\omega) = 1$ (*cf.* **[FS3]**), so that (perhaps after a compact perturbation) $\mathcal{M}$ is a 1-manifold with a single boundary point corresponding to the unique reducible self-dual connection. The component of $\mathcal{M}$ containing the reducible connection must then have a noncompact end. This indicates that a self-dual connection "pops off" the end **[T1**,§10]. That is, there is a sequence of connections in $\mathcal{M}$ which limits to the 'union' of a nontrivial self-dual connection C over $\Sigma \times \mathbb{R}$ which is asymptotically trivial near $+\infty$ and is asymptotically a flat connection α' near $-\infty$ together with a self-dual connection over W which is asymptotic to α'. Since C is self-dual and nontrivial, its charge $\frac{1}{8\pi^2}\int_{\Sigma\times\mathbb{R}} \mathrm{Tr}(F_C \wedge F_C) > 0$. Now for any asymptotically trivial connection A in E over W we have $\frac{1}{8\pi^2}\int_W \mathrm{Tr}(F_A \wedge F_A) = e^2/(pq(pqk-1)) = 1/(pq(pqk-1))$. Then $0 < -\mathrm{CS}(\theta, \alpha') = \frac{1}{8\pi^2}\int_{\Sigma\times\mathbb{R}} \mathrm{Tr}(F_C \wedge F_C) \leq 1/(pq(pqk-1))$, so that $-\mathrm{CS}(\theta,\alpha') = 1/(pq(pqk-1))$.

Our result of Corollary 5.2 was expanded by Furuta **[Fu2]**, who showed that Θ_3^H is infinitely generated. Below is the proof of Furuta's result which we gave in **[FS6]** using the τ-invariant. (Furuta's proof is similar.)

THEOREM 5.3 (FURUTA). *Let p and q be pairwise relatively prime integers. The collection of homology 3-spheres $\{\Sigma(p,q,pqk-1) | k \geq 1\}$ are linearly independent over $\mathbf{Z}$ in Θ_3^H.*

PROOF: Fix $k \geq 2$ and suppose that $\Sigma(p,q,pqk-1) = \sum_{j=1}^k n_j\Sigma(p,q,pqj-1)$ in Θ_3^H, where $n_j \in \mathbf{Z}$ and $n_k \leq 0$. Then there is a cobordism Y between $\Sigma(p,q,pqk-1)$ and the disjoint union $\coprod_{j=1}^k n_j\Sigma(p,q,pqj-1)$ with Y having the cohomology of a $(1+\sum|n_j|)$-punctured 4-sphere. Now cap off the $-n_k$ copies of $-\Sigma(p,q,pqk-1)$ by adjoining to Y the positive definite manifolds $-N_k$ bounded by $-\Sigma(p,q,pqk-1)$, where N_k denotes the canonical resolution. Let V be the resulting positive definite 4-manifold, and, as usual, let W denote the mapping cylinder of the orbit map $\Sigma(p,q,pqk-1) \to S^2$. Finally, let $X = W \cup_{\Sigma(p,q,pqk-1)} V$ and consider the $SO(3)$ V-bundle $E = L \oplus \mathbf{R}$ over the positive definite orbifold X, where the Euler class of L comes from the dual of ω. For any asymptotically trivial connection A in E over X we have $\frac{1}{8\pi^2}\int_X \mathrm{Tr}(F_A \wedge F_A) = 1/(pq(pqk-1))$. The moduli space $\mathcal{M}$ of asymptotically trivial self-dual connections in E has dimension $R(p,q,pqk-1;1) = 1$, so that (perhaps after a compact perturbation) there is a component of $\mathcal{M}$ which is an arc with one endpoint corresponding to the reducible self-dual connection. As in the argument above, the noncompactness of $\mathcal{M}$ implies that there is a nontrivial self-dual connection C over $Y = \pm\Sigma(p,q,pqj-1) \times \mathbf{R}$, for some $j < k$, which is asymptotically trivial near $+\infty$ and is asymptotically a flat connection α' near $-\infty$. Also, there is a self-dual connection B over X which is asymptotically flat at the ends of X; so $\frac{1}{8\pi^2}\int_X \mathrm{Tr}(F_B \wedge F_B) \geq 0$. However, $\frac{1}{pq(pqk-1)} = \frac{1}{8\pi^2}\int_X \mathrm{Tr}(F_A \wedge F_A) = \frac{1}{8\pi^2}\int_Y \mathrm{Tr}(F_C \wedge F_C) + \frac{1}{8\pi^2}\int_{\hat{X}} Tr(F_B \wedge F_B) \geq \tau(\Sigma(p,q,pqj-1)) + \frac{1}{8\pi^2}\int_X \mathrm{Tr}(F_B \wedge F_B) \geq \frac{1}{pq(pqj-1)}$, a contradiction.

Other non-cobordism relationships can be detected by the explicit computations of $\tau(\Sigma(p,q,r))$. For example, $\tau(\Sigma(2,3,7)) = 25/42$ and $\tau(-\Sigma(2,3,7)) = 4 - 121/42 = 47/42$. Thus, the proof of Theorem 5.3 shows that $\Sigma(2,3,5)$ is not a multiple of $\Sigma(2,3,7)$ in Θ_3^H. Further uses of these τ invariants will be given in §7.

6. η and ρ-Invariants

Atiyah, Patodi, and Singer introduced in [**APS1-3**] a real-valued invariant for flat connections in trivialized bundles over odd-dimensional manifolds. These invariants arose from their study of index theorems for manifolds with boundary. In this section we shall describe these invariants and show how they have been used in low dimensional topology.

Let B be a self-adjoint elliptic operator on a compact manifold M. The eigenvalues λ of B are real and discrete. Atiyah, Patodi, and Singer [**APS1**] define the function

$$\eta(s) = \sum_{\lambda \neq 0} (\mathrm{sign}\lambda)|\lambda|^{-s}.$$

In [**APS1**] it is shown that $\eta(s)$ has a finite value at $s = 0$. Note that for a finite-dimensional operator B, the η-function evaluated at 0, $\eta(0)$, counts the difference between the number of positive and negative eigenvalues of B.

Now let Σ be a 3-manifold with a flat connection a in a trivialized $U(n)$-bundle over Σ. Let Ω^p_{ad} denote the space $L^p_k(\Omega(\Sigma) \otimes \mathfrak{u}_n)$, and consider the self-adjoint elliptic operator

$$B_a = \star d_a - d_a \star : \Omega^0_{\mathrm{ad}} \oplus \Omega^2_{\mathrm{ad}} \to \Omega^0_{\mathrm{ad}} \oplus \Omega^2_{\mathrm{ad}}.$$

Since the operator B_a involves the $\star$ operator, it depends upon a Riemannian metric on Σ and changes sign when the orientation is changed. The importance of the η-invariant to low-dimensional topology is due to the role that the η-invariant of B_a plays in the computation of twisted signatures of 4-manifolds with boundary.

Let X be a 4-manifold with boundary Σ and let $\beta : \pi_1(X) \to U(n)$ be a unitary representation of its fundamental group. This defines a flat vector bundle V_β over X, or equivalently a local coefficient system. The cohomology groups $H^*(X; V_\beta)$ and $H^*(X, \Sigma; V_\beta)$ have a natural pairing into $\mathbf{C}$ given by cup product, the inner product on V_β, and the evaluation of the top cycle of X mod Σ. This induces a nondegenerate form on $\hat{H}^*(X; V_\beta)$, the image of the relative cohomology in the absolute cohomology. On $\hat{H}^2(X; V_\beta)$ this form is Hermitian and the signature of this form is denoted by $\operatorname{sign}_\beta(X)$.

Assume that the metric on Σ is extended to a metric on X which is a product near Σ and that the restriction of β to Σ is a. It is shown in [**APS3**] that

$$\operatorname{sign}_\beta(X) = n \int_X L(p_1) - \eta_a(0)$$

where $L(p_1)$ is the Hirzebruch L-polynomial of X. Thus $\eta_a(0)$ may be viewed as a signature defect. However, it depends on the choice of a Riemannian metric on Σ. To resolve this dependency, Atiyah, Patodi, and Singer define the reduced η-function by

$$\rho_a(s) = \eta_a(s) - \eta_\theta(s)$$

where θ is the trivial $U(n)$ connection. An application of the above signature formula to $\Sigma \times I$ shows that $\rho_a(0)$ is independent of the Riemannian metric on Σ and is a diffeomorphism invariant of Σ and a. It is denoted by $\rho_a(\Sigma)$. Furthermore, if $\Sigma = \partial X$ with a extending to a flat unitary connection β over X, then

$$(6.1) \qquad \rho_a(\Sigma) = n \cdot \operatorname{sign}(X) - \operatorname{sign}_\beta(X)$$

The ρ_a-invariants were made important in low dimensional topology via the Casson-Gordon invariants for knots [**CG**]. We shall next discuss these invariants and indicate how gauge theory can enter into their considerations.

A smooth knot K in S^3 is called *slice* if there is a smooth 2-disk $D \subset B^4$ with $K = \partial D$. Oriented knots K_0, K_1 are *cobordant* if there is a smoothly embedded oriented annulus C in $S^3 \times I$ with $\partial C = K_1 \times \{1\} - K_0 \times \{0\}$. Addition of cobordism classes of oriented knots is given by connected sum, resulting in the knot cobordism group Θ_1^3 in which slice knots represent the trivial element.

Suppose the knot $K \subset S^3$ bounds an oriented surface $F \subset S^3$. Thicken F to an embedding $F \times I \subset S^3$. Then given $x, y \in H_1(F)$, let $\lambda(x, y) =$ linking number of $x \times 0$ and $y \times 1$. This defines the *Seifert form*, a bilinear form $\lambda : H_1(F) \times H_1(F) \to \mathbf{Z}$, such that $\lambda(x, y) - \lambda(y, x)$ is the intersection number of x and y. It is called *null-cobordant* if it vanishes on a subgroup of $H_1(F)$ of dimension $\frac{1}{2} \dim H_1(F))$. J.

Levine has proved (in all dimensions) that if K is slice, then any Seifert form for K is null-cobordant. Such a knot is called *algebraically slice*. Furthermore, in higher (odd) dimensions the analogous condition is necessary and sufficient for K to be slice.

A knot K is called a *ribbon* knot if it bounds an immersed disc (ribbon) in S^3 each of whose singularities consists of two sheets intersecting in an arc which is interior to one of the sheets. Ribbon knots are slice, for one can push the interior of the ribbon into B^4 and then deform slightly a neighborhood of each arc. An old problem of Fox, which is still unresolved, asks whether every slice knot is ribbon. In [**CG**], Casson and Gordon present an invariant for detecting when an algebraically slice knot fails to be ribbon and a modified version of this invariant that detects when it fails to be slice. We discuss these ribbon invariants and indicate how gauge theory makes them slice invariants.

Let L be the double branched covering of the knot K in S^3. If K is slice, then the double covering of B^4 branched over the slicing disc is a 4-manifold W with $\tilde{H}_*(W;\mathbb{Q}) = 0$, and if the image of $H_1(L)$ in $H_1(W)$ has order m, then $|H_1(L)| = m^2$. Furthermore, if the slicing disc is obtained by deforming a ribbon, then $\pi_1(L)$ surjects onto $\pi_1(W)$.

Let $\chi : H_1(L) \to U(1)$ be a representation with image the m-th roots of unity, $\mathbb{C}_m$. Then χ is induced by a map $L \to K(\mathbb{C}_m, 1)$. Since $\Omega^3 K(\mathbb{C}_m, 1)$ is torsion, rL bounds a compact 4-manifold W over $K(\mathbb{C}_m, 1)$, for some $r > 0$. Thus the representation χ factors through $H_1(W)$ and induces a flat $U(1)$ bundle V_χ over W. Then the Casson-Gordon invariant $\sigma(K,\chi)$ is defined by

$$\sigma(K,\chi) = \frac{1}{r}(\mathrm{sign}(W) - \mathrm{sign}_\chi(W)). \tag{6.2}$$

Hence it follows from (6.1) that $\sigma(K,\chi) = \rho_\chi(L)$.

Casson-Gordon invariants were originally applied to those knots K in S^3 whose double branched covering is a lens space $L = L(p,q)$. This contains the collection of 2-bridge knots. If a knot K in this class is ribbon, then $\pi_1(L) = \mathbb{Z}_{m^2}$ and $\pi_1(W) = \mathbb{Z}_m$. Using W to compute $\sigma(K,\chi)$, one gets that $H_0(W;V_\chi) = 0$ if χ is non-trivial, and since the m-fold covering of W is simply-connected, $H_1(W;V_\chi) = H_3(W;V_\chi) = 0$ (this is where the ribbon assumption is used). Also the Euler characteristic of W with V_χ coefficients is the same as its Euler characteristic with $\mathbb{Z}$ coefficients, namely 1; so that $H_2(W;V_\chi)$ has dimension 1. Thus $\sigma(K,\chi) = \pm 1$. Calculations then show that there are algebraically slice 2-bridge knots K for which there is a χ with $\sigma(K,\chi) \neq \pm 1$; hence they are not ribbon. Casson-Gordon then proceed in [**CG**] to refine these invariants. By considering infinite cyclic coverings they define invariants of (K,χ) that show that these K are also not slice in the case that m is a prime power order.

At this point gauge theory can enter the picture to show that in fact if K is slice, then $\sigma(K,\chi) = \pm 1$ for any m. This was done in [**FS4**] as follows. Consider the orbifold X =cone$(L) \cup_L W$. The bundle V_χ over W extends as a flat V-bundle over X. The flat connection determined by V_χ is both self-dual and anti-self-dual. Now consider the moduli spaces $\mathcal{M}_+$ and $\mathcal{M}_-$ of self-dual and anti-self-dual connections in the $SO(3)$ V-bundle $E_\chi = V_\chi \oplus \mathbb{R}$. Each contains the reducible flat connection determined by χ and is compact (since these flat connections are representations into a compact group). Using the index theorem one computes that the formal dimension of $\mathcal{M}_\pm$ is $-2 \pm \sigma(K,\chi)$. As in the proof of Theorem 5.1, this is an odd integer. If $\dim\mathcal{M}_\pm > 0$, then a perturbation of the equations has a moduli space that is a compact manifold of dimension $\dim\mathcal{M}_\pm$ with an odd number of singularities of the form a cone on a complex projective space, and as in (5.1), an odd number of complex projective spaces cannot bound in $\mathcal{B}$. Thus the formal dimensions of both $\mathcal{M}_\pm$, are negative, and so $\sigma(K,\chi) = \pm 1$.

This program was extended by G. Matic in **[Ma]** and independently D. Ruberman in **[R]**. They show that if L is a rational homology sphere which has an integral homology sphere as a finite cover, and if L bounds a rationally acyclic 4-manifold W which has a character χ with image $\mathbb{C}_m$, then the same conclusion holds. The main new ingredient is the gauge theory for manifolds with ends developed by C. Taubes in **[T1]**. Rather than coning off the boundary, the idea is to add an open collar to W^4 and consider self-dual and anti-self-dual connections that are asymptotically flat. With the results of **[T1]** in place, the proof is formally the same.

Note that the only ρ_α-invariants that were used in the above set-up were those associated with representations $\alpha : \pi_1(L) \to U(2)$ that factored through a finite group. In general, there are many irreducible representations. What role do these invariants play in the study of Θ^3_1? We should keep this question in mind during the next few sections, where these irreducible representations play an essential role.

7. An Integer Instanton Invariant

Let Σ be a homology 3-sphere, and let $\alpha' : \pi_1(\Sigma) \to SO(3)$ be an irreducible (i.e. non-trivial) representation, which corresponds to a flat connection in the trivial bundle over Σ. Fix a trivialization of the $SO(3)$-bundle over Σ; this gives us a fixed trivial connection θ. The Chern-Simons invariant $\mathrm{CS}(\theta,\alpha')$ of α' is the Pontryagin charge of a connection on the trivial bundle over $\Sigma \times \mathbb{R}$ that tends asymptotically to α' near $+\infty$ and to θ near $-\infty$. (Recall from §3 that this calculation takes place in $\mathcal{A}_\Sigma$, not in $\mathcal{B}_\Sigma$.) Let $\alpha'(\Sigma)$ denote that connection gauge equivalent to α' with $-\mathrm{CS}(\theta,\alpha'(\Sigma)) \in [0,4)$.

We can associate an integer to the representation α' as follows. Let $\mathcal{M}(\alpha'(\Sigma),\theta)$ denote the moduli space of self-dual connections over $\Sigma\times\mathbb{R}$ whose Pontryagin charge is $-\mathrm{CS}(\theta,\alpha'(\Sigma))$ and that are asymptotically $\alpha'(\Sigma)$ near $-\infty$ and θ near $+\infty$. Let $\dim\mathcal{M}(\alpha'(\Sigma),\theta)$ denote its virtual dimension as predicted by the index theorem and define

$$I(\alpha') = \dim\mathcal{M}(\alpha'(\Sigma),\theta) + h_{\alpha'}$$

where $h_{\alpha'}$ is the sum of the dimensions of $H^i(\Sigma; \mathbf{V}_{\alpha'})$, $i = 0, 1$. This definition does not depend on our original trivialization θ, for if θ is changed by a gauge equivalence, then $\alpha'(\Sigma)$ will change by the same gauge equivalence. When considered with appropriate boundary conditions (*c.f.* [**APS1**],[**T1**]), the self-duality operator (whose index gives the formal dimension of the above moduli spaces) restricts to the boundary of a smooth 4-manifold as a self-adjoint elliptic operator. Let $\eta(s)$ denote its η-function. The Atiyah-Patodi-Singer index theorem can be used to compute $I(\alpha')$:

$$I(\alpha') = \int_{\Sigma\times\mathbb{R}} \hat{A}(\Sigma\times\mathbb{R})ch(V_-)ch(\mathfrak{g}) - \frac{1}{2}(h_\theta + \eta_\theta(0)) + \frac{1}{2}(h_{\alpha'(\Sigma)} + \eta_{\alpha'(\Sigma)}(0))$$

where the forms $\hat{A}(\Sigma\times\mathbb{R})$ and $ch(V_-)$ are computed from the Riemannian connection on $\Sigma\times\mathbb{R}$ (choose a product metric, say) and $\mathfrak{g}$ is the $SO(3)$ bundle over $\Sigma\times\mathbb{R}$ carrying the connection $\alpha'(\Sigma)$. Recalling that $\rho_{\alpha'} = \eta_{\alpha'}(0) - \eta_\theta(0)$ and noting that $h_\theta = 3$ we have

$$\text{(7.1)} \qquad I(\alpha') = \int_{\Sigma\times\mathbb{R}} \hat{A}(\Sigma\times\mathbb{R})ch(V_-)ch(\mathfrak{g}) - \frac{3}{2} + \frac{h_{a_\alpha(\Sigma)}}{2} + \frac{\rho_{a_\alpha(\Sigma)}}{2}.$$

The integral term of (7.1) is

$$\int_{\Sigma\times\mathbb{R}} \hat{A}(\Sigma\times\mathbb{R})ch(V_-)ch(\mathfrak{g}) = 2\int_{\Sigma\times\mathbb{R}} p_1(\mathfrak{g}) + \frac{3}{2}\int_{\Sigma\times\mathbb{R}}(\mathcal{L} - \mathcal{E})$$

where $\mathcal{L}$ and $\mathcal{E}$ are the L-polynomial and the Euler form of $\Sigma\times\mathbb{R}$. Since $\mathfrak{g}$ is our $SO(3)$-bundle over $\Sigma\times\mathbb{R}$ with connection $\alpha'(\Sigma)$, we have $\int_{\Sigma\times\mathbb{R}} p_1(\mathfrak{g}) = -\mathrm{CS}(\theta, \alpha'(\Sigma))$. We then get as in [APS1,§4] that the integral term is

$$\text{(7.2)} \quad -2\mathrm{CS}(\theta,\alpha'(\Sigma)) - \frac{3}{2}[\chi(\Sigma\times\mathbb{R}) - \sigma(\Sigma\times\mathbb{R})] + \frac{3}{2}(\eta_\theta(0) - \eta_\theta(0)) = -2\mathrm{CS}(\theta, \alpha'(\Sigma)).$$

Combining (7.1) and (7.2) we have

$$\text{(7.3)} \qquad I(\alpha') = -2\mathrm{CS}(\theta, \alpha'(\Sigma)) - \frac{3}{2} + \frac{h_{\alpha'(\Sigma)}}{2} + \frac{\rho_{\alpha'(\Sigma)}}{2} \in \mathbb{Z}.$$

We now have a relationship between the Chern-Simons invariants introduced in §3 and the ρ_α-invariants of the self-duality operator, namely

$$4\mathrm{CS}(\alpha') \equiv \rho_{\alpha'} + h_{\alpha'} - 3 \mod 2\mathbb{Z}.$$

Like the Chern-Simons invariant, the integer invariant $I(\alpha')$ is useful in detecting when homology 3-spheres are not homology cobordant.

In [**FS6**] we computed $I(\alpha')$ for $\alpha' \in \mathcal{R}(\Sigma(p,q,r))$. We now indicate the key ideas behind the computation. As in §5 we have the integer

$$(7.4)\quad R(a_1,\ldots,a_n;e\cdot\omega)=\frac{2}{a}-3+m+\sum_{i=1}^{m}\frac{2}{a_i}\sum_{k=1}^{a_i-1}\cot\left(\frac{\pi ak}{a_i^2}\right)\cot\left(\frac{\pi k}{a_i}\right)\sin^2\left(\frac{\pi ek}{a_i}\right)$$

($a = a_1,\ldots,a_n$), which is the virtual dimension of the moduli space of asymptotically trivial self-dual connections in the $SO(3)$ V-bundle $L_{e\cdot\omega}\oplus\mathbb{R}$ over W. It follows from Proposition 4.1 that $h_{\alpha'}=0$ for each nontrivial representation of a Brieskorn sphere. Combining (7.4) with the choices of e that arise from representations of $\pi_1(\Sigma(p,q,r))$ (see §4) and Theorem 4.2 yields the computation of $I(\alpha')$; for $-\mathrm{CS}(\theta,\alpha'(\Sigma))=\frac{e^2}{pqr}-4k\in[0,4)$. Then $I(\alpha')=R(p,q,r;e)-8k$.

THEOREM 7.5 (*c.f.*[**FS6**]). *Suppose $R(a_1,\ldots,a_n;\omega)\geq 1$. If $\Sigma(a_1,\ldots,a_n)$ is homology cobordant to a homology 3-sphere Σ, then*
  (1) *$\tau(\Sigma)\leq\tau(\Sigma(a_1,\ldots,a_n))=\frac{1}{a}$, and*
  (2) *$1\leq I(\alpha')\leq R(a_1,\ldots,a_n;\omega)$ for some representation $\alpha'\in\mathcal{R}(\Sigma)$ with $0\leq -CS(\alpha')\leq\frac{1}{a}$.*

Furthermore, $\Sigma(a_1,\ldots,a_n)$ is not homology cobordant via a simply-connected homology cobordism to any other homology sphere.

PROOF: Let X be the union of the mapping cylinder W of $\Sigma(a_1,\ldots,a_n)$ and the homology cobordism, and consider the $SO(3)$ V-bundle $E=L\oplus\mathbb{R}$ over x, where L is the $SO(2)$ V-bundle whose Euler class comes from the dual of the preferred class $\omega\in H_2(W;\mathbb{Z})$. Let $\mathcal{M}$ denote the moduli space of asymptotically trivial self-dual connections on E with Pontryagin charge $\frac{1}{a}$. We now apply the ideas of the proof of Theorem 5.3 to $\mathcal{M}$.

The moduli space $\mathcal{M}$ contains a single reducible connection, and so is nonempty. Thus $\mathcal{M}$ is a manifold of odd dimension $R(a_1,\ldots,a_n;\omega)\geq 1$, and has one singularity, a cone on a complex projective space. As in (5.1) this complex projective space cannot be null-cobordant inside $\mathcal{B}_X$. Therefore $\mathcal{M}$ has an 'asymptotic end' as in (5.3). Since the smallest Pontryagin number of a bundle on the suspension of a lens space $L(a_i,b_i)$ which admits a self-dual (V-) connection is $\frac{4}{a_i}>\frac{1}{a}$, no instanton can pop off at a cone point. This means that there is a nontrivial flat $SO(3)$ connection α' on Σ and a sequence of connections in $\mathcal{M}$ which limits to the 'union' of self-dual connections B over X and C over $\Sigma\times\mathbb{R}$, where B tends asymptotically to α' and C is asymptotic to α' at $-\infty$ and to θ at $+\infty$. The Pontryagin charge of C is $-\mathrm{CS}(\theta,\alpha')$; hence $0\leq-\mathrm{CS}(\theta,\alpha')\leq\frac{1}{a}$. Notice that this also means that $\alpha'(\Sigma)=\alpha'$. Now if we apply the very same argument to W rather than to X we will find a flat connection β' on $\Sigma(a_1,\ldots,a_n)$ with $0\leq-\mathrm{CS}(\theta,\beta')\leq\frac{1}{a}$. Since the Chern-Simons invariants of flat connections on $\Sigma(a_1,\ldots,a_n)$ are all of the form $\frac{e^2}{a}$, this proves the assertion (1).

To prove assertion (2), note that by the translational invariance of the self-duality equation on $\Sigma \times \mathbb{R}$ we get $1 \leq \mathcal{M}_\Sigma(\alpha', \theta)$ for the component of the moduli space containing the connection C. Furthermore, if $\mathcal{M}_X(\alpha')$ is the component of the moduli space containing B, then by the index theorem of [**APS1**]

$$R(a_1, \ldots, a_n; \omega) = \dim\mathcal{M} = \dim\mathcal{M}_X(\alpha') + \dim\mathcal{M}_\Sigma(\alpha', \theta) + h_{\alpha'}$$
$$\geq \dim\mathcal{M}_\Sigma(\alpha', \theta) + h_{\alpha'} = I(\alpha') \geq 1.$$

The last statement follows as in Proposition 1.7 of [**T1**]. Let U be a simply-connected homology cobordism from $\Sigma(a_1, \ldots, a_n)$ to Σ, and let V be the simply-connected homology cobordism from $\Sigma(a_1, \ldots, a_n)$ to itself obtained by doubling U along Σ. We obtain a reducible (asymptotically trivial) self-dual connection on the bundle $E = L_e \oplus \mathbb{R}$ over the union $\tilde{X}$ of the mapping cylinder W of $\Sigma(a_1, \ldots, a_n)$ with infinitely many copies of V adjoined. Let $\mathcal{M}$ be the moduli space of asymptotically trivial self-dual connections on E. Now, since X has a simply-connected end, $\mathcal{M}$ is compact ([**T1**]). As above, since $\dim \mathcal{M} = R(a_1, \ldots, a_n) > 0$, we can cut down to obtain a compact moduli space with one end point, a contradiction.

For example, $\tau(2,7,15) = \tau(2,3,35)$ and $R(2,3,35;\omega) = 1$. However, for the unique α in $\mathcal{R}(\Sigma(2,7,15))$ with $\mathrm{CS}(\alpha) = \tau(\Sigma(2,7,15))$, we have $I(\alpha) = -5$; so that $\Sigma(2,7,15)$ is not homology cobordant to $\Sigma(2,3,35)$.

8. Instanton Homology and an Extension

We now return to our discussion in §2 of Casson's invariant and Taubes' interpretation in terms of gauge theory. In the situation where all nontrivial representations of $\pi_1(\Sigma)$ in $SU(2)$ are nondegenerate, recall that $\lambda(\Sigma) = \frac{1}{2}\sum_{\alpha \in \mathcal{R}^*(\Sigma)}(-1)^{\mathrm{SF}(\alpha)}$ where $\mathrm{SF}(\alpha)$ is the spectral flow of the operator D_b of §2 as b varies from θ to α. Equivalently we can restate this for $SO(3)$ representations with the orientation conventions we have been using, *i.e.* $\lambda(\Sigma) = \frac{1}{2}\sum_{\alpha' \in \mathcal{R}^*(\Sigma)}(-1)^{\mathrm{SF}(\alpha', \theta)}$ where $\mathrm{SF}(\alpha', \theta)$ is the spectral flow of D_b (defined now for $SO(3)$ connections) as b varies from α' to the trivial $SO(3)$ connection. Floer interpreted this sum as an Euler characteristic. The idea is as follows.

If a flat $SO(3)$ connection is changed by a gauge transformation g, its spectral flow to a fixed trivial connection changes by $8 \cdot \deg(g)$. Similarly, if the choice of trivial connection is changed, the spectral flow again changes by a multiple of 8. This means that $\mathrm{SF}(\alpha', \theta)$ is well-defined on $\mathcal{R}^*(\Sigma)$ as an integer mod 8. Floer defines chain groups $R_n(\Sigma)$ indexed by $\mathbb{Z}_8$ as

$$R_n(\Sigma) = \mathbb{Z}\langle \alpha' \in \mathcal{R}^*(\Sigma) | \mathrm{SF}(\alpha', \theta) \equiv n \pmod 8 \rangle$$

(so $\lambda(\Sigma) = \frac{1}{2}\sum_{n=0}^{7}(-1)^n \mathrm{rank}(R_n(\Sigma))$). The boundary operator is given by

$$\partial\alpha' = \sum \#\mathcal{M}^1(\alpha', \beta') \cdot \beta'$$

where $\mathcal{M}^1(\alpha',\beta')$ is the union of 1-dimensional components of the moduli space of self-dual connections on $\Sigma\times\mathbb{R}$ which tend asymptotically to α' as $t\to-\infty$ and to β' as $t\to+\infty$, and "#"denotes a count with signs. Floer [**F**] shows that this defines a chain complex. Its homology is the instanton homology $I_*(\Sigma)$ graded by $\mathbb{Z}_8$, and again can be defined even when there are nondegenerate representations of $\pi_1(\Sigma)$. Since flat connections are the critical points of the Chern-Simons function, and since it can be shown that the gradient trajectories of the Chern-Simons function are exactly the finite-action self-dual connections on $\Sigma\times\mathbb{R}$, this theory becomes quite analogous to Witten's calculation of the homology of a manifold from a Morse function [**W**]. Instanton homology is a very important new invariant and deserves a much more complete description than we have given here. We refer the reader to the original source [**F**] and to the excellent survey papers [**At**] and [**Br**].

In [**FS8**] we show how to extend this theory to one with integer grading. For simplicity continue the assumption that all nontrivial ($SO(3)$) representations of $\pi_1(\Sigma)$ are nondegenerate, and also assume that there is no nontrivial representation with Chern-Simons invariant congruent to 0 mod 4. (These restrictions are satisfied by Brieskorn spheres and in general can be removed.) As in the previous section, we fix a trivial $SO(3)$ connection θ over Σ.

For any integer m, let $\mathcal{R}_m(\Sigma)$ be the free abelian group generated by the $\alpha'\in\mathcal{R}^*(\Sigma)$ with $I(\alpha')=m$. (Since we are assuming that all these representations are nondegenerate, $h_{\alpha'}=0$ here.) For $n\in\mathbb{Z}_8$ and $s\in\mathbb{Z}$ with $s\equiv n \pmod 8$, define the free abelian groups

$$F_sR_n(\Sigma)=\sum_{j\geq 0}\mathcal{R}_{s+8j}(\Sigma).$$

Then

$$\cdots\subset F_{s+8}R_n(\Sigma)\subset F_sR_n(\Sigma)\subset F_{s-8}R_n(\Sigma)\subset\cdots\subset R_n(\Sigma)$$

is a finite length decreasing filtration of $R_n(\Sigma)$. Furthermore, it can be shown (using the fact that the Chern-Simons functional is non-increasing along gradient trajectories) that Floer's boundary operator $\partial: F_sR_n(\Sigma)\to F_{s-1}R_{n-1}(\Sigma)$ preserves the filtration. Thus Floer's $\mathbb{Z}_8$-graded complex $(R_*(\Sigma),\partial)$ has a decreasing bounded filtration $(F_sR_*(\Sigma),\partial)$. For $n\in\mathbb{Z}_8$ and $s\equiv n \pmod 8$ let $\mathcal{I}_{s,n}(\Sigma)$ denote the homology of the complex

$$\cdots\xrightarrow{\partial}F_{s+1}R_{n+1}(\Sigma)\xrightarrow{\partial}F_sR_n(\Sigma)\xrightarrow{\partial}F_{s-1}R_{n-1}(\Sigma)\xrightarrow{\partial}\cdots.$$

We then have a bounded filtration on $I_n(\Sigma)$ defined by

$$F_sI_n(\Sigma)=\mathrm{im}[\mathcal{I}_{s,n}(\Sigma)\to I_n(\Sigma)]$$

with

$$\cdots\subset F_{s+8}I_n(\Sigma)\subset F_sI_n(\Sigma)\subset F_{s-8}I_n(\Sigma)\subset\cdots\subset I_n(\Sigma).$$

As usual, there is a homology spectral sequence:

THEOREM 8.1 [**FS8**]. *There is a convergent E^1-spectral sequence $(E^r_{s,n}(\Sigma), d^r)$ such that for $n \in \mathbb{Z}_8$ and $s \in \mathbb{Z}$ with $s \equiv n \pmod 8$ we have*

$$E^\infty_{s,n}(\Sigma) \cong F_s I_n(\Sigma)/F_{s+8} I_n(\Sigma)$$

Furthermore, the groups $E^r_{s,n}(\Sigma)$ are topological invariants.

In particular for $s \in \mathbb{Z}$ the groups $\tilde{I}_s(\Sigma) = E^1_{s,n}(\Sigma)$ give an integer-graded version of instanton homology. As in §7 this theory is easily seen to be independent of our choice of bundle-trivialization. See [**FS8**] for more details.

In [**FS6**] we gave an algorithm for computing $I_*(\Sigma(p,q,r))$. We list some examples. The groups I_i are free over $\mathbb{Z}$ and vanish for odd i, so we denote the instanton homology $I_*(\Sigma(p,q,r))$ of $\Sigma(p,q,r)$ as an ordered 4-tuple (f_0, f_1, f_2, f_3) where f_i is the rank of $I_{2i+1}(\Sigma(p,q,r))$.

$$I_*(\Sigma(2,3,6k\pm1)) = \begin{cases} (\frac{k\mp1}{2}, \frac{k\pm1}{2}, \frac{k\mp1}{2}, \frac{k\pm1}{2}) & \text{for } k \text{ odd} \\ (\frac{k}{2}, \frac{k}{2}, \frac{k}{2}, \frac{k}{2}) & \text{for } k \text{ even} \end{cases}$$

$$I_*(\Sigma(2,5,10k\pm1)) = \begin{cases} (\frac{3k\mp1}{2}, \frac{3k\pm1}{2}, \frac{3k\mp1}{2}, \frac{3k\pm1}{2}) & \text{for } k \text{ odd} \\ (\frac{3k}{2}, \frac{3k}{2}, \frac{3k}{2}, \frac{3k}{2}) & \text{for } k \text{ even} \end{cases}$$

$$I_*(\Sigma(2,5,10k\pm3)) = \begin{cases} (\frac{3k\pm1}{2}, \frac{3k\pm1}{2}, \frac{3k\pm1}{2}, \frac{3k\pm1}{2}) & \text{for } k \text{ odd} \\ (\frac{3k\pm2}{2}, \frac{3k}{2}, \frac{3k\pm2}{2}, \frac{3k}{2}) & \text{for } k \text{ even} \end{cases}$$

$$I_*(\Sigma(2,7,14k\pm1)) = \begin{cases} (3k\mp1, 3k\pm1, 3k\mp1, 3k\pm1) & \text{for } k \text{ odd} \\ (3k, 3k, 3k, 3k) & \text{for } k \text{ even} \end{cases}$$

$$I_*(\Sigma(2,7,14k\pm3)) = (3k, 3k\pm1, 3k, 3k\pm1)$$

$$I_*(\Sigma(2,7,14k\pm5)) = (3k\pm1, 3k\pm1, 3k\pm1, 3k\pm1)$$

$$I_*(\Sigma(3,4,12k\pm1)) = \begin{cases} (\frac{5k\mp1}{2}, \frac{5k\pm1}{2}, \frac{5k\mp1}{2}, \frac{5k\pm1}{2}) & \text{for } k \text{ odd} \\ (\frac{5k}{2}, \frac{5k}{2}, \frac{5k}{2}, \frac{5k}{2}) & \text{for } k \text{ even} \end{cases}$$

$$I_*(\Sigma(3,4,12k-5)) = \begin{cases} (\frac{5k-1}{2}, \frac{5k-3}{2}, \frac{5k-1}{2}, \frac{5k-3}{2}) & \text{for } k \text{ odd} \\ (\frac{5k-2}{2}, \frac{5k-2}{2}, \frac{5k-2}{2}, \frac{5k-2}{2}) & \text{for } k \text{ even} \end{cases}$$

$$I_*(\Sigma(3,5,15k\pm2)) = \begin{cases} (4k, 4k\pm1, 4k, 4k\pm1) & \text{for } k \text{ odd} \\ (4k\pm1, 4k, 4k\pm1, 4k\mp1) & \text{for } k \text{ even} \end{cases}$$

We conclude by listing some explicit computations of the Poincaré-Laurent polynomials $p(a_1, a_2, a_3)(t)$ of the homology groups $\tilde{I}_*(\Sigma(p,q,r))$.

$p(2,3,5) = t + t^5$
$p(2,3,11) = t + t^3 + t^5 + t^7$
$p(2,3,17) = t + t^3 + 2t^5 + t^7 + t^9$

$p(2,3,23) = t + 2t^3 + 2t^5 + 2t^7 + t^9$
$p(2,3,29) = t + t^3 + 3t^5 + 2t^7 + 2t^9 + t^{11}$
$p(2,3,35) = t + 2t^3 + 3t^5 + 3t^7 + 2t^9 + t^{11}$
$p(2,3,41) = t + t^3 + 4t^5 + 3t^7 + 3t^9 + 2t^{11}$
$p(2,3,47) = t + 2t^3 + 3t^5 + 4t^7 + 3t^9 + 2t^{11} + t^{13}$
$p(2,3,53) = t + t^3 + 4t^5 + 4t^7 + 4t^9 + 3t^{11} + t^{13}$
$p(2,3,59) = t + 2t^3 + 3t^5 + 5t^7 + 4t^9 + 3t^{11} + 2t^{13}$
$p(2,3,65) = t + t^3 + 4t^5 + 5t^7 + 5t^9 + 4t^{11} + 2t^{13}$
$p(2,3,71) = t + 2t^3 + 3t^5 + 5t^7 + 5t^9 + 4t^{11} + 3t^{13} + t^{15}$
$p(2,3,7) = t^{-1} + t^3$
$p(2,3,13) = t^{-1} + t + t^3 + t^5$
$p(2,3,19) = 2t^{-1} + t + 2t^3 + t^5$
$p(2,3,25) = 2t^{-1} + 2t + 2t^3 + 2t^5$
$p(2,3,31) = 2t^{-1} + 2t + 3t^3 + 2t^5 + t^7$
$p(2,3,37) = t^{-1} + 3t + 3t^3 + 2t^5 + 2t^7$
$p(2,3,37) = 2t^{-1} + 3t + 4t^3 + 3t^5 + 2t^7$
$p(2,3,43) = t^{-1} + 4t + 4t^3 + 4t^5 + 3t^7$
$p(2,3,49) = 2t^{-1} + 3t + 5t^3 + 4t^5 + 3t^7 + t^9$
$p(3,4,13) = t^{-5} + 2t^{-3} + 3t^{-1} + 2t + 2t^3$
$p(4,5,21) = 4t^{-9} + 5t^{-7} + 8t^{-5} + 6t^{-3} + 5t^{-1} + t + t^3$
$p(5,6,31) = 2t^{-15} + 10t^{-13} + 11t^{-11} + 15t^{-9} + 13t^{-7} + 8t^{-5} + 5t^{-3} + 4t^{-1} + t + t^3$

For $\alpha' \in \mathcal{R}(\Sigma(2,3,6k-1))$, $\mathrm{CS}(\alpha') \equiv \frac{e^2}{36k-6} \mod 4$ where $e \equiv 1 \mod 6$. Choosing the largest e and computing $R(2,3,6k-1; e\cdot\omega)$, it can be shown that for a given any positive integer N, there is a K with $\tilde{I}_{2N-1}(\Sigma(2,3,6K-1)) \neq 0$. Similarly, it can be shown that for any negative integer N, there is a K with $\tilde{I}_{2N-1}(\Sigma(K,K+1,K(K+1)+1)) \neq 0$.

9. Homotopy K3 Surfaces Containing $\Sigma(2,3,7)$

In this section we shall describe an application of the technology which we have discussed in previous sections to the problem of computing the Donaldson invariants of smooth 4-manifolds homotopy equivalent to the K3-surface (*i.e.* homotopy $K3$-surfaces). The Donaldson invariant may be briefly described as follows. Given a smooth simply-connected 4-manifold, M, for a generic metric g on M, the moduli space, $\mathcal{M}_{k,M}(g) \subset \mathcal{B}_{k,M}$ of anti-self-dual $SU(2)$-connections ($\star F_A = -F_A$) with $c_2 = k$ is a manifold of dimension $8k - 3(1+b_2^+)$. (It may have singular points if $b_2^+ = 0$.) If b_2^+ is odd, then this dimension is even, say $2d$. Donaldson has defined a homomorphism $\mu : H_2(M;\mathbf{Z}) \to H^2(\mathcal{B}_{k,M};\mathbf{Z})$ (see [**D4**]) and has shown that if $k > \frac{3}{4}(1+b_2^+)$ then there is a well-defined pairing $q(z_1,\ldots,z_d) = \mu(z_1) \cup \ldots \mu(z_d)[\mathcal{M}_{k,M}(g)]$. This defines Donaldson's invariant $q : \oplus_d H^2(M;\mathbf{Z}) \to \mathbf{Z}$. Our theorem is:

THEOREM 9.1 [**FS7**]. *If the Brieskorn sphere $\Sigma(2,3,7)$ is embedded in a 4-manifold which is homotopy equivalent to a $K3$-surface, then there are classes $z_1, \ldots, z_{10} \in H_2(M;\mathbf{Z})$ such that $q(z_1, \ldots, z_{10}) \equiv 1 \pmod 2$*

This result plays a role in an important result of S. Akbulut.

THEOREM 9.2 [**Ak**]. *There is a compact contractible 4-manifold W which has a fake relative smooth structure.*

To prove this Akbulut gives a construction which produces a homotopy K3 surface M containing $\Sigma(2,3,7)$ and a compact, contractible 4-manifold W together with a self-diffeomorphism f of ∂W which extends to a self-homeomorphism of W such that either

(1) all of the Donaldson polynomial invariants of M are trivial, or
(2) there is no self-diffeomorphism of W extending f.

Theorem 9.1 then implies that (1) is false; so Akbulut's construction gives (2).

The proof of Theorem 9.1 proceeds via a degeneration argument. The intersection form form of a $K3$-surface is $2E_8 \oplus 3H$, and $\Sigma(2,3,7)$ embeds in the standard $K3$-surface, splitting it into two submanifolds, X and Y, where X has intersection form $E_8 \oplus H$ and boundary $\Sigma(2,3,7)$ and Y has intersection form $E_8 \oplus 2H$ and boundary $-\Sigma(2,3,7)$. Using the work of Donaldson [**D2**], it is not difficult to see that if $\Sigma(2,3,7)$ embeds in any homotopy $K3$-surface, M, then it splits M as $X \cup Y$ in a similar manner. The idea is then to study the effect of letting the metric g degenerate to a metric which stretches a tube $\Sigma(2,3,7) \times (-1,1)$ in M until it has infinite length.

For the homology classes $z_1, \ldots, z_{10}$, we choose classes $z_1, \ldots, z_4 \in H_2(X;\mathbf{Z}) = E_8 \oplus H$ such that $z_1, z_2 \in E_8$ satisfy $z_1^2 = z_2^2 = 2$ and $z_1 \cdot z_2 = 1$, and choose $z_3, z_4 \in H$ with $z_3^2 = z_4^2 = 0$ and $z_3 \cdot z_4 = 1$. Similarly choose $z_5, \ldots, z_{10} \in H_2(Y;\mathbf{Z}) = E_8 \oplus 2H$ such that z_5, z_6 form a pair in E_8 and z_7, z_8 and z_9, z_{10} form pairs in the two copies of H. Suppose for a moment that $q(z_1, \ldots, z_{10}) \neq 0$ and follow the above-described degeneration of metrics. In the process M is pulled apart into noncompact manifolds X_+ and Y_- ($X_+ = X \cup \mathbf{R}^+$, and $Y_- = \mathbf{R}^- \cup Y$). A finite-action anti-self-dual connection on X_+ or Y_- is asymptotically flat, and the techniques of §4 can be used to show that $\Sigma(2,3,7)$ has only two gauge equivalence classes of nontrivial flat connections.

If $q(z_1, \ldots, z_{10}) \neq 0$ the degeneration process will produce nontrivial "relative" Donaldson invariants which can be described roughly as $q_X(z_1, \ldots, z_4) = \sum_\rho \mu(z_1) \cup \cdots \cup \mu(z_4)[\mathcal{M}_X^8(\rho)]$, where ρ is a flat connection on $\Sigma(2,3,7)$, and $q_Y(z_5, \ldots, z_{10}) = \sum_\rho \mu(z_5) \cup \cdots \cup \mu(z_{10})[\mathcal{M}_Y^{12}(\rho)]$. One can now make counting arguments (see [**FS7**]) to the effect that only one possible ρ can appear in the above sums, and furthermore, for this ρ we have

$$q(z_1, \ldots, z_{10}) = q_X(z_1, \ldots, z_4)(\rho) \cdot q_Y(z_5, \ldots, z_{10})(\rho).$$

The proof is now concluded by constructing the moduli spaces $\mathcal{M}_X^8(\rho)$ and $\mathcal{M}_Y^{12}(\rho)$ whose appropriate relative invariants are odd. The philosophy is to apply the proofs of Donaldson's Theorems B and C of [**D2**]. These theorems prove that there are no *closed* 4-manifolds which have the intersection forms of X or Y. Applying the proofs to X and Y, rather than contradictions we obtain information about the asymptotic behavior of certain moduli spaces. As in the proof of (5.3) we get anti-self-dual connections popping off the ends of $\mathcal{M}_{2,X}(\theta)$ and $\mathcal{M}_{3,Y}(\theta)$ which leave us the correct moduli spaces over X and Y (see [**FS7**]).

REFERENCES

[**Ak**] S.Akbulut, *A fake compact contractible 4-manifold*, preprint, 1989.

[**AM**] S. Akbulut and J. McCarthy, *Casson's invariant for oriented homology 3-spheres, an exposition*, "Mathematical Notes, Vol.36," Princeton University Press, 1990.

[**At**] M.F. Atiyah, *New invariants of 3 and 4 dimensional manifolds, in 'The Mathematical Heritage of Hermann Weyl'*, Proceedings Symposia Pure Math **48** (1988).

[**APS1**] M.F. Atiyah, V.K. Patodi, and I.M. Singer, *Spectral asymmetry and Riemannian geometry. I.*, Math. Proc. Camb. Phil. Soc. **77** (1975), 43 -69.

[**APS2**] ———, *Spectral asymmetry and Riemannian geometry. II.*, Math. Proc. Camb. Phil. Soc. **78** (1975), 405–432.

[**APS3**] ———, *Spectral asymmetry and Riemannian geometry. III.*, Math. Proc. Camb. Phil. Soc. **79** (1976), 71 -99.

[**B**] E. Brieskorn, *Beispiele zur Differetial topologie von singularities*, Inv. Math. **2** (1966), 1–14.

[**BO**] S.Bauer and C. Okonek, *The algebraic geometry of representation spaces associated to Seifert fibered homology 3-spheres*, preprint.

[**Br**] P. Braam, *Floer homology groups for homology three-spheres*, University of Utrecht Preprint, 1988.

[**BZ**] G. Burde and H. Zieschang, "Knots," de Gruyter Studies in Math. 5, Walter de Gruyter & Co., New York, 1985.

[**CG**] A. Casson and C. Gordon, *Cobordism of Classical knots*, in "A la Recherche de la Topologie Perdue," Progress in Mathematics, Birkhäuser, Boston, MA, 1986.

[**D1**] S. Donaldson, *An application of gauge theory to the topology of 4-manifolds*, J. Diff. Geom. **18** (1983), 269–316.

[**D2**] ———, *Connections, cohomology, and the intersection form of 4-manifolds*, J. Diff. Geom. **24** (1986), 275–341.

[**D3**] ———, *The orientation of Yang-Mills moduli spaces and 4-dimensional topology*, J. Diff. Geom. **26** (1987), 397–428.

[**D4**] ———, *Polynomial invariants for smooth 4-manifolds*, preprint, 1987.

[**FS1**] R. Fintushel and R. Stern, *An exotic free involution on S^4*, Annals of Math. **113** (1981), 357–365.

[**FS2**] ———, *Seifert-fibered 3-manifolds and nonorientable 4-manifolds*, Contemporary Math. **20** (1983), 103–119.

[**FS3**] ———, *Pseudofree orbifolds*, Annals of Math. **122** (1985), 335 -364.

[**FS4**] ———, *Rational cobordisms of spherical space forms*, Topology **26** (1987), 385–393.

[**FS5**] ———, *Definite 4-manifolds*, J. Diff. Geom. **28** (1988), 133–141.

[FS6] ——————, *Instanton homology of Seifert fibered homology three spheres*, to appear, Proc. London Math. Soc. (1990).

[FS7] ——————, *Homotopy K3-surfaces containing* $\Sigma(2,3,7)$, preprint.

[FS8] ——————, *Integer graded instanton homology groups for homology 3-spheres*, preprint.

[F] A. Floer, *An instanton invariant for 3-manifolds*, Commun. Math. Phys. **118** (1988), 215–240.

[FT] M. Freedman and L. Taylor, *Λ-splitting 4-manifolds*, Topology **16** (1977), 181–184.

[Fu1] M. Furuta, *The homology cobordism group of homology 3-spheres*, preprint.

[Fu2] ————, *Homology cobordism group of homology 3-spheres*, University of Tokyo preprint, 1987.

[FuS] M. Furuta and B. Steer, preprint.

[GL] C. Gordon and J. Luecke, *Knots are determined by their complements*, 1988 preprint.

[GS] D. Galewski and R.Stern, *Classification of simplicial triangulations of topological manifolds*, Annals of Math. **111** (1980), 1–34.

[HH] F. Hirzebruch and H. Hopf, *Felder von Flachenelementen in 4-dimensionalen Mannigfaltigkeiten*, Math. Annalen **136** (1958), 156–172.

[HLM] H. Hilden, M. Lozano and J. Montesinos, *Universal knots*, Topology **24** (1985), 499–504.

[KK1] P. Kirk and E. Klassen, *Representation spaces of Seifert fibered homology spheres*, preprint.

[KK2] ——————, *Chern-Simons invariants of 3-manifolds and representation spaces of knot groups*, to appear, Math. Annalen.

[Ma] G. Matic, *Rational homology cobordisms*, J. Diff. Geom. **28** (1988), 277–307.

[Mat] T. Matumoto, *Triangulations of manifolds*, A.M.S. Proc. of Symp. in Pure Math. **32** (1977), 3–6.

[My] R. Myers, *Involutions on homology 3-spheres*, Pacific J. Math. **94** (1981), 379–384.

[NR] W. Neumann and F. Raymond, *Seifert manifolds, μ-invariant, and orientation reversing maps*, Springer Lecture Notes **664** (1977), 163–196.

[R] D. Ruberman, *Rational homology cobordisms of rational space forms*, Topology **27** (1988), 401–414.

[T1] C. Taubes, *Gauge theory on asymptotically periodic 4-manifolds*, J.Diff.Geom. **25** (1987), 363–430.

[T2] ————, *Casson's invariant and gauge theory*, Harvard preprint, 1988.

Department of Mathematics, Michigan State University, East Lansing, Michigan 48824
Department of Mathematics, University of California, Irvine, California 92717

On the Floer Homology of Seifert fibered Homology 3–Spheres

CHRISTIAN OKONEK
Mathematisches Institut der Universität Bonn
Wegelerstr.10
D–5300 Bonn 1, F.R.G.

1. Introduction

This note contains a somewhat expanded version of my talk given at the Durham Geometry Symposium. In this talk, which was based on a joint paper with S.Bauer [B/O], I tried to explain how some fairly remote results in algebraic geometry — like e.g. Deligne's solution of the Weil conjectures — can be used to calculate the instanton homology of Seifert fibered homology 3–spheres.

Instanton homology groups $I_n(\Sigma), n \in \mathbf{Z}/_8$ have recently been defined by A.Floer for every oriented 3–dimensional $\mathbf{Z}$–homology sphere Σ[F].
They provide an important link between the geometry of 3– and 4–dimensional manifolds [A1]. In fact, Donaldson has shown that his polynomial invariants γ_X^k for a 4–manifold X [D2], which can be decomposed along a homology sphere Σ into two pieces factor through the Floer homology of Σ [A1]. A nice introduction to the Donaldson–Floer theory can be found in Atiyah's survey articles [A1],[A2].

The definition of the groups $I_n(\Sigma)$ uses gauge theoretic constructions in dimension 3 and 4, so that it is difficult to compute them for a general homology sphere. For the subclass of Seifert fibered homology spheres however, Fintushel and Stern have found a way to calculate the instanton homology provided that a certain conjecture holds [F/S2].
This conjecture, which I will explain in a moment, has recently been settled by Kirk and Klassen [K/K], so that the Fintushel–Stern program can be carried out. What it comes down to is to understand the representation spaces

$$R^*(\Sigma) = \mathrm{Hom}^*(\pi_1(\Sigma), SU(2))/_{conj.}$$

of conjugacy classes of irreducible $SU(2)$–representations of the fundamental group $\pi_1(\Sigma)$.
The Floer homology of a Seifert fibered homology sphere Σ is determined by the Betti numbers of the components of $R^*(\Sigma)$ and certain (explicitely computable) integers associated to each component [F/S2].

One way to study these representation spaces $R^*(\Sigma)$ is to identify them — via Donaldson's solution of the Kobayashi–Hitchin conjecture [D1] — with moduli spaces of stable vector bundles over certain algebraic surfaces [O/V2]. In this way they become complex projective varieties which turn out to be smooth and rational. Moreover, these moduli spaces come with a stratification whose individual strata can be described in terms of secant varieties of rational normal curves [B/O].
Using a trick due to Deligne–Illusie [D/I], one can assume that everything is defined over an algebraic number ring, so that the Weil conjectures can be applied. The explicit geometric description of the strata of the moduli spaces then allows to determine the Zeta functions associated to the components and thereby to compute their Betti numbers.
The final result is an algorithm which, in principle, could be implemented on a computer.

2. Floer Homology

I recall very briefly the idea of the definition of Floer's instanton homology groups; a much more detailed description can be found in Floer's paper [F] and the survey article [BR].
Let Σ be an oriented $\mathbf{Z}$–homology sphere of dimension 3. Consider the space $\mathcal{B}$ of gauge equivalence classes of $SU(2)$–connections on the trivial $SU(2)$–bundle over Σ. The map $A \longmapsto F_A$, sending a connection A to its curvature, defines a natural 1–Form F on $\mathcal{B}$ which is locally exact. More precisely, F is — up to a constant — the differential of the Chern–Simons function

$$f : \mathcal{B} \longrightarrow \mathbf{R}/_{4\mathbf{Z}}.$$

This function associates to a connection A the integral

$$\frac{1}{4\pi^2} \int_{\Sigma \times [0,1]} tr(F_{A_t} \wedge F_{A_t})$$

where $A_t := (1-t)A + t\theta$ is a path of connections from A to the trivial connection θ, thought of as a connection on $\Sigma \times [0,1]$.
The critical set of the Chern–Simons function, i.e. the zeros of F, can be identified — via the monodromy representation — with the space

$$R(\Sigma) = \mathrm{Hom}(\pi_1(\Sigma), SU(2))/_{conj.}$$

of conjugacy classes of $SU(2)$–representations of $\pi_1(\Sigma)$.
Suppose now that all non–trivial critical points of f are non–degenerate (if not one uses a suitable Fredholm perturbation); this means that $R(\Sigma)$ is finite and that the Hessian of f (considered as an operator on the tangent space) is an isomorphism at every critical point $\alpha = [A]$ in $R^*(\Sigma) = R(\Sigma)\setminus\{[\theta]\}$.
Let $S(\theta, \alpha) \in \mathbf{Z}/_8$ denote the spectral flow associated to a path of connections

from the trivial connection to A [BR]. Floer's instanton homology $I_*(\Sigma)$ is then defined as the homology of the following chain complex (R_*, ∂):
The nth chain groupR_n of $R_* = \bigoplus_{n \in \mathbf{Z}/8} R_n$ is the free abelian group generated by the elements $\alpha \in R^*(\Sigma)$ with $S(\theta, \alpha) = n$.
The boundary operator $\partial : R_n \longrightarrow R_{n-1}$ is given by

$$\partial\alpha = \sum_{\substack{\beta \in R^*(\Sigma) \\ S(\theta,\beta)=n-1}} m(\alpha,\beta)\beta;$$

here $m(\alpha, \beta)$ is the number of oriented 1-dimensional components of the moduli space $\mathcal{M}_\Sigma(\alpha, \beta)$ of self-dual connections (relative to a product metric) on $\Sigma \times \mathbf{R}$, which are asymptotic to α and β for t tending to $\pm\infty$ [F]. Floer shows that $\partial^2 = 0$ and proves that the homology of the instanton chain complex (R_*, ∂) is independent of the various choices (metric, perturbation).

3. Seifert fibered Homology 3-Spheres

Let $\pi : \Sigma \longrightarrow \Sigma/_{S^1}$ be a Seifert fibration of a homology 3-sphere Σ with n exceptional orbits $\pi^{-1}(x_i), i = 1, \ldots, n$ of multiplicities $a_1, \ldots, a_n$. The multiplicities are necessarily pairwise relatively prime and the orbit space $\Sigma/_{S^1}$ is homeomorphic to $(S^2; (x_1, a_1), \ldots, (x_n, a_n))$ [N/R].
Conversely, given n pairwise coprime integers $a_i \geq 2$, there exists a Seifert fibered homology 3-sphere Σ with these multiplicities; its diffeomorphism type is determined by $\underline{a} = (a_1, \ldots, a_n)$ [N/R].
Denote such a homology sphere by $\Sigma(\underline{a}) = \Sigma(a_1, \ldots, a_n)$. I will always assume that the multiplicities are indexed in such a way that at most a_1 is even. The links of certain Brieskorn complete intersections provide standard models of Seifert fibered homology spheres [N/R]. The fundamental group of $\Sigma(\underline{a})$ has the following representation [F/S2]:
Let $a := a_1 \cdot \ldots \cdot a_n$ and choose integers $b, b_i, i = 1, \ldots, n$ with $a\left(-b + \sum_{i=1}^{n} \frac{b_i}{a_i}\right) = 1$. Then

$$\pi_1(\Sigma(\underline{a})) \cong \left\langle t_1, \ldots, t_n, h \mid h \text{ central}, \ t_i^{a_i} = h^{-b_i}, t_1 \cdot \ldots \cdot t_n = h^b \right\rangle.$$

If $n \geq 3$, then this group is infinite with center $\langle h \rangle \cong \mathbf{Z}$, except for $\pi_1(\Sigma(2,3,5)) \cong SL(2, \mathbf{F}_5)$ with center $\langle h \rangle \cong \mathbf{Z}/2$. The quotient $\pi_1(\Sigma(\underline{a}))/_{\langle h \rangle}$ is isomorphic to the 2-orbifold fundamental group $\pi_1^{orb}(\Sigma(\underline{a})/S^1)$ of the decomposition surface [F/S2]. It is isomorphic to a cocompact Fuchsian group of genus 0 with representation

$$\langle t_1, \ldots, t_n \mid t_i^{a_i} = 1, t_1 \cdot \ldots \cdot t_n = 1 \rangle.$$

4. Representation Spaces of Seifert fibered Homology Spheres

Fix a Seifert fibered homology sphere $\Sigma(\underline{a}) = \Sigma(a_1, \ldots, a_n)$ with $n \geq 3$ exceptional fibers and consider a representation

$$\alpha : \pi_1(\Sigma(\underline{a})) \longrightarrow SU(2).$$

If α is irreducible, then the generator h of the center must be mapped to $\pm 1 \in SU(2)$; the images $\alpha(t_i)$ of the remaining generators are conjugate to diagonal matrices

$$\alpha(t_i) \sim \begin{pmatrix} \omega_i^{l_i} & \\ & \omega_i^{-l_i} \end{pmatrix}$$

with $\omega_i = \exp\left(\frac{2\pi\sqrt{-1}}{a_i}\right)$ and certain numbers $l_i \in \mathbf{Z}/a_i$. These numbers $(\pm l_1, \ldots, \pm l_n)$ are the rotation numbers of the representation α.

Proposition 1 ([F/S2],[B/O]) *The representation $R^*(\Sigma(\underline{a}))$ of a Seifert fibered homology sphere is a closed differentiable manifold with several components. The rotation numbers of an irreducible representation α are invariants of the connected component of α in $R^*(\Sigma(\underline{a}))$. A component with rotation numbers $(\pm l_1, \ldots, \pm l_n)$ has dimension $2(m-3)$, where $m = \sharp\{i \mid 2l_i \neq 0\}$.*

Furthermore, there exists at most one component in $R^*(\Sigma(\underline{a}))$ realizing a given set of rotation numbers [B/O].

In the special case of 3 exceptional orbits, i.e. for Brieskorn spheres $\Sigma(a_1, a_2, a_3)$, the associated representation spaces are finite.
Fintushel and Stern have shown that the number of elements in $R^*(\Sigma(a_1, a_2, a_3))$ is equal to $-\frac{1}{4}$ times the signature of the Milnor fiber of the corresponding singularities [F/S2].
The Casson invariant of a Brieskorn sphere is therefore equal to $\frac{1}{8}$ times the signature of its Milnor fiber.
The latter result has recently been generalized by Neumann and Wahl [N/W].

5. The Instanton Chain Complex of Seifert Spheres

In this section I recall the relevant results of Fintushel and Stern's paper [F/S2]. Again, fix a Seifert fibered homology sphere $\Sigma(\underline{a}) = \Sigma(a_1, \ldots, a_n), n \geq 3$. For any integer e let

$$R(\underline{a}, e) = \frac{2e^2}{a} - 3 + m + \sum_{i=1}^{n} \frac{2}{a_i} \sum_{k=1}^{a_i - 1} \cot\left(\frac{\pi a k}{a_i^2}\right) \cot\left(\frac{\pi k}{a_i}\right) \sin^2\left(\frac{\pi e k}{a_i}\right),$$

where $m = \sharp\{i \mid e \not\equiv 0 (\mathrm{mod}\ a_i)\}$. This number is the virtual dimension of a certain moduli space of instantons; it is always odd [F/S1].

Theorem 1 ([F/S2]) *The spectral flow $S(\theta,\alpha)$ of an element $\alpha \in R^*(\Sigma(\underline{a}))$ with rotation numbers $(\pm l_1,\dots,\pm l_n)$ is given by*

$$S(\theta,\alpha) = -R(\underline{a},e) - 3 (mod\, 8),$$

if e satisfies $e \equiv \sum_{i=1}^{n} l_i \frac{a}{a_i} (mod\, 2a)$.

If now every representation $\alpha \in R^*(\Sigma(\underline{a}))$ is non-degenerate, then the grading in the instanton chain complex is always even, so that the boundary operator must vanish. This assumption holds if $n \equiv 3$, i.e. for Brieskorn spheres $\Sigma(a_1,a_2,a_3)$. Explicit examples of instanton homology groups $I_*(\Sigma(a_1,a_2,a_3))$ can be found in [F/S2].

In the general case $n > 3$ the elements of $R^*(\Sigma(\underline{a}))$ are usually degenerate, so that one has to perturb the Chern–Simons function. Fintushel and Stern use a Morse function on $R^*(\Sigma(\underline{a}))$ to produce a perturbation with non-degenerate critical points.

Theorem 2 ([F/S2]) *Let $g : R^*(\Sigma(\underline{a}))_\alpha \longrightarrow \mathbf{R}$ a Morse function on the component of α in $R^*(\Sigma(\underline{a}))$. The critical points of g are basis elements of the instanton chain complex.*

A critical point $\beta \in R^*(\Sigma(\underline{a}))_\alpha$ of g with Morse index $\mu_g(\beta)$ has grading $S(\theta,\alpha) + \mu_g(\beta)$.

In order to make explicit computations possible Fintushel and Stern show how $R^*(\Sigma(\underline{a}))$ can be described as a configuration space of certain linkages in S^3 [F/S2]. With this method they find copies of S^2 as 2–dimensional components. On the basis of these examples they make the following

Conjecture *Every component of $R^*(\Sigma(\underline{a}))$ admits Morse functions with only even index critical points.*

Note that this conjecture implies that the instanton chain complex of $\Sigma(\underline{a})$ is concentrated in even dimensions, the boundary operator vanishes and the Floer homology can be read off from the rotation numbers and the Betti numbers of the components of $R^*(\Sigma(\underline{a}))$.

6. The Algebraic Geometry of the Representation Spaces

$R^*(\Sigma(\underline{a}))$ Every representation $\alpha : \pi_1(\Sigma(\underline{a})) \longrightarrow SU(2)$ induces a representation $\alpha : \pi_1^{orb}(\Sigma(\underline{a})/S^1) \longrightarrow PU(2)$ s.t. the following diagram commutes:

$$\begin{array}{ccc} \pi_1(\Sigma(\underline{a})) & \xrightarrow{\alpha} & SU(2) \\ \downarrow /\langle h \rangle & & \downarrow /\mathbf{Z}/2 \\ \pi_1^{orb}(\Sigma(\underline{a})/S^1) & \xrightarrow{\bar{\alpha}} & PU(2). \end{array}$$

This correspondence yields an identification of $R^*(\Sigma(\underline{a}))$ with the representation space

$$\mathrm{Hom}^*\left(\pi_1^{orb}\left(\Sigma(\underline{a})/S^1\right), PU(2)\right)/_{conj.}$$

of the orbifold fundamental group in $PU(2)$. Recall that

$$\pi_1^{orb}\left(\Sigma(\underline{a})/S^1\right) \cong \langle t_1, \ldots, t_n \mid t_i^{a_i} = 1, t_1 \cdot \ldots \cdot t_n = 1 \rangle .$$

In the sequel I shall denote this group by $\Gamma(\underline{a})$.

Theorem 3 ([B/O]) *The representation space $Hom^*(\Gamma(\underline{a}), PU(2))/_{conj.}$ admits the structure a of smooth complex projective variety whose components are rational.*

The proof has two essential steps:

i) Interpretation of $\mathrm{Hom}^*(\Gamma(\underline{a}), PU(2))/_{conj.}$ as moduli space of stable vector bundles:
Consider a rational elliptic surface over $\mathbf{P}^1$ — defined by a generic pencil of plane cubic curves — and perform logarithmic transformations of multiplicities $a_1, \ldots, a_n$ along smooth fibers over $x_1, \ldots, x_n \in \mathbf{P}^1$ [B/P/V]. The resulting elliptic surfaces $X(\underline{a}) = X(a_1, \ldots, a_n)$ over $\mathbf{P}^1$ are algebraic with fundamental group $\pi_1(X(\underline{a})) \cong \Gamma(\underline{a})$ [U].
(The algebraic structure of $X(\underline{a})$ depends on the choices which are involved in the logarithmic transformations, but the C^∞–type does not [U]). Choose a (sufficiently nice) ample divisor $H = H(\underline{a})$ on $X = X(\underline{a})$ and let $\mathcal{M}_X^H(c_1, c_2)$ denote the moduli space of H–stable rank–2 bundles over X with Chern classes c_1, c_2 [O/S/S].
Using Donaldson's solution of the Kobayashi–Hitchin conjecture [D1] and some simple arguments [O/V1] one obtains an identification of
$\mathrm{Hom}^*(\pi_1(X(\underline{a})), PU(2))/_{conj.}$ with the differentiable space underlying the disjoint union $\mathcal{M}_X^H(0,0) \amalg \mathcal{M}_X^H(K,0)$. Here $K = -c_1(X)$ is the canonical class of X.

ii) Description of the moduli space of stable vector bundles:
The moduli spaces $\mathcal{M}_X^H(0,0)$ and $\mathcal{M}_X^H(K,0)$ can be handled by similar methods; but a little trick allows to avoid computing the latter. Indeed, the homomorphism

$$\tau : \Gamma(2a_1, a_2, \ldots, a_n) \longrightarrow \Gamma(a_1, \ldots, a_n)$$

sending a generator t_i of $\Gamma(2a_1, a_2, \ldots, a_n)$ to the corresponding generator in $\Gamma(a_1, a_2, \ldots, a_n)$ induces an isomorphism

$$\tau^* : \mathcal{M}_X^H(0,0) \amalg \mathcal{M}_X^H(K,0) \longrightarrow \mathcal{M}_{\hat{X}}^{\hat{H}}(0,0)$$

with the moduli space $\mathcal{M}_{\hat{X}}^{\hat{H}}(0,0)$ of stable bundles over an elliptic surface $\hat{X}$ of type $X(2a_1, a_2, \ldots, a_n)$.
Consider now a stable 2–bundle $\mathcal{E}$ over X with trivial Chern classes. $\mathcal{E}$ admits a unique representation as an extension

$$(*) \quad 0 \longrightarrow \mathcal{O}(-D) \longrightarrow \mathcal{E} \longrightarrow \mathcal{O}(D) \longrightarrow 0$$

of line bundles, where $D \sim dF + \sum_{i=1}^{n} d_i F_i$ is a vertical divisor with $0 \leq d_i < a_i$. Here F denotes a generic fiber of X over $\mathbf{P}^1$, $a_i F_i \sim F, i = 1, \ldots, n$ the n multiple fibers over the points $x_1, \ldots x_n \in \mathbf{P}^1$.
The line bundles $\mathcal{O}(D)$ which occur in such extensions form a finite subset $I \subset NS(X)$ in the Neron–Severi group of X.
Conversely, for every line bundle $\mathcal{O}(D) \in I$ and every (non-trivial) extension class $[\epsilon] \in \mathbf{P}\left(\mathrm{Ext}^1\left(\mathcal{O}(D), \mathcal{O}(-D)\right)\right)$ one obtains a simple 2-bundle $\mathcal{E}$ given by the extension

$$\epsilon: \quad 0 \longrightarrow \mathcal{O}(-D) \longrightarrow \mathcal{E} \longrightarrow \mathcal{O}(D) \longrightarrow 0.$$

Denote by $\mathcal{M}_X^s(0,0)$ the moduli space of simple 2-bundles over X with trivial Chern classes. This is a locally (but not globally) Hausdorff complex space containing $\mathcal{M}_X^H(0,0)$ as (Hausdorff) open subspace [O/V2]. Let $\mathbf{P}(D) = \mathbf{P}\left(\mathrm{Ext}^1\left(\mathcal{O}(D), \mathcal{O}(-D)\right)\right)$ and define Zariski open subsets $U(D) \subset \mathbf{P}(D)$ by $U(D) = \mathbf{P}(D) \cap \mathcal{M}_X^H(0,0)$. Then one has a cartesian diagram

$$\begin{array}{ccc} \coprod_I \mathbf{P}(D) & \hookrightarrow & \mathcal{M}_X^s(0,0) \\ \cup & & \cup \\ \coprod_I U(D) & \xrightarrow{\cong} & \mathcal{M}_X^H(0,0) \end{array}$$

and a stratification $\coprod_I U(D)$ of $\mathcal{M}_X^H(0,0)$ by locally closed subspaces $U(D)$, each sitting as a Zariski–open subset in its 'own' projective space. In order to prove the smoothness of $\mathcal{M}_X^H(0,0)$ as a complex algebraic variety, one shows that the coefficients d_i of a divisor D determine the rotation numbers of the component which contains $U(D)$.
More precisely, if a representation $\alpha : \pi_1(X(\underline{a})) \to SU(2)$ corresponds to a vector bundle $\mathcal{E}$ given by a class $[\epsilon] \in U(D)$ with $D \sim dF + \sum_{i=1}^{n} d_i F_i$, then α has the rotation numbers $(\pm l_1, \ldots, \pm l_n)$.
The rationality of all components of $\mathcal{M}_X^H(0,0)$ now follows immediately.

The next point is to understand the projective varieties $\mathbf{P}(D) \backslash U(D)$ parametrizing (simple but) unstable bundles.
A bundle $\mathcal{E}$ given by $[\epsilon] \in \mathbf{P}(D)$ is unstable if and only if X contains a vertical curve E with $\mathcal{O}(D - E) \in I$ s.t. $[\epsilon]$ is contained in the projective kernel $\mathbf{P}(\mathrm{Ker}(\bullet E))$ of the multiplication map

$$\bullet E \quad : \quad \mathrm{Ext}^1\left(\mathcal{O}(D), \mathcal{O}(-D)\right) \longrightarrow \mathrm{Ext}^1\left(\mathcal{O}(D-E), \mathcal{O}(-D)\right).$$

These kernels form a projective bundle $\mathbf{P}(\mathrm{Ker}(\bullet \mid E \mid))$ over the linear system $\mid E \mid$ which admits a natural map

$$\psi_{|E|} : \mathbf{P}(\mathrm{Ker}(\bullet \mid E \mid)) \longrightarrow \mathbf{P}(D)$$

to the projective space $\mathbf{P}(D)$. The image of $\psi_{|E|}$ is the subvariety $\mathrm{Dest}(\mid E \mid)$ of bundles which are destabilized by curves in $\mid E \mid$.

Recall that the join $W_1 * \ldots * W_k$ of subvarieties $W_1, \ldots, W_k \subset \mathbf{P}^N$ is the smallest closed subvariety of $\mathbf{P}^N$ containing $\mathrm{span}(w_1, \ldots, w_k)$ for every tuple $(w_1, \ldots, w_k) \in W_1 \times \ldots \times W_k$. The secant variety $\mathrm{Sec}_k(W) = W * \ldots * W$ (k times) is a particular case.

Proposition 2 ([B/O]) *Suppose that* $\dim \mathbf{P}(D) > 0$. *There exists a natural embedding of* $(\mathbf{P}^1; x_1, \ldots, x_n)$ *as a rational normal curve* $N(D) \subset \mathbf{P}(D)$ *with marked points* $\overline{x}_j$, *such that the following holds:*

i) Every destabilizing subvariety $Dest(\mid E \mid)$ *of* $\mathbf{P}(D)$ *is a join* $Sec_k(N(D)) * \{\overline{x}_{i_1}\} * \cdots * \{\overline{x}_{i_l}\}$ *of a secant variety of* $N(D)$ *and some of the points* $\overline{x}_j$.

ii) $\mathbf{P}(D) \backslash U(D)$ *is a finite union of destabilizing subvarieties* $Dest(\mid E \mid)$.

iii) The intersection $Dest(\mid E \mid) \cap Dest(\mid E' \mid)$ *of two destabilizing subvarieties in* $\mathbf{P}(D)$ *is a finite union of other destabilizing subvarieties.*

The following picture illustrates a typical 2–dimensional situation

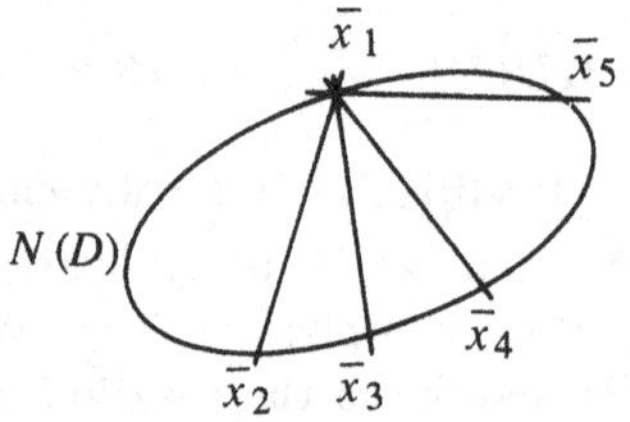

7. The Betti Numbers of the Moduli Spaces

The way in which the stratification of $\mathcal{M} = \mathcal{M}_X^H(0,0)$ has been defined makes it difficult to describe the normal bundles of the various strata.
To circumvent this difficulty one can use an approach which has been applied by Harder and Narasimhan in a similar situation [H/N]. Their idea was to calculate the Betti numbers of a moduli space by 'counting points over finite fields and then use the Weil conjectures'. In the situation at hand the space $\mathcal{M}$ is not a priori defined over a number ring. However, one can find an extension of $\mathcal{M}$ over the spectrum of a subring $A \subset \mathbf{C}$ and a closed point in Spec (A) with residue field $\mathbf{F}_q$ of positive characteristic p, so that there exists a good prime $l \neq p$ with

$$H^i(\mathcal{M}; \mathcal{Q}_l) \cong H^i_{\acute{e}t}(\overline{\mathcal{M}}_{\overline{\mathbf{F}}_q}; \mathcal{Q}_l)$$

for all i.
Consider now the Zeta function of $\overline{\mathcal{M}}_{\overline{\mathbf{F}}_q}$,

$$Z(\overline{\mathcal{M}}_{\overline{\mathbf{F}}_q}, t) = \exp\left(\sum_{k \geq 1} \frac{v_k(\overline{\mathcal{M}}_{\overline{\mathbf{F}}_q})}{k} t^k\right),$$

where $v_k(\overline{\mathcal{M}}_{\overline{\mathbf{F}}_q})$ counts the number of points in $\mathcal{M}_{\mathbf{F}_{q^k}}$. By Deligne's solution of the Weil conjectures $Z(\overline{\mathcal{M}}_{\overline{\mathbf{F}}_q}, t)$ can be written in the form

$$Z(\overline{\mathcal{M}}_{\overline{\mathbf{F}}_q}, t) = \frac{P_1 \cdot \ldots \cdot P_{2n-1}}{P_0 \cdot \ldots \cdot P_{2n}}$$

with polynomials P_i of degree $\dim H^i_{\acute{e}t}(\overline{\mathcal{M}}_{\overline{\mathbf{F}}_q}; \mathcal{Q}_l)$, whose zeros all have absolute value $p^{-\frac{i}{2}}$ [D]. In our situation we have to determine the Zeta functions of the strata $U(D)$ or, equivalently, of the varieties $\mathbf{P}(D)\backslash U(D)$ of unstable bundles. Using the explicit description of these varieties as joins over secant varieties of rational normal curves, it is possible to calculate these Zeta functions. One finds that associated to each component of $\mathcal{M}$, there are natural numbers $b_0 = 1, b_2, \ldots, b_{2(m-3)}$, so that the counting function v_k of the component has the form

$$v_k = \sum_{i=0}^{2(m-3)} b_{2i}(q^k)^i.$$

Of course, this implies that the Zeta function of this component has the product decomposition

$$Z(t) = \Pi_{i=0}^{2(m-3)}(1 - q^i t)^{-b_{2i}}.$$

More precisely:

Theorem 4 ([B/O]) *The odd Betti numbers of the moduli space $\mathcal{M}^H_X(0,0)$ are zero. The even Betti numbers can be determined by a numerical algorithm.*

Explicit formulas can be found in [B/O].

8. Final Remarks

As I already mentioned in the introduction, the conjecture of Fintushel and Stern has been shown to be true. In fact, there are at least three different (announcements of) proofs.

The first one — by Kirk and Klassen [K/K] — uses the concept of linkages in S^3 to construct directly a Morse function with only even index critical points.

A second proof has been announced by Furuta and Steer [F/ST]. Their starting point is the observation that the representation space $\mathrm{Hom}^*\left(\pi_1^{orb}(\Sigma(\underline{a})/S^1), PU(2)\right)/_{conj.}$ can be interpreted as moduli space of equivariant Yang–Mills connections over a suitable covering surface of $\Sigma(\underline{a})/S^1$. Extending the Atiyah–Bott method [A/B] to this equivariant setting they give formulas for the Poincaré polynomials of the instanton homology.

There is still another interpretation of $\mathrm{Hom}^*\left(\pi_1^{orb}(\Sigma(\underline{a})/S^1), PU(2)\right)/_{conj.}$, going back to Mehta and Seshadri [M/S]. These authors show that representation spaces of cocompact Fuchsian groups can be identified with moduli spaces of parabolic vector bundles on marked Riemann surfaces.

Using this description of $R^*(\Sigma(\underline{a}))$ Bauer [B] gives a third proof of the conjecture. Finally, I like to mention another point of view which might be interesting. The fundamental group $\pi_1(\Sigma(\underline{a}))$ of a Seifert fibered sphere is isomorphic to the local fundamental group of a corresponding Brieskorn complete intersection singularity. Thus unitary representations of this group should give rise to reflexive modules over the local ring of such singularities; equivalently, they should define vector bundles over their minimal resolutions. It might be useful to consider the relevant singularities as group quotients of cone singularities [P].

References

[A1] **Atiyah,M.F.(1988)** New invariants of 3– and 4–dimensional manifolds. In: The Mathematical Heritage of Herman Weyl, Proc.Symp.Pure Math.48, Amer. Math. Soc..

[A2] **Atiyah,M.F.(1989)** Topological quantum field theory. Preprint, Oxford.

[A/B] **Atiyah,M.F.,Bott,R.(1982)** The Yang–Mills equations over Riemann surfaces. Phil.Trans.R.Soc. London A, 308,523–615.

[B/P/V] **Barth,W.,Peters,Ch.,Van de Ven,A. (1985)** Compact complex surfaces. Erg.der Math.(3)4. Springer: Berlin, Heidelberg, New York.

[BA] **Bauer,S.(1989)** Parabolic bundles, elliptic surfaces and $SU(2)$–representation spaces of genus zero Fuchsian groups. Preprint 67, MPI Bonn.

[B/O] **Bauer,S.,Okonek,Ch.(1990)** The algebraic geometry of representation spaces associated to Seifert fibered homology 3–spheres. Math.Ann..

[B] **Braam,P.(1988)** Floer homology groups for homology three–spheres. Preprint, Univ.of Utrecht.

[D] **Deligne,P.(1974)** La conjecture de Weil,I.Publ.Math.IHES 43,273–307.

[D/I] **Deligne,P.,Illusie,L.(1987)** Relèvements modulo p^2 et décomposition du complex de de Rham. Invent.math. 89,247–270.

[D1] **Donaldson,S.K.(1985)** Anti–self–dual Yang–Mills connections over complex algebraic surfaces and stable vector bundles. Proc.London Math.Soc.(3), 50,1–26.

[D2] **Donaldson.S.K.(to appear)** Polynomial invariants for smooth four–manifolds. Topology.

[F/S1] **Fintushel,R.,Stern,R.(1985)** Pseudofree orbifolds. Ann.of Math. 122,335–364.

[F/S2] **Fintushel,R.,Stern,R.(1988)** Instanton homology of Seifert fibered homology 3–spheres. Preprint, Univ.of Utah, Salt Lake City.

[F] **Floer,A.(1988)** An instanton invariant for 3–manifolds. Commun.Math.Phys. 118,215–240.

[F/ST] **Furuta,M.,Steer,B.(1989)** The moduli spaces of flat connections on certain 3–manifolds. Preprint, Oxford.

[H/N] **Harder,G.,Narasimhan,M.S.(1975)** On the cohomology groups of moduli spaces of vector bundles over curves. Math.Ann. 212,215–248.

[K/K] **Kirk,P.A.,Klassen,E.P.(1989)** Representation spaces of Seifert fibered homology spheres. Preprint, California Institute of Tech., Pasadena.

[M/S] **Mehta,V.B.,Seshadri,C.S.(1980)** Moduli of vector bundles on curves with parabolic structure. Math.Ann. 248,205–239.

[N/R] **Neumann,W.,Raymond,F.(1978)** Seifert manifolds, plumbing, μ–invariant and orientation reversing maps. In: Algebraic and Geometric Topology, Springer LNM 664, 162–194.

[N/W] **Neumann,W.,Wahl,J.(to appear)** Casson invariants of links of singularities. Comment.Math.Helv..

[O/S/S] **Okonek,Ch.,Schneider,M., Spindler,H.(1980)** Vector bundles on complex projective spaces. Progress in Mathematics 3, Birkhäuser, Boston.

[O/V1] **Okonek,Ch.,Van de Ven,A.(1989)** Γ–type invariants associated to $PU(2)$–bundles and the differentiable structure of Barlow's surface. Invent.math., 95,601–614.

[O/V2] **Okonek,Ch.,Van de Ven,A.(1989)** Instantons, stable bundles and C^∞–structures on algebraic surfaces. Preprint, Bonn.

[P] **Pinkham,H.(1977)** Normal surface singularities with $\mathbf{C}^*$–action. Math.Ann. 227,183–193.

[U] **Ue,M.(1986)** On the diffeomorphism types of elliptic surfaces with multiple fibres. Invent.math. 84,633–643.

$\mathbf{Z}_a$-invariant SU(2) instantons over the Four Sphere

MIKIO FURUTA
The University of Tokyo, Hongo Tokyo 113, Japan and
The Mathematical Institute , Oxford

1. INTRODUCTION

The purpose of this article is to give a classification of invariant SU(2)-instantons on S^4 for some equivariant SU(2)-bundles over S^4 and to give some applications. We use the term *instanton* for anti-self-dual connection here. It is well known that the moduli spaces of SU(2)-instantons on the standard four sphere S^4 are smooth manifolds. When a group Γ acts on S^4 isometrically and P is a Γ-invariant SU(2)-bundle, the moduli space $M(P)$ of Γ-invariant instantons on P is defined as the quotient of the space of Γ-invariant instantons divided by the Γ-equivariant gauge transformations. Then $M(P)$ has a natural smooth structure as well. Instantons on S^4 are classified by the ADHM-construction [2] or the monad description [5], so, in principle, we have a description of invariant instantons. On the other hand, for some group actions the invariant instantons have some geometric interpretation: M. F. Atiyah pointed out as an important example that when Γ is the rotations around S^2 in S^4, the Γ-invariant instantons are interpreted as hyperbolic monopoles [1]. In this article we consider subgroups of the maximal torus of SO(4) as Γ and Γ-equivariant SU(2)-bundles over S^4 for which the moduli spaces of invariant instantons are one-dimensional, in particular when Γ is a finite cyclic group $\mathbf{Z}_a$ of order a.

A crucial observation is as follows. Suppose Γ is a cyclic group and the Γ-action is semifree with fixed point set $\{0, \infty\}$. The quotient S^3/Γ is called a lens space. Because $(S^3/\Gamma) \times \mathbf{R}$ is conformally equivalent to $(S^4 \setminus \{0, \infty\})/\Gamma$, a Γ-invariant instanton on S^4 induces an instanton on $(S^3/\Gamma) \times \mathbf{R}$. Conversely Uhlenbeck's removable singularity theorem [19] implies that an instanton on $(S^3/\Gamma) \times \mathbf{R}$ with L^2-bounded curvature comes from a Γ-invariant instanton on S^4.

In Section 2 a classification of invariant instantons is described for some equivariant bundles over S^4. This is a special case considered by D. M. Austin [3] and Y. Hashimoto and the author [13]. (The results of [13] are an extension of

---

Supported by the U.K. Science and Engineering Research Council

Hashimoto's MSc thesis (University of Tokyo, 1987)). As a byproduct the Euler numbers of the moduli spaces of SU(2)-instantons on S^4 are given [10]. In Section 3 an analogue of Floer's instanton homology group is defined for lens spaces with odd order fundamental groups [12]. Using the classification in Section 2 and the above correspondence between invariant instantons on S^4 and instantons on $(S^3/\Gamma) \times \mathbf{R}$, these groups are described explicitly. In Section 4 an application to cobordisms among lens spaces is explained; this is an extension of an argument in [11].

Acknowledgement. The author is grateful to M. Crabb for his reading through the manuscript and all his comments.

2. Description of one-dimensional moduli space

2.1 $S^1 \times S^1$-invariant instantons.

Let $T = S^1 \times S^1$ be a maximal torus of SO(4). Then topological isomorphism classes of T-equivariant SO(3)-bundles over S^4 with negative first Pontrjagin class are parametrized by $\{(k_1, k_2) : k_1, k_2 \in \mathbf{N}\}$. We write $\bar{P}(k_1, k_2)$ for the T-bundle corresponding to (k_1, k_2). Then $\bar{P}(k_1, k_2)$ is characterized by

(i) The T-action on $\bar{P}(k_1, k_2)_\infty$ (resp. $\bar{P}(k_1, k_2)_0$) is given by the conjugacy class of a homomorphism $f : T \to \mathrm{SO}(3)$ defined by

$$f(t_1, t_2) = \begin{pmatrix} \cos\theta & -\sin\theta & 0 \\ \sin\theta & \cos\theta & 0 \\ 0 & 0 & 1 \end{pmatrix},$$

where $e^{i\theta} = t_1^{k_1} t_2^{k_2}$ (resp. $t_1^{k_1} t_2^{-k_2}$). We call f the isotropy representation at ∞ (resp. 0).

(ii) $p_1(\bar{P}(k_1, k_2))[S^4] = -4k_1k_2$.

In this subsection we consider T-invariant instantons on $\bar{P}(k_1, k_2)$. Taking a double covering $\tilde{T}$ of T, if necessary, we may consider $\tilde{T}$-invariant instantons on a $\tilde{T}$-equivariant SU(2)-bundle $P(k_1, k_2)$ instead of $\bar{P}(k_1, k_2)$. (The choice of the double covering depends on k_1 and k_2.) Note that $c_2(P(k_1, k_2))[S^4] = k_1k_2$.

The ADHM-construction reduces classification of SU(2)-instantons to that of certain holomorphic $\mathrm{SL}_2(\mathbf{C})$-bundles over $\mathbf{P}^3$. Moreover S. K. Donaldson reduced the classification to that of holomorphic $\mathrm{SL}_2(\mathbf{C})$-bundles over $\mathbf{P}^2 = \mathbf{C}^2 \cup l_\infty$ which are trivial over l_∞. A pair of an SU(2)-instanton and a base point of the SU(2)-bundle at ∞ corresponds to a pair of a holomorphic $\mathrm{SL}_2(\mathbf{C})$-bundle and a holomorphic trivialization of the bundle on l_∞.

By considering this procedure equivariantly, we can reduce the classification of $\tilde{T}$-invariant SU(2)-instantons to that of $\tilde{T}$-equivariant holomorphic SL_2-bundles over $\mathbf{P}^2$ which are trivial on l_∞. Here we regard T as a subgroup of $\mathrm{U}(2) = \mathrm{SO}(4) \cap \mathrm{GL}_2(\mathbf{C})$ which acts on $\mathbf{P}^2$ naturally.

Since the $\tilde{T}$-action preserves the holomorphic structure of the bundle, it can be extended to an action of the complexification $\tilde{T}^{\mathbf{C}}$. Because $\mathbf{P}^2$ has a dense $\tilde{T}^{\mathbf{C}}$-orbit, it would be expected that a small number of data should classify $\tilde{T}^{\mathbf{C}}$-equivariant holomorphic bundles. In fact such equivariant $SL_2(\mathbf{C})$-bundles were classified by T. Kaneyama [17]. (Kaneyama assumed that the $\tilde{T}^{\mathbf{C}}$-acton is algebraic. This is shown by, for instance, looking at a $\tilde{T}^{\mathbf{C}}$-action on *monads*, which we shall mention later.)

In our case they are parametrized by $(k_1, k_2) \in \mathbf{N}$ and are expressed as $E(k_1, k_2)$ in the following exact sequence.

$$0 \to \mathcal{O}(-k_1 - k_2) \xrightarrow{\varphi} \mathcal{O} \oplus \mathcal{O}(-k_2) \oplus \mathcal{O}(-k_1) \to E(k_1, k_2) \to 0,$$

where $\varphi(f) = (z_0^{k_1+k_2} f, z_1^{k_1} f, z_2^{k_2} f)$. (We write (z_0, z_1, z_2) for the homogeneous coordinate of $\mathbf{P}^2$.) Let $t \in \tilde{T}^{\mathbf{C}}$ be a lift of $(t_1, t_2) \in T^{\mathbf{C}} = \mathbf{C}^* \times \mathbf{C}^*$. Then the action of t on $E(k_1, k_2)$ is induced from the actions on $\mathcal{O}(-k_1 - k_2)$, $\mathcal{O}$, $\mathcal{O}(-k_2)$ and $\mathcal{O}(-k_1)$ which are defined below.

$$\begin{aligned}
t \cdot f(z_0, z_1, z_2) &= t_1^{-k_1/2} t_2^{-k_2/2} f(z_0, t_1^{-1} z_1, t_2^{-1} z_2) && f \in \mathcal{O}(-k_1 - k_2) \\
t \cdot f(z_0, z_1, z_2) &= t_1^{-k_1/2} t_2^{-k_2/2} f(z_0, t_1^{-1} z_1, t_2^{-1} z_2) && f \in \mathcal{O} \\
t \cdot f(z_0, z_1, z_2) &= t_1^{k_1/2} t_2^{-k_2/2} f(z_0, t_1^{-1} z_1, t_2^{-1} z_2) && f \in \mathcal{O}(-k_2) \\
t \cdot f(z_0, z_1, z_2) &= t_1^{-k_1/2} t_2^{k_2/2} f(z_0, t_1^{-1} z_1, t_2^{-1} z_2) && f \in \mathcal{O}(-k_1)
\end{aligned}$$

The $\tilde{T}$-invariant instantons on the $\tilde{T}$-equivariant SU(2)-bundle $P(k_1, k_2)$ over S^4 correspond to the $\tilde{T}^{\mathbf{C}}$-equivariant $SL_2(\mathbf{C})$-bundle $E(k_1, k_2)$. Note that the base points of an SU(2)-bundle at ∞ are parametrized by SU(2) and the trivializations of a holomorphic bundle on l_∞ are, if any, parametrized by $SL_2(\mathbf{C})$. So, in general, one holomorphic bundle gives rise to a family of (non-based) instantons parametrized by $SL_2(\mathbf{C})/SU(2)$. But we have now group symmetries. If we consider the base points and the trivializations compatible with the $\tilde{T}$-action and the $\tilde{T}^{\mathbf{C}}$-action, then they are parametrized by U(1) and $\mathbf{C}^*$ respectively. Hence one $\tilde{T}^{\mathbf{C}}$-equivariant holomorphic bundle gives rise to a family of $\tilde{T}$-invariant instantons parametrized by $\mathbf{C}^*/U(1) = \mathbf{R}_+$. So we have

THEOREM 1 [3,13]. $M(P(k_1, k_2)) \cong \mathbf{R}_+$.

Note that the dilation $r : S^4 \to S^4$ $(r \in \mathbf{R}_+)$ $r(x) = rx$, $(x \in \mathbf{R}^4 \cup \{\infty\} = S^4)$ induces a free $\mathbf{R}_+$-action on $M(P(k_1, k_2))$. Theorem 1 says that $M(P(k_1, k_2))$ consists of exactly one orbit.

In particular the dimension of $M(P(k_1, k_2))$ is one. This can also be shown from the Atiyah-Bott-Lefschetz formula. Let P_k be an SU(2)-bundle over S^4 such that $c_2(P_k)[S^4] = k$ and M_k be the moduli space of instantons on P_k. The SO(5)-action on S^4 cannot be lifted to P_k if $k \neq 0$, but there is an SO(5)-action on M_k because the action can be lifted up to gauge transformations. Forgetting the T-action we can think of $M(P(k_1, k_2))$ as a submanifold of M_k for $k = k_1 k_2$. (Since every non-trivial instanton on S^4 is irreducible, the map $M(P(k_1, k_2)) \to M_k$ is injective. Its image is contained in the fixed point set M_k^T.) For $[A] \in M(P(k_1, k_2))$ the tangent space $(TM_k)_{[A]}$ is a T-module. From the Atiyah-Bott-Lefschetz formula we have

LEMMA 1 [12].

$$(TM_k)_{[A]} = -1 + \sum a_{ij} s_1^i s_2^j,$$
$$a_{ij} = |\{\varepsilon \in \{\pm 1\} : |i| \leq k_1 - \frac{1+\varepsilon}{2}, |j| \leq k_2 - \frac{1-\varepsilon}{2}\}|,$$

where s_1 and s_2 are the components of $(TS^4)_0$ as complex representation spaces of T: $(TS^4)_0 = s_1 + s_2$. (We fix an identification $S^4 = \mathbf{C}^2 \cup \{\infty\}$.)

The constant term of the above two-variable Laurent polynomial is one. This gives the dimension of $M(P(k_1, k_2))$.

We already gave an expression of the holomorphic $\tilde{T}$-bundle corresponding to an element of $M(P(k_1, k_2))$. But to consider $\mathbf{Z}_a$-invariant instantons later it is convenient to give the monad description.

We recall the monad description [5]. Let V be a k-dimensional complex vector space and W a 2-dimensional complex vector space. Suppose four linear maps α_1, α_2, a, b are given, where $b : W \to V$, $\alpha_1, \alpha_2 : V \to V$ and $a : V \to W$. When they satisfy

$$[\alpha_1, \alpha_2] + ba = 0,$$

one has maps A_Z and B_Z below parametrized by $Z = (z_0, z_1, z_2) \in \mathbf{C}^3 \setminus \{0\}$:

$$0 \longrightarrow V \xrightarrow{A_Z} V \oplus V \oplus W \xrightarrow{B_Z} V \longrightarrow 0,$$

$$A_Z = \begin{pmatrix} z_0\alpha_1 + z_1 I_k \\ z_0\alpha_2 + z_2 I_k \\ z_0 a \end{pmatrix}, \qquad B_Z = (-z_0\alpha_2 - z_2 I_k, z_0\alpha_1 + z_1 I_k, z_0 b).$$

They satisfy $B_Z A_Z = 0$. Moreover if A_Z is injective and B_Z is surjective for every Z, one has a holomorphic bundle $\coprod_{[Z]\in \mathbf{P}^2} \mathrm{Ker}\, B_Z / \mathrm{Im}\, A_Z$ over $\mathbf{P}^2$. A trivialization on l_∞ corresponds to an isomorphism $W \cong \mathbf{C}^2$.

For example $P(3,2)$ is given by

$$W = \langle f_{-\frac{3}{2},-1}, f_{\frac{3}{2},1} \rangle,$$
$$V = \langle e_{-1,\frac{1}{2}}, e_{0,\frac{1}{2}}, e_{1,\frac{1}{2}}, e_{-1,-\frac{1}{2}}, e_{0,-\frac{1}{2}}, e_{1,-\frac{1}{2}} \rangle$$

and α_1, α_2, a, b defined below.

$$\begin{array}{ccccccccc}
 & & e_{-1,\frac{1}{2}} & \xrightarrow{\alpha_1} & e_{0,\frac{1}{2}} & \xrightarrow{\alpha_1} & e_{1,\frac{1}{2}} & \xrightarrow{a} & f_{\frac{3}{2},1} \\
 & & {\scriptstyle\alpha_2}\uparrow & & {\scriptstyle\alpha_2}\uparrow & & {\scriptstyle\alpha_2}\uparrow & & \\
f_{-\frac{3}{2},-1} & \xrightarrow{b} & e_{-1,-\frac{1}{2}} & \xrightarrow{\alpha_1} & e_{0,-\frac{1}{2}} & \xrightarrow{\alpha_1} & e_{1,-\frac{1}{2}} & &
\end{array}$$

Other matrix elements of α_1, α_2, a and b are defined to be zero. Here we define a $\tilde{T}$-action on W and V by

$$t \cdot e_{\mu\nu} = t_1^{\mu} t_2^{\nu} e_{\mu\nu}, \qquad t \cdot f_{\mu\nu} = t_1^{\mu} t_2^{\nu} f_{\mu\nu},$$

where $t \in \tilde{T}$ is a lift of $(t_1, t_2) \in T$. Then these data define a $\tilde{T}$-equivariant holomorphic $\mathrm{SL}_2(\mathbf{C})$-bundle. In general the monad for $P(k_1, k_2)$ is given in a similar way.

To recover an instanton from a holomorphic bundle, one has to solve a (finite dimensional) variational problem [5]: find a hermitian metric on V such that $\| \alpha_1 \|^2 + \| \alpha_2 \|^2 + \| a \|^2 + \| b \|^2$ attains the minimum. (The metric of W is fixed because W is associated to the fibre of the SU(2)-bundle at ∞.) Then a solution gives rise to a set of data for the ADHM-construction from which one can find a connection form. For $P(k_1, k_2)$ the variational problem is reduced to an equation which is similar to *Kirchhoff's law*. Firstly let us write down solutions for $P(3, 2)$. Let $r \in \mathbf{R}_+$ be a parameter.

$$\begin{array}{ccccccccc}
 & & e_{-1,\frac{1}{2}} & \xrightarrow{\;r\;} & e_{0,\frac{1}{2}} & \xrightarrow{\frac{1+\sqrt{5}}{2}r} & e_{1,\frac{1}{2}} & \xrightarrow{\frac{3+\sqrt{5}}{2}r} & f_{\frac{3}{2},1} \\
 & & {\scriptstyle r}\uparrow & & {\scriptstyle \frac{-1+\sqrt{5}}{2}r}\uparrow & & {\scriptstyle r}\uparrow & & \\
f_{-\frac{3}{2},-1} & \xrightarrow[\frac{3+\sqrt{5}}{2}r]{} & e_{-1,-\frac{1}{2}} & \xrightarrow[\frac{1+\sqrt{5}}{2}r]{} & e_{0,-\frac{1}{2}} & \xrightarrow[r]{} & e_{1,-\frac{1}{2}} & &
\end{array}$$

Here a metric on V is given so that $\{e_{\mu\nu}\}$ is an orthonormal basis, and the positive number written at each arrow describes the square norm of a matrix element of α_1, α_2, a and b for this basis. In general, if we call these positive numbers *flows*, the equation is

(i) At each vertex the sum of the entering flows is zero. (In the above example we have, for instance, $r + (1+\sqrt{5})r/2 - (3+\sqrt{5})r/2 = 0$ at the vertex $e_{1,\frac{1}{2}}$.)

(ii) At each unit square the two products of the flows corresponding to $\alpha_1\alpha_2$ and $\alpha_2\alpha_1$ agree.

(Every monad describing $P(k_1, k_2)$ satisfies (ii). So (i) is the essential equation.) The author does not know how to solve this equation for general $P(k_1, k_2)$.

2.2 S^1-invariant instantons.

If one takes a sufficiently complicated subgroup of T, one could expect that its fixed point set in M_k is the same as the fixed point set for the T-action. For a natural number p let T_p be the subgroup $\{(t, t^p) : t \in S^1\}$ of T. We show that T_p satisfies this property if p is odd and larger than k. The following argument is outlined in [10]. For a fixed point $[A] \in M_k^{T_p}$ there is a unique lift of the T_p-action to an action on the adjoint bundle $ad\, P_k = P_k \times_{ad} \mathfrak{su}(2)$ preserving A. Suppose that the isotropy representation of the T_p-action on $(ad\, P_k)$ at ∞ (resp. 0) is

$$(t, t^p) \mapsto \begin{pmatrix} \cos\theta & -\sin\theta & 0 \\ \sin\theta & \cos\theta & 0 \\ 0 & 0 & 1 \end{pmatrix},$$

where $e^{i\theta} = t^{l_\infty}$ (resp. $e^{i\theta} = t^{l_0}$). Then one can use the Atiyah-Bott-Lefschetz formula to calculate the T_p-action on $(TM_k)_{[A]}$. The result is

$$Tr(t|(TM_k)_{[A]}) = -\frac{1+t^{1+p}}{(1-t)(1-t^p)}(t^{l_0}+t^{-l_0}+1) - \frac{1+t^{1-p}}{(1-t)(1-t^{-p})}(t^{l_\infty}+t^{-l_\infty}+1).$$

The right hand side must be a Laurent polynomial in t with non-negative coefficients and its value at $t=1$ must be equal to $\dim M_k = 8k-3$. From these requirements we can easily show, replacing l_∞ and l_0 by $-l_\infty$ and $-l_0$ if necessary, that $l_\infty = k_1 + pk_2$, $l_0 = k_1 - pk_2$ and $k = k_1k_2$ for some $k_1, k_2 \in \mathbf{N}$. (Looking at the value at 1 we have $2(l_\infty^2 - l_0^2)/p - 3 = 8k-3$, i. e. $l_\infty^2 - l_0^2 = 4pk$. Looking at the value at $e^{2\pi i/r}$ we have $l_0 \equiv \pm l_\infty \bmod p$. Note that we assumed that p is odd.)
Then the T_p-action on $ad\, P$ is isomorphic to a restriction of the T-action on $ad\, P(k_1, k_2$
Moreover if p is larger than k, then the constant term of the Laurent polynomial is one. Hence the dimension of $M_k^{T_p}$ is one. Since $\mathbf{R}_+$ acts freely on $M_k^{T_p}$, it must be a disjoint union of copies of $\mathbf{R}_+$. Because T is commutative, the T-action on M_k preserves $M_k^{T_p}$ as a set. Since an action of a compact group on $\mathbf{R}_+$ must be trivial, the T-action on $M_k^{T_p}$ must be trivial, so we have $M_k^{T_p} = M_k^T$.
We know from Theorem 1 that M_k^T is a disjoint union of $\mathbf{R}_+$'s and the number of components is equal to the number $d(k)$ of positive divisors of k. As an application we find the Euler characteristic number of M_k.

THEOREM 2. $\chi(M_k) = d(k)$.

Here $\chi(M_k) = \sum_i (-1)^i \dim H^i(M_k, \mathbf{R})$.

PROOF: Let X be $M_k \setminus M_k^T$. Then we have shown $X^{T_p} = \emptyset$. The T_p-action on X may not be free. So the quotient space X/T_p is not smooth in general, but is an orbifold, or a V-manifold. The projection map $X \to X/T_p$ is not in general a circle bundle, but is a circle bundle in the category of orbifolds (a circle V-bundle). Then we still have a version of the Thom-Gysin exact sequence for the de Rham cohomology groups:

$$\to H^{i-2}(X/T_p, \mathbf{R}) \to H^i(X/T_p, \mathbf{R}) \to H^i(X, \mathbf{R}) \to,$$

where $H^*(X, \mathbf{R})$ is the de Rham cohomology of X and $H^*(X/T_p, \mathbf{R})$ is here defined as the cohomology of the chain complex of smooth differential forms on X which are *basic* for the T_p-action. Then if $H^*(X, \mathbf{R})$ and $H^*(X/T_p, \mathbf{R})$ are finite dimensional, we obtain that $\chi(X) = \chi(X/T_p) - \chi(X/T_p) = 0$ and

$$\chi(M_k) = \chi(M_k^T) + \chi(X) = d(k) + 0 = d(k).$$

It suffices to see the finiteness of dim $H^*(M_k, \mathbf{R})$. This follows from the fact that the moduli space of based instantons $\tilde{M}_k$ is an SO(3)-bundle over M_k and that $\tilde{M}_k$ is a quasi-affine variety [5]. (In fact Hashimoto gave an explicit embedding of $\tilde{M}_k$ into an affine space by identifying $\tilde{M}_k$ with a space of representations of a certain algebra [14].) ∎

The above argument does not give which Betti number is not zero. But the following facts are known.

(i) J. Hurtubise showed that the first Betti number of M_k is zero. In fact $\pi_1(M_k) = 0$ if k is odd and $\pi_1(M_k) = \mathbf{Z}_2$ if even [15].

(ii) Y. Kamiyama showed that the second Betti number of M_k is also zero. In fact $H^2(M_k, \mathbf{Z}_p) = 0$ for a prime number p larger than k [16].

2.3 $\mathbf{Z}_a$-invariant instantons. Let V_1 and V_2 be two faithful complex 1-dimensional representation spaces of $\mathbf{Z}_a$. We think of $(V_1 \oplus V_2) \cup \{\infty\}$ as S^4 with a $\mathbf{Z}_a$-action. For simplicity we assume from now on that a is odd. (When a is even, the argument is parallel except that we need a double covering of $\mathbf{Z}_a$. For the details see [3,13].) Let L be the kernel of the map $\mathbf{Z} \times \mathbf{Z} \to \mathrm{Hom}(\mathbf{Z}_a, S^1)$ defined by $(i, j) \mapsto V_1^{\otimes i} \otimes V_2^{\otimes j}$. Then $\mathrm{Hom}(\mathbf{Z}_a, S^1)$ can be identified with lattice points on the torus $(\mathbf{R} \times \mathbf{R})/L$. A $\mathbf{Z}_a$-equivariant SU(2)-bundle P over S^4 is specified by the $\mathbf{Z}_a$-actions on $(P)_\infty$ and $(P)_0$ and the second Chern class $c_2(P)$. A $\mathbf{Z}_a$-action on $(P)_\infty$ or $(P)_0$ corresponds to a conjugacy class of a homomorphism from $\mathbf{Z}_a$ to SU(2), i. e. the isotropy representation. These three satisfy a compatibility condition. Note that the set $\mathrm{Hom}(\mathbf{Z}_a, \mathrm{SU}(2))/\mathrm{conj.}$ can be identified with the unordered pairs $\{f_1, f_2\}$ such that $f_1 + f_2 \equiv 0 \bmod L$. The compatibility condition is described as follows [12]. Let $\{f_0, -f_0\}$ and $\{f_\infty, -f_\infty\}$ be the pairs corresponding to the isotropy representations at 0 and ∞. Take one of the rectangles on $\mathbf{R} \times \mathbf{R}$ which satisfy.

(i) The four edges are parallel to $\mathbf{R} \times \{0\}$ or $\{0\} \times \mathbf{R}$, so the vertices are written as (x_1, y_1), (x_1, y_2), (x_2, y_1), (x_2, y_2) for $x_1 < x_2$ and $y_1 < y_2$. Moreover x_1, x_2, y_1 and y_2 are integers.

(ii) The projection of the two-point set $\{(x_1, y_2), (x_2, y_1)\}$ to $(\mathbf{R} \times \mathbf{R})/L$ is equal to $\{f_0, -f_0\}$ and the projection of $\{(x_1, y_1), (x_2, y_2)\}$ is $\{f_\infty, -f_\infty\}$.

Then the area $(x_2 - x_1)(y_2 - y_1)$ of the rectangle is well defined mod a and the compatibility condition is

$$c_2(P)[S^4] \equiv (x_2 - x_1)(y_2 - y_1) \bmod a.$$

In fact the $\tilde{T}$-equivariant bundle $P(x_2 - x_1, y_2 - y_1)$ can be regarded as a $\mathbf{Z}_a$-equivariant bundle P by restriction of the action, which satisfies $c_2(P) = (x_2 -$

$x_1)(y_2 - y_1)$ and has the required isotropy representations at 0 and ∞. (Here, since a is odd, we have a unique lift $\mathbf{Z}_a \to \tilde{T}$ of $\mathbf{Z}_a \subset T$.)
If we identify $\mathrm{Hom}(\mathbf{Z}_a, \mathrm{SU}(2))/\mathrm{conj.}$ with the set of isomorphism classes of flat SU(2)-connections on the lens space $S^3/\mathbf{Z}_a$, then the above formula could give the Chern-Simons invariants of these flat connections.
The $\mathbf{Z}_a$-equivariant bundles with one-dimensional moduli spaces are classified by using Lemma 1.

THEOREM 3 [3,13,12]. *Let $\{f_0, -f_0\}$ and $\{f_\infty, -f_\infty\}$ be any two unordered pairs corresponding to two elements of* $\mathrm{Hom}(\mathbf{Z}_a, \mathrm{SU}(2))/conj.$ *If the projection of a rectangle on* $\mathbf{R}\times\mathbf{R}$ *satisfying (i) and (ii) above to* $(\mathbf{R}\times\mathbf{R})/L$ *does not have self-intersection except vertices, then* $P(x_2 - x_1, y_2 - y_1)$ *with the restricted* $\mathbf{Z}_a$*-action has a one-dimensional moduli space of invariant instantons. Conversely all* $\mathbf{Z}_a$*-equivariant* SU(2)*-bundles with one-dimensional moduli spaces are given in this way. Moreover for each such* $\mathbf{Z}_a$*-equivariant* SU(2)*-bundle, the one-dimensional moduli space is diffeomorphic to* $\mathbf{R}_+$.

PROOF: Suppose that a rectangle with vertices $\{(x_i, y_j) : i, j = 1, 2\}$ satisfies the above assumption. From Lemma 1 it is equivalent to say that the number of identity representations in a $\mathbf{Z}_a$-module $(TM_k)_{[A]}$ is one for $[A] \in M(P)$, where P is $P(x_2 - x_1, y_2 - y_1)$ with the restricted $\mathbf{Z}_a$-action. Hence $M(P)$ is one-dimensional. Conversely if the moduli space for a $\mathbf{Z}_a$-equivariant SU(2)-bundle P is non-empty and one-dimensional, then, as in the previous subsection, the $\mathbf{Z}_a$-action can be extended to a $\tilde{T}$-action for a double cover $\tilde{T}$ of T and $\mathbf{Z}_a$-invariant instantons are actually $\tilde{T}$-invariant. Hence we can use the classification of $\tilde{T}$-invariant instantons to obtain the result. ∎

The monad description of $P(x_2 - x_1, y_2 - y_1)$ can be explained using the corresponding rectangle: the space V which appears in the monad is spanned by vectors corresponding to unit squares $\{(z_1, z_2), (z_1 + 1, z_2), (z_1, z_2 + 1), (z_1 + 1, z_2 + 1)\}$ $(z_1, z_2 \in \mathbf{Z})$ sitting in the rectangle. The maps α_1 and α_2 could be seen as 'flows' on the rectangle from the left to the right and from the bottom to the top respectively. The space W can be understood as a vector space spanned by two vectors corresponding to the two-point set $\{f_\infty, -f_\infty\}$. The maps a and b are local 'flows' around these two points.
When a rectangle on $(\mathbf{R} \times \mathbf{R})/L$ has self-intersection only on an edge, Lemma 1 can be used to see that the dimension of the corresponding moduli space is three. In fact in this case the 'flow' obtains one more (complex) dimensional freedom at the tangential edge. Similarly suppose that two rectangles corresponding to one-dimensional moduli spaces have intersection only on their edges and that the two vertices of the one rectangle for the isotropy representation at ∞ are equal to the two vertices of the other for the isotropy representation at 0, then the union of the two rectangles could be used to construct a three-dimensional moduli space. These pictures could give the monad description of all three-dimensional moduli spaces

to show that the the possibility of the diffeomorphism type of three-dimensional moduli space is only $(S^2 - \{ n\text{-points} \}) \times \mathbf{R}_+$, where $n = 0, 1, 2$ [3].
In general it is convenient to describe a $\mathbf{Z}_a$-equivariant monad as a collection of finite dimensional vector spaces assigned to each unit square of $(\mathbf{R} \times \mathbf{R})/L$ and $\{f_\infty, -f_\infty\}$ together with four 'flows'. The dimension of the vector space assigned to f_∞ or $-f_\infty$ should be one.

When we fix the vector spaces, we can construct a deformation of a monad by deforming the 'flows', which gives a family of instantons. Take an oriented closed path on the torus $(\mathbf{R} \times \mathbf{R})/L$ which lies in $(\mathbf{Z} \times \mathbf{R})/L \cup (\mathbf{R} \times \mathbf{Z})/L$. Then for each non-zero complex number c new flows are defined as follows.

(i) In the place where the flow does not go across the path, the new flow is the same as before.

(ii) To give the new flow, each component of flow (which is a homomorphism between vector spaces) is multiplied by c^i when it goes across the path, where i is the multiplicity of the intersection between the path and the flow.

(When the path goes through f_∞ or $-f_\infty$, we can arrange the new flow so that, for instance, the local flow a is the same as before and b is multiplied by c^i as above.) Since a torus has two closed paths homologically independent of each other, we can construct a family of $\mathbf{Z}_a$-equivariant holomorphic $\mathrm{SL}_2(\mathbf{C})$-bundles parametrized by $\mathbf{C}^* \times \mathbf{C}^*$, which gives a map from $\mathbf{C}^* \times \mathbf{C}^*$ to the moduli space divided by dilations. If P is a $\mathbf{Z}_a$-equivariant bundle such that $M(P)/\mathbf{R}_+$ is compact, then the image of this map should have a compact closure.
A similar idea is used in [13] to classify P such that $M(P)/\mathbf{R}_+$ is compact: the cases we have described ($M(P) \cong \mathbf{R}_+$ and $M(P) \cong S^2 \times \mathbf{R}_+$) turn out to be the only possibilities.
It is now well known that an end of a moduli space corresponds to a splitting of the bundle associated with 'bubbles'. Therefore by collecting our results, the following criterion can be shown.

THEOREM 4 [3,13]. *A $\mathbf{Z}_a$-equivariant* SU(2)*-bundle allows an invariant instanton on it, if and only if it is isomorphic to a connected sum of finitely many $\tilde{T}$-equivariant* SU(2)*-bundles with non-negative second Chern class. The connected sum is constructed by gluing neighbourhoods of* 0 *and* ∞ *together.*

To glue $P(k_1, k_2)$ and $P(l_1, l_2)$ at 0 and ∞, one needs the condition that the isotropy representation of $P(k_1, k_2)$ at 0 is isomorphic to that of $P(l_1, l_2)$ at ∞ if restricted to $\mathbf{Z}_a$. (In [13] the above theorem is shown under the assumption $a > c_2(P)[S^4]$. Austin proved it in general by using a different argument in [3].)

EXAMPLE 1. Let V_0 be the standard complex one-dimensional representation space of $\mathbf{Z}_a$. Let l_1 and l_2 be natural numbers coprime to each other, $V_1 = V_0^{\otimes l_2}$ and $V_2 = V_0^{\otimes -l_1}$. Suppose that a is sufficiently large compared with l_1 and l_2. (The precise condition will be given soon later.) Then $P(l_1, l_2)$ has the following properties.

(i) $P(l_1, l_2)$ with the restricted $\mathbf{Z}_a$-action has a one-dimensional moduli space.

The isotropy representation at ∞ is trivial and that at 0 is $V_0^{\otimes l_1 l_2} \otimes V_0^{\otimes -l_1 l_2}$.

(ii) Let P be a $\mathbf{Z}_a$-equivariant SU(2)-bundle over S^4 such that the isotropy representation at 0 is the same as that of $P(l_1, l_2)$. Suppose that there exists at least one invariant instanton on P and $c_2(P)[S^4] \leq c_2(P(l_1, l_2))[S^4](= l_1 l_2)$. Then $P \cong P(l_1, l_2)$.

PROOF: (i) If we substitute $s_1 = V_1$ and $s_2 = V_2$ in the Laurent polynomial given in Lemma 1, then the number of identity representations of $\mathbf{Z}_a$ in it is equal to the dimension. It turns out to be one if a is sufficiently large so that the conditions $xl_2 - yl_1 \equiv 0 \bmod a$, $|k_1| \leq l_1$ and $|k_2| \leq l_2$ for integers x and y imply $x = l_1$, $y = l_2$ or $x = -l_1$, $y = -l_2$.
(ii) From Theorem 4 we may assume that $P = P(k_1, k_2)$ for some k_1, k_2 and the isotropy representations of $P(k_1, k_2)$ and $P(l_1, l_2)$ at 0 agree. Therefore if a is sufficiently large so that the conditions $k_1 l_2 + k_2 l_1 \equiv \pm 2 l_1 l_2 \bmod a$ and $k_1 k_2 \leq l_1 l_2$ imply $k_1 = l_1$ and $k_2 = l_2$, then we obtain the result. ∎

We use this example in Section 4.

3. An analogue of Floer's instanton homology for lens spaces

For an oriented homology 3-sphere Σ, A. Floer defined the instanton homology groups $HI_*(\Sigma)$ using instantons on $\Sigma \times \mathbf{R}$ [9]. For closed 4-manifold or V-manifolds, it has been important to consider some cohomology classes on moduli spaces of instantons [6,7]. For instance Donaldson's polynomial invariant is defined by using certain cohomology classes. For 4-manifolds with boundaries, R. Fintushel and R. Stern used some $\mathbf{Z}_2$-cohomology classes introduced by Donaldson [6] to compute the polynomial invariants valued in the instanton homology groups in a special case [8]. However it has not been made clear how these $\mathbf{Z}_2$-cohomology classes are related to the instanton homology in a general context. In this section we use one of the $\mathbf{Z}_2$-cohomology classes to define an analogue of the instanton homology groups for lens spaces with odd order fundamental groups, as an attempt to understand this cohomology class. For the details see [12].

A lens space is not a homology 3-sphere. It is only a rational homology 3-sphere and *every* flat SU(2)-connection is reducible. When the instanton homology is defined, a reducible flat connection gives rise to a difficulty which ought to be solved in itself: a reducible connection has a non-trivial symmetry and it causes a quotient singularity in the space of connections. We do not deal with this problem here. However when one uses a cohomology class of degree one, as we shall see, it is rather easier to evaluate the class on certain moduli spaces if a flat connection has exactly one-dimensional symmetry.

Recall that the instanton homology groups are defined by using a chain complex (C, ∂) under a certain transversality assumption.

(i) C is spanned by the classes of irreducible flat SU(2)-connections on Σ.

(ii) The matrix element of ∂ for irreducible flat connections A_1 and A_2 is given by counting the number (with sign) of the components of a one-dimensional moduli space of instantons on $\Sigma \times \mathbf{R}$ which connect A_1 to A_2.

For a lens space $S^3/\mathbf{Z}_a$ with a odd, firstly we define (C', ∂') similarly.

(i)' C' as a $\mathbf{Z}_2$-vector space spanned by the classes of non-trivial (reducible) flat SU(2)-connections on $S^3/\mathbf{Z}_a$.

(ii)' The matrix element of ∂' for non-trivial flat connections A_1 and A_2 is given by counting the number (up to mod 2) of the components of a one-dimensional moduli space of instantons on $(S^3/\mathbf{Z}_a) \times \mathbf{R}$ which connect A_1 to A_2.

While Floer showed $\partial^2 = 0$, there is no reason for the square of ∂' to be zero as we shall see. Recall that the proof of $\partial^2 = 0$ depends on the two facts below.

(iii) The matrix element of ∂^2 for A_1 and A_3 is given by counting the number (with sign) of the ends of a two-dimensional moduli space of instantons on $\Sigma \times \mathbf{R}$ which connect A_1 to A_3.

(iv) The two-dimensional moduli space has a free $\mathbf{R}$ action and the quotient is one-dimensional. Hence the number of ends (with sign) is zero.

The reason why the two dimensional moduli space comes in is explained as follows [9]. Suppose we are given a non-empty one-dimensional moduli space of instantons which connect A_1 to A_2, and similarly A_2 to A_3. Then one can construct an end of a moduli space $M(A_1, A_3)$ of instantons which connect A_1 to A_3. When A_2 is irreducible, the dimension of the moduli space is two, which is the sum of the dimensions of two moduli spaces. However if A_2 is a non-trivial reducible connection, then $M(A_1, A_3)$ becomes three-dimensional, where the extra one dimension comes from the dimension of the symmetry of A_2. In this case the number of ends $M(A_1, A_3)/\mathbf{R}$ is not necessary zero (mod 2) since its dimension is two.

The idea to define an analogue of instanton homology groups is as follows.

(iv)' In the above situation suppose A_2 is a non-trivial reducible flat connection and suppose we have a $\mathbf{Z}_2$-cohomology class u of degree one. An end of $M(A_1, A_3)/\mathbf{R}$ is diffeomorphic to $S^1 \times (0, 1)$. (Here S^1 is the symmetry of A_2 which gives an extra parameter in gluing two connections.) Then the number of ends such that $u[S^1] = 1$ should be zero mod 2, since it is the evaluation of u by the boundary of truncated $M(A_1, A_3)/\mathbf{R}$.

We define u by using the Dirac operator $D(A_1, A_3)$ twisted by the bundle on which A_1 and A_3 are connected. If the numerical index of $D(A_1, A_3)$ is even, then the determinant line bundle for the family of the Dirac operators descends to a real line bundle on $M(A_1, A_3)$ [6]. Then u is defined as its first Stiefel-Whitney class. Let $D(A_1, A_2)$ and $D(A_2, A_3)$ be similar twisted Dirac operators. Then we have

$$\operatorname{ind} D(A_1, A_3) = \operatorname{ind} D(A_1, A_2) + \operatorname{ind} D(A_2, A_3)$$

(when Σ is a lens space). Hence if $\operatorname{ind} D(A_1, A_3)$ is even, then the parities of $\operatorname{ind} D(A_1, A_2)$ and $\operatorname{ind} D(A_2, A_3)$ agree and moreover we can show that it is also equal to $u[S^1]$ [12 Proposition 3.2].

Now we define a map $\partial'' : C' \to C'$ as follows.

(ii)" The matrix element of ∂'' for A_1 and A_2 is given by the number (up to mod 2) of components of a one-dimensional moduli space of instantons which connect A_1 to A_2 such that the associated twisted Dirac operator has odd index.

Then the argument in (iv)' implies $\partial''^2 = 0$. An analogue of the instanton homology is defined as the homology group of (C', ∂'').

REMARK. We can introduce a $\mathbf{Z}_8$-grading for (C', ∂'') [12].

Recall that instantons on $(S^3/\mathbf{Z}_a) \times \mathbf{R}$ with L^2-bounded curvatures can be regarded as $\mathbf{Z}_a$-invariant instantons on S^4. We can use the classification of $\mathbf{Z}_a$-equivariant SU(2)-bundles which have one-dimensional moduli spaces of invariant instantons to describe the boundary map ∂'' explicitly. In addition to Theorem 3 we only have to calculate the index of the twisted Dirac operator up to mod 2. It can be shown that the index is equal to the second Chern number mod 2. Hence we obtain the following description.

PROPOSITION 1. [12 Theorem 4.2]

(i) *C' is a $\mathbf{Z}_2$-vector space spanned by the pairs $\{f, -f\}$ ($f \neq 0$) of lattice points of $(\mathbf{R} \times \mathbf{R})/L$.*

(ii) *The matrix elements of ∂'' correspond to rectangles on $(\mathbf{R} \times \mathbf{R})/L$ without self-intersection such that the centre of the rectangles are of the form $(a/2, a/2)$ mod L*

4. COBORDISMS AMONG LENS SPACES

One could regard a moduli space M of instantons on a 4-manifold X as a cobordism between its ends and its singularities. When both can be described by some topological data of X, each cohomology class of M of degree dim $M - 1$ gives rise to a certain equation for the data. Such an idea was first developed by Donaldson [4] and subsequently used by Fintushel and Stern [7] and T. Lawson [18] for V-manifolds.
Suppose given a sequence of instantons. Then an end of the moduli space of instantons corresponds to divergence of their curvatures at some points on the 4-manifold. On a smooth 4-manifold the divergence could be captured by instantons on S^4 [4]. On the other hand the divergence on a V-manifold could be understood by using instanton on S^4/Γ, where Γ is a finite subgroup of SO(4).
Lawson used a certain non-existence result of invariant instantons on S^4 to see compactness of some moduli spaces for V-manifolds and used it to obtain some results in topology [18].
In this section we use Example 1 in Section 2.3 to give an application to topology. Let l_1 and l_2 be natural numbers coprime to each other and a be an odd number sufficiently large compared with l_1 and l_2. Let us identify S^3 with the unit sphere of the $\mathbf{Z}_a$-module $V_0^{\otimes l_2} \oplus V_0^{\otimes -l_1}$, where V_0 is the standard one-dimensional representation. We write $L(a; l_1, l_2)$ for the quotient $S^3/\mathbf{Z}_a$.

THEOREM 5. *Suppose X be an oriented closed 4-dimensional V-manifold which satisfies*

(i) $\pi_1(X) = 1$. *(The fundamental group of the underlying space of X, not the orbifold fundamental group of X.)*
(ii) $H_2(X, \mathbf{Q}) = 0$.
(iii) *There is a singular point p whose neighbourhood is of the form $cL(a; l_1, l_2)$ (the cone on $L(a; l_1, l_2)$).*

Then there is a singular point ($\neq p$) whose neighbourhood is of the form S^3/Γ with $|\Gamma| \geq a/(l_1 l_2)$.

SKETCH OF A PROOF: Let us identify S^4 with $(V_0^{\otimes l_2} \oplus V_0^{\otimes -l_1}) \cup \{\infty\}$. Then we can think of X as a 'connected sum' of X and $S^4/\mathbf{Z}_a$ at p and ∞. Let P be an SU(2)-V-bundle defined by a 'connected sum' of $X \times \mathrm{SU}(2)$ and $P(l_1, l_2)/\mathbf{Z}_a$. (We can take the connected sum because the $\mathbf{Z}_a$-action on $(P(l_1, l_2))_\infty$ is trivial.)
Let $M(P)$ be the moduli space of instantons on P. Then, after deforming slightly if necessary, $M(P)$ has a structure of a smooth one-dimensional manifold. Using the one-dimensional moduli space of invariant instanton on $P(l_1, l_2)$, we can construct an end of $M(P)$ diffeomorphic to an interval [4,6,11]. The property of $P(l_1, l_2)$ showed in Example 1 (ii) says that this is the only end where curvature of instantons diverges at p. Since the number of ends of a one-dimensional manifold is even, there must be an end where curvature of instantons diverges at some other point q. The amount of L^2-norm of the curvature concentrated on q is equal to or less than the total amount of the L^2-norm. When we write the neighbourhood of q as $c(S^3/\Gamma)$, then this inequality implies $|\Gamma| \geq a/(l_1 l_2)$. ∎

COROLLARY 1. *Let Σ be the connected sum $L(a; l_1, l_2)\#(\#_{i=1}^n S^3/\mathbf{Z}_{a_i})$, where $S^3/\mathbf{Z}_{a_i}$ is a lens space with fundamental group $\mathbf{Z}_{a_i}$. Suppose that a is an odd number sufficiently large compared with l_1 and l_2, and that $a_i < a/(l_1 l_2)$ for $i = 1, \ldots, n$. Then Σ cannot be smoothly embedded in S^4.*

PROOF: Suppose there is an embedding $\Sigma \subset S^4$. Then S^4 is divided into two pieces. From one of them a counterexample of the previous theorem can be constructed. ∎

Other applications of $P(l_1, l_2)$ are given in [11] when $l_1 = l_2 = 1$.
We remark that, using an argument similar to the above, one could conversely show the existence of some invariant instantons on S^4 *topologically* without appealing to any classification. The simplest example would be to show that M_1 is non-empty: let P be an SU(2)-bundle over $\mathbf{P}^2$ with $c_2(P)[\mathbf{P}^2] = 1$, then the moduli space $M(P)$ of instantons on P cannot be compact because a singular point of $M(P)$ requires an end of $M(P)$, which implies $M_1 \neq \emptyset$. In order to consider invariant instantons on S^4 one could use, instead of $\mathbf{P}^2$, the quotient of a weighted S^1-action on $S^5 = S(\mathbf{C}^3)$, which is a rational $\mathbf{P}^2$ with (at most) three singular points of the form $c(S^3/\mathbf{Z}_a)$. A similar argument was used by Fintushel and Stern for another direction [7].

References

1. M. F. Atiyah, *Magnetic monopoles in hyperbolic spaces*, Vector bundles on algebraic varieties, Tata Institute of Fundamental Research, Bombay, 1984.
2. M. F. Atiyah. V. G. Drinfeld, N. J. Hitchin and Yu. I. Manin, *Constructions of instantons*, Phys. Lett. **65 A** (1978), 185-187.
3. D. M. Austin, SO(3)-*Instantons on* $L(p,q) \times \mathbf{R}$, preprint.
4. S. K. Donaldson, *An application of gauge theory of 4-dimensional topology*, J. Differential Geometry **18** (1983), 279-315.
5. S. K. Donaldson, *Instantons and geometric invariant theory*, Commun. Math. Phys. **93** (1984), 453-461.
6. S. K. Donaldson, *Connections cohomology and the intersection forms of 4-manifolds*, J. Differential Geometry **24** (1986), 295-341.
7. R. Fintushel and R. Stern, *Pseudofree orbifolds*, Ann. of Math. **122** (1985), 335-364.
8. R. Fintushel and R. Stern, *Homotopy K3 surfaces containing* $\Sigma(2,3,7)$, preprint.
9. A. Floer, *An instanton invariant for 3-manifolds*, Commun. Math. Phys. **118** (1988), 215-240.
10. M. Furuta, *Euler number of moduli spaces of instantons*, Proc. Japan Acad. **63, Ser. A** (1987), 266-267.
11. M. Furuta, *On self-dual pseudo-connections on some orbifolds*, preprint.
12. M. Furuta, *An analogue of Floer homology for lens spaces*, preprint.
13. M. Furuta and Y. Hashimoto, *Invariant instantons on* S^4, preprint.
14. Y. Hashimoto, *Instantons and representations of an associative algebra*, preprint, University of Tokyo.
15. J. Hurtubise, *Instantons and jumping lines*, Commun. Math. Phys. **105** (1986), 107-122.
16. Y. Kamiyama, *The 2 dimensional cohomology group of moduli space of instantons*, preprint, University of Tokyo.
17. T. Kaneyama, *On equivariant vector bundles on an almost homogeneous variety*, Nagoya Math. J. **57** (1975), 65-86.
18. T. Lawson, *Compactness results for orbifold instantons*, Math. Z. **200** (1988), 123-140.
19. K. K. Uhlenbeck, *Removable singularities in Yang-Mills fields*, Commun, Math. Phys. **83** (1982), 11-30.

PART 3

DIFFERENTIAL GEOMETRY AND MATHEMATICAL PHYSICS

Differential geometry has proved to be a natural setting for large parts of mathematical physics and conversely mathematical physics has provided a supply of new ideas and problems for differential geometers. Perhaps the best known example of this two-way interaction is given by the instanton solutions of the Yang-Mills equations – first noted by physicists but now playing an important role in several areas of mathematics. There are, however, many other very interesting special equations and theories; some of them, like the "monopole" equations, close relatives of the Yang-Mills instantons and some rather different. In this section we have a number of papers which exemplify this rich interaction.

The paper of Manton describes the Skyrme model, which is of practical interest in physics and is also very attractive mathematically. The theory appears to offer a number of challenging problems in the calculus of variations; being a variant of the well-known harmonic map theory in which the energy integrand is modified by a quartic term. The intriguing scheme described by Manton, relating the Skyrme model to instantons, also displays very well the beautiful classical geometry involved in the explicit description of the instanton solutions.

An important line of research on instantons, going back to the seminal paper of Atiyah and Jones [AJ], bears on the limiting behaviour of the homotopy and homology groups of the instanton moduli spaces over S^4, for large Chern numbers. The "Atiyah-Jones conjecture" suggests that these agree with the homotopy and homology groups of the third loop space of the structure group. From quite different directions, Taubes and Kirwan have made important advances on this problem recently. Analogous problems and results apply to the moduli spaces of monopoles. The paper of Cohen and Jones below describes more refined results in this direction, giving a complete description of the homology of all the monopole moduli spaces in terms of the braid groups (which also enter into the Jones theory of link invariants, as described in Atiyah's lecture in Durham).

Next we have two papers on the oldest branch of differential geometry—the geometry of submanifolds. The article of Hartley and Tucker develops a general framework for dealing with variational problems for submanifolds, more complicated than the simple minimal surface problem. A well-known instance of this kind of theory in the mathematical literature is the work initiated by Willmore, for surfaces in $\mathbf{R}^3$. The paper of Burstall gives a fine illustration of the application of the holomorphic geometry of the Penrose twistor space to minimal surfaces.

The papers of Tod and Wood are quite closely related. Both consider special differential geometric structures in 3-dimensions, with particular reference to Thurston's homogeneous geometries, which have to do with the space of geodesics in a 3-manifold (the mini-twistor space, in Tod's terminology). These structures seem to have a good deal of potential, posing many natural questions (for example the existence of Einstein-Weyl structures on general 3-manifolds) and offering scope for significant interactions with 4-dimensional geometry. Note that there are similarities between the discussion in the last section of Wood's paper, on the passage from a Seifert 3-manifold to an elliptic surface, and the technique described in the paper of Okonek for studying the representations of the fundamental group of a Seifert manifold.

[AJ] Atiyah, M.F. and Jones, J.D.S. *Topological aspects of Yang-Mills theory* Commun. Math. Phys. **61** (1978) 97-118

Skyrme Fields and Instantons

N.S. MANTON

Department of Applied Mathematics and Theoretical Physics
University of Cambridge
Silver Street, Cambridge CB3 9EW
England

ABSTRACT The first part of this paper is a brief review of the Skyrme model, and some of the mathematical problems it raises. The second part is a summary of the proposal by M.F. Atiyah and the author to derive families of Skyrme fields from Yang-Mills instantons.

1 THE SKYRME MODEL

Hadronic physics at modest energies (a few GeV) is concerned with the interactions of nucleons (protons and neutrons) and of pions. About 30 years ago, Skyrme suggested a model for these particles which is still useful (Skyrme, 1962), despite the fact that the particles are now believed to be bound states of quarks. In the Skyrme model only the pion field appears, and the nucleons are quantum states of a classical soliton solution of the pion field equations, known as the Skyrmion.

Nucleons have baryon number 1, their antiparticles have baryon number -1, and pions have baryon number 0. In any physical process the total baryon number is unchanged. In the Skyrme model, a field configuration has a conserved integral topological charge which Skyrme identified with the baryon number. The Skyrmion has charge 1, and there is a similar solution with charge -1.

Skyrme's pion field is a scalar field U taking values in $SU(2)$. I shall mainly consider fields at a given time, and not discuss dynamics much. In this case, U is a map from physical space $\mathbf{R}^3$ to $SU(2)$. The uniform field $U = 1$ represents the vacuum, and all field configurations are assumed to be asymptotically like the vacuum so $U(\mathbf{x}) \to 1$ as $|\mathbf{x}| \to \infty$. Space may therefore be compactified to a 3-sphere of infinite radius. Let $S^3(R)$ denote a 3-sphere with its standard metric and radius R. $SU(2)$ with its standard metric is $S^3(1)$. U is effectively a map

$$U : S^3(\infty) \to S^3(1) . \tag{1}$$

Its degree, $\deg U$, is a topological invariant and an integer. Skyrme identified $\deg U$ with the baryon number.

U maps an infinitesimal sphere of radius ϵ, centred at $\mathbf{x}$, to a neighbourhood of $U(\mathbf{x})$. To lowest order in ϵ, this neighbourhood is an ellipsoid, with principal axes $\mu_1\epsilon, \mu_2\epsilon$ and $\mu_3\epsilon$, say. The energy density at $\mathbf{x}$, proposed by Skyrme, is (Manton, 1987)

$$e(\mathbf{x}) = \mu_1^2 + \mu_2^2 + \mu_3^2 + \mu_1^2\mu_2^2 + \mu_2^2\mu_3^2 + \mu_3^2\mu_1^2 \tag{2}$$

and the total field energy is

$$E = \int_{\mathbf{R}^3} e(\mathbf{x})\, d^3x\,. \tag{3}$$

Note the following about this energy expression:
1) E is the potential energy of the Skyrme field at a given time . Using Lorentz invariance one can obtain the kinetic energy and hence the Lagrangian for dynamical fields. The kinetic energy expression defines a metric on the function space of static fields.
2) The vacuum field $U = 1$ has zero energy, and $\deg U = 0$.
3) Presumably, finite energy implies that $U(\mathbf{x}) \to$ const as $|\mathbf{x}| \to \infty$, but this may not have been rigorously proved.
4) E is in dimensionless form. The energy unit and length unit are determined from experimental properties of hadrons.
5) The symmetries of E are the Euclidean group of $\mathbf{R}^3$, and the $O(4)$ group of $S^3(1)$. The latter is the "chiral symmetry" group. The choice of a vacuum $U = 1$ breaks this down to $O(3)$ which is "isospin symmetry".

It follows immediately from (2) that

$$e(\mathbf{x}) \geq 6\mu_1\mu_2\mu_3\,, \tag{4}$$

and since $\mu_1\mu_2\mu_3$ is the modulus of the Jacobian of the map U, and the volume of $S^3(1)$ is $2\pi^2$, the energy satisfies the inequality (Fadeev, 1976)

$$E \geq 12\pi^2\,|\deg U|\,. \tag{5}$$

Let E_n denote the infimum of the energy for fields of degree n. The symmetry $U \to U^{-1}$ (the inverse in $SU(2)$) changes the sign of $\deg U$, so $E_n = E_{-n}$. It has been shown (Castillejo and Kugler, 1987) that $E_n < E_l + E_{n-l}$ for any integer l not equal to 0 or n, and Esteban has shown, assuming this inequality, that the infimum is attained for each integer n by a smooth field whose energy is concentrated in a single region of space (Esteban, 1986). The physical meaning of the strict inequality is that there are attractive forces in the Skyrme model. It is easy to prove that $E_n \leq E_l + E_{n-l}$ by considering fields of degrees l and $n-l$ glued together at a large separation, but the strict inequality is less obvious.

The vacuum is the lowest energy field of degree 0. The lowest energy field of degree 1 is not known for certain, but physicists have assumed, and numerical evidence makes it likely, that it is spherically symmetric. The lowest energy spherically symmetric field is known as the Skyrmion, and its standard form is

$$U(\mathbf{x}) = \cos f(r)1 + i \sin f(r)\hat{\mathbf{x}} \cdot \sigma \,. \tag{6}$$

Here $r = |\mathbf{x}|, \hat{\mathbf{x}} = \mathbf{x}/r$ and σ^1, σ^2 and σ^3 are the Pauli matrices. The profile function $f(r)$ has been determined numerically, by solving the variational equation for f obtained from the energy functional. The boundary conditions are $f(0) = \pi$ and $f(r) \to 0$ as $r \to \infty$, so U is continuous at the origin and $\deg U = 1$. The energy of the Skyrmion is $1.231\ldots \times 12\pi^2$ (Adkins, Nappi and Witten, 1983; Jackson and Rho, 1983). A six-parameter family of Skyrmions is generated from (6) by the action of symmetries. The centre can be moved to an arbitrary point, and the orientation changed by conjugating U with some fixed element of $SU(2)$. Replacing f by $-f$ gives the anti-Skyrmion with the same energy but degree -1. The physical nucleons and anti-nucleons are obtained by promoting the position and orientation collective coordinates to dynamical variables and quantizing these. The quantum states are characterized by their momentum, their spin and their isospin. The nucleons have spin $\frac{1}{2}$ and isospin $\frac{1}{2}$, with the isospin "up" for the proton and "down" for the neutron.

Skyrme's motivation for choosing the energy density (2) was to have a simple generalization of the harmonic map energy density $\mu_1^2 + \mu_2^2 + \mu_3^2$ whose variational equations had non-trivial solutions in $\mathbf{R}^3$. Skyrme's energy density is geometrically natural in three dimensions, and one may use it to define the energy of maps from any 3-dimensional Riemannian manifold M to another Riemannian manifold N. In this general context one may ask: (i) Is the infimum of the energy for maps in each homotopy class attained by some smooth map that satisfies the variational equations? (ii) Are there saddle-point solutions, i.e. non-minimal solutions of the variational equations? (iii) Is the Skyrme energy functional in some sense a Morse function? These questions are open.

Some explicit (numerical) solutions to the equations have been found for special geometries. For example, solutions of all degrees are known for maps $U : S^3(R) \to S^3(1)$, where R is finite (Jackson, Manton and Wirzba, 1989). Most of these are saddle points, and most do not have good limiting behaviour as $R \to \infty$. Other solutions, representing Skyrme crystals, are known for maps from a flat 3-torus to $S^3(1)$ (Kugler and Shtrikman, 1988; Castillejo et al, 1988). On the other hand, no solutions other than the vacuum are known for maps of degree zero from $\mathbf{R}^3$ to $S^3(1)$, despite an attempt to find a saddle-point solution (Bagger, Goldstein and Soldate, 1985).

A more detailed problem is the following. It is fairly easy to understand that the energy bound $12\pi^2$ is exceeded by the Skyrmion because $\mathbf{R}^3$ and $S^3(1)$ are not isometric (Manton, 1987). More generally, there is a topological lower bound on the energy for maps $U : M \to N$ which cannot be attained unless U is an isometry. The problem is to find a stronger lower bound in the case that M and N are geometrically distinct and there are no isometric maps between them.

For maps of degree 2 from $\mathbf{R}^3$ to $S^3(1)$, two solutions of the Skyrme field equations are known. The first has the spherically symmetric form (6), but with $f(0) = 2\pi$. The energy is $1.83\ldots \times 24\pi^2$, which is greater than that of two well-separated Skyrmions (Jackson and Rho, 1983). It is a saddle-point of the energy functional and has six unstable modes as well as six zero modes (Wirzba and Bang, 1989). The other solution is axisymmetric and has energy $1.18\ldots \times 24\pi^2$, which is less than that of two well-separated Skyrmions (Kopeliovich and Shtern, 1987; Verbaarschot, 1987; Schramm, Dothan and Biedenharn, 1988). The energy density is concentrated in a toroidal region. It is likely that this solution is the lowest energy Skyrme field of degree 2, and it has eight zero modes.

One of the central problems in hadronic physics is to understand the interaction of two nucleons at low energy. It is known experimentally that there is one bound state of a proton and neutron - the deuteron - and there is a wealth of scattering data. Much can be described with semi-phenomenological nucleon-nucleon potential models, but there is no deep understanding of these. It is not yet possible to calculate low energy phenomena using QCD, the theory of quarks and their interactions. It is therefore a challenge to see if the Skyrme model can describe them. In principle, one should treat the Skyrme model as a quantum field theory and restrict attention to the sector where the fields have degree 2. In practice, this leads to all sorts of conceptual and computational difficulties. Instead, one may try to select a finite dimensional submanifold of Skyrme fields, whose coordinates are the physically relevant degrees of freedom at low energy, i.e. collective coordinates, and one should quantize.

The simplest version of this idea is to quantize the eight collective coordinates of the orbit of the lowest energy degree 2 solution (Braaten and Carson, 1988). One of the quantum states is qualitatively like the deuteron. However, to describe the deuteron quantitatively and to describe nucleon-nucleon scattering one needs at least 12 collective coordinates since two well-separated Skyrmions have 6 collective coordinates each, namely their positions and orientations. A candidate for a 12-dimensional set of Skyrme fields of degree 2 is the unstable manifold of the orbit of the spherically symmetric solution (Manton, 1988). This manifold has not been investigated in detail, but it probably includes well-separated Skyrmions in all positions and orientations, as well as the orbit of lowest energy fields. It is an open problem to determine numerically which fields lie on this manifold, and to ascertain whether the manifold is smooth at the lowest energy fields or has cusps there. It needs to be smooth to give a physically sensible model.

2 SKYRME FIELDS FROM INSTANTONS

One natural way to obtain static Skyrme fields is as the holonomy of instantons (Atiyah and Manton, 1989).

Suppose A_μ is any $SU(2)$ Yang-Mills gauge potential in (Euclidean) $\mathbf{R}^4$ with finite action and 2nd Chern class k. In a suitable gauge, $A_\mu(x)$ decays faster than $|x|^{-1}$ as $|x| \to \infty$. Let the time-lines in $\mathbf{R}^4$ denote the lines parallel to the time axis. They are labelled by points of $\mathbf{R}^3$. Let $U(\mathbf{x})$ be the holonomy of A_μ along the time-line labelled by $\mathbf{x}$. Formally

$$U(\mathbf{x}) = \mathcal{P} \exp - \int_{-\infty}^{\infty} A_\tau(\mathbf{x}, \tau)\, d\tau \tag{7}$$

where τ is the (Euclidean) time-cordinate and $\mathcal{P}$ denotes path ordering. U takes values in $SU(2)$, and hence may be regarded as a Skyrme field in $\mathbf{R}^3$.

U is unchanged under a large class of gauge transformations, but to ensure a completely gauge invariant definition of U one should regard A_μ as defined on $\mathbf{R}^4$ conformally compactified to S^4, and U as the holonomy along a circle on S^4 which starts and ends at the point at infinity. In most cases, this is equivalent to closing the contour in (7) with a semi-circle at infinity. U is then well-defined up to conjugation by a fixed ($\mathbf{x}$-independent) element of $SU(2)$. Also, in this way, one ensures that $U(\mathbf{x}) \to 1$ as $|\mathbf{x}| \to \infty$. It is a basic topological fact that the field U, regarded as a Skyrme field, has degree k.

The moduli space of k-instantons (anti-self-dual Yang-Mills fields of 2nd Chern class k), M_k, is an $8k$-dimensional connected manifold (not $8k$-3 because one allows conjugation by fixed elements of $SU(2)$) (Atiyah, 1979). The holonomies have one dimension less, as a time-translation doesn't affect them. These instantons therefore generate a connected $(8k-1)$-dimensional manifold of Skyrme fields of degree k, $\tilde{M}_k = M_k/\mathbf{R}$.

There is no precise relationship between the anti-self-duality equations for Yang-Mills fields in $\mathbf{R}^4$ and the Skyrme equations in $\mathbf{R}^3$, so it is not surprising that none of the instanton-generated Skyrme fields are solutions of the Skyrme equations. However, some are good approximations to solutions discussed in Sect. 1, and $\tilde{M}_k$ smoothly interpolates between these approximate solutions. In fact, the coordinates of $\tilde{M}_k$ seem to correspond well to the collective coordinates of Skyrme fields relevant to k-nucleon physics at low energy. The symmetry group acting on M_k is the product of $SO(3)$ (the adjoint action of $SU(2)$ on U) and the 15-dimensional conformal group of $\mathbf{R}^4$. The $SO(3)$ survives the holonomy construction as the isospin symmetry of Skyrme fields, but the conformal group is broken down to the Euclidean group of $\mathbf{R}^3$ and dilations.

The 1-instantons generate a 7-dimensional set of spherically symmetric Skyrme fields. The seven coordinates define the centre, the orientation and the scale size. In the standard position, the Skyrme field is of the form (6), with

$$f(r) = \pi\left[1 - \left(1 + \frac{\lambda^2}{r^2}\right)^{-\frac{1}{2}}\right]. \tag{8}$$

λ is the scale parameter. For this simple profile, the minimal value of the Skyrme energy is $1.24\ldots \times 12\pi^2$ when $\lambda^2 = 2.11\ldots$, which exceeds the energy of the Skyrmion by less than 1%.

The 2-instantons generate a 15-dimensional manifold of Skyrme fields. Some of these 2-instantons may be identified with two well-separated 1-instantons, and as the time-separation tends to infinity, the resulting Skyrme field tends to a product of two Skyrme fields of degree 1. 14 of the 15 dimensions are accounted for by the positions, orientations and scales of these two degree 1 fields, and the last is the time separation of the instantons which has little effect in the limit. If the spatial separation is also large, then the time-separation may be continuously increased from $-\infty$ to ∞. The effect is to reverse the order of the product of the Skyrme fields. A particularly symmetric configuration can occur when the time-separation is zero.

2-instantons with rotational symmetry about the time-line $\mathbf{x} = 0$ generate Skyrme fields of the form (6). The profile function $f(r)$ is quite complicated in general. However, the 2-instantons with time reversal symmetry which correspond to two well-separated single instantons of the same scale size λ (they have the same orientation because of the symmetry) give, in the limit of infinite separation, the simple profile

$$f(r) = 2\pi\left[1 - \left(1 + \frac{\lambda^2}{r^2}\right)^{-\frac{1}{2}}\right]. \tag{9}$$

The Skyrme energy is minimized when $\lambda^2 = 2.62\ldots$ and then $E = 1.86\ldots \times 24\pi^2$. This is probably the lowest energy Skyrme field of degree 2 with $SO(3)$ symmetry that is generated from instantons. Its energy again exceeds the energy of the $SO(3)$ symmetric solution of the Skyrme equations by about 1%.

A general formula for 2-instantons is known (Jackiw, Nohl and Rebbi, 1977). Hartshorne has given a geometric characterisation of 2-instantons, and shown that all can be expressed in this form (Hartshorne, 1978). These instantons are obtained from an $SU(2)$ matrix U_0 and a potential

$$\rho(x) = \frac{\lambda_1}{(x-X_1)^2} + \frac{\lambda_2}{(x-X_2)^2} + \frac{\lambda_3}{(x-X_3)^2} \tag{10}$$

where X_1, X_2 and X_3 are distinct points in $\mathbf{R}^4$ (poles), and λ_1, λ_2 and λ_3 are positive constants (weights). $(x-X_i)^2$ denotes the square of the Euclidean distance from x to X_i. In terms of U_0 and ρ the time-component of the gauge potential is

$$A_\tau = U_0\left(\frac{i}{2}\frac{\nabla\rho}{\rho}\cdot\sigma\right)U_0^{-1}\,. \tag{11}$$

An arbitrary U_0 is necessary to obtain the full 16-dimensional moduli space of instantons. The recipe for obtaining the associated Skyrme fields is to take the formula (7) and multiply by -1. The factor -1 comes from closing the contour.

The formula (11) depends on 17 parameters (the pole positions, ratios of the weights, and U_0), but, as shown by Jackiw et al., there is a 1-parameter family of changes to the poles and weights whose effect is simply a gauge transformation. This may be described geometrically as follows, according to Hartshorne. Suppose for the moment that X_1, X_2 and X_3 are not collinear. Then associated with the poles and weights are two coplanar conics in $\mathbf{R}^4$. (See Figure). Let A_1, A_2 and A_3 be the interior points on the sides of the triangle $X_1\,X_2\,X_3$, defined by

$$\frac{X_1A_3}{A_3X_2} = \frac{\lambda_1}{\lambda_2}\,, \quad \frac{X_2A_1}{A_1X_3} = \frac{\lambda_2}{\lambda_3}\,, \quad \frac{X_3A_2}{A_2X_1} = \frac{\lambda_3}{\lambda_1}\,. \tag{12}$$

The first conic is the unique ellipse which is tangent to the sides of the triangle at A_1, A_2 and A_3. The existence of the ellipse follows, by the converse of Brianchon's theorem, from the concurrency of the lines $A_1\,X_1$, $A_2\,X_2$ and $A_3\,X_3$, and this in turn follows, by Ceva's theorem, because

$$\frac{X_1\,A_3\cdot X_2\,A_1\cdot X_3\,A_2}{A_3\,X_2\cdot A_1\,X_3\cdot A_2\,X_1} = 1\,. \tag{13}$$

The second conic is the circumcircle of the triangle $X_1\,X_2\,X_3$.

Now, we have a pair of conics with a triangle $X_1\,X_2\,X_3$ circumscribing one and inscribed in the other. By Poncelet's theorem, there is a porism (a one-parameter family) of such triangles. A second triangle $X_1'X_2'X_3'$ is shown in the Figure, tangent to the ellipse at A_1', A_2' and A_3'. Each triangle of the porism has associated poles (the vertices) and weights (defined up to an irrelevant multiplicative constant by the analogue of formulae (12)), but they all give the same instanton, up to gauge tranformations. The pair of conics is the gauge invariant data which defines the instanton.

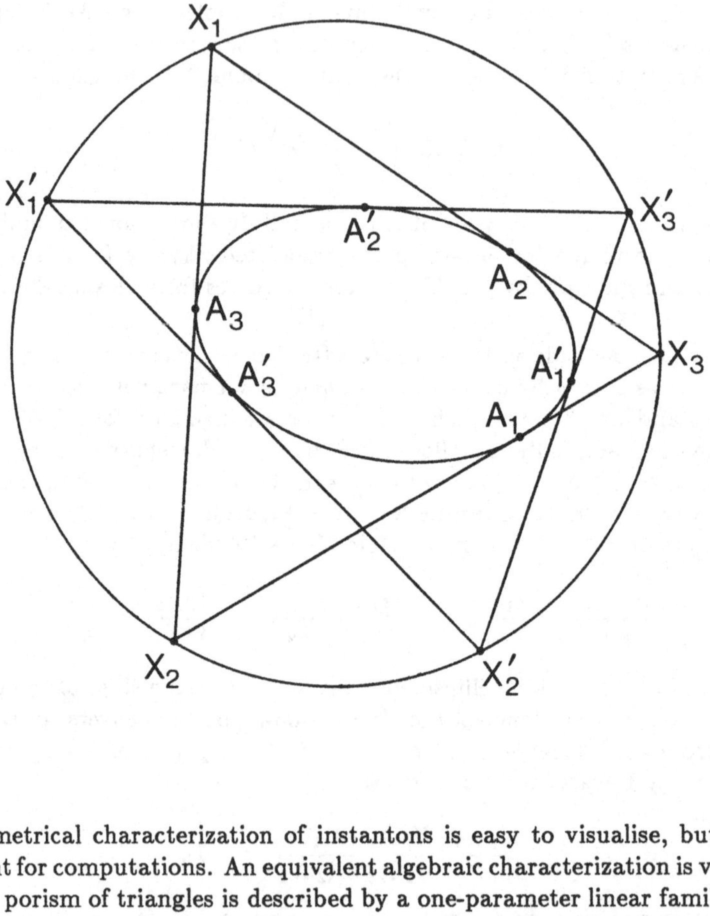

This geometrical characterization of instantons is easy to visualise, but not very convenient for computations. An equivalent algebraic characterization is very useful. Here, the porism of triangles is described by a one-parameter linear family of cubic equations. Let t denote the (real) rational coordinate along the circle $t = \tan \frac{1}{2}\theta$, where θ is an angular coordinate. Suppose that the vertices of one triangle of the porism have coordinates t_1, t_2 and t_3. Associated with the triangle is the cubic equation

$$p(t) \equiv (t - t_1)\,(t - t_2)\,(t - t_3) = 0\,. \tag{14}$$

Associated with a second triangle of the porism, with vertices t'_1, t'_2 and t'_3 is the cubic equation

$$p'(t) \equiv (t - t'_1)(t - t'_2)(t - t'_3) = 0 . \tag{15}$$

It is a remarkable fact that any triangle of the porism is associated with a cubic equation of the form

$$\mu\, p(t) + \mu' p'(t) = 0 . \tag{16}$$

The porism is therefore given by a (projective) line of cubic equations, with inhomogeneous parameter μ'/μ. The same characterization of the porism as a line of cubics also applies when the circle degenerates to a line. In this case, t is simply a linear coordinate.

2-instantons are rather well understood, but the evaluation of the associated Skyrme fields involves computing the holonomy. In practice, this means integrating the ordinary differential equation

$$\frac{d\Psi}{d\tau} = C(\tau)\Psi \tag{17}$$

along a time-line. C is a 2×2 matrix of rational functions of τ which depend on the instanton parameters as well as on $\mathbf{x}$, and Ψ is a 2-component vector. Eq. (17) is of Fuchsian type, and since the integral is from $-\infty$ to $+\infty$ one may complete the path of integration with a large semi-circle in the complex τ-plane. The holonomy is therefore a monodromy of the operator C. Since the problem is non-abelian, the monodromy cannot be calculated by simply adding residues. It would be very interesting if these monodromies could be determined without numerically integrating (17).

So far, it has only been possible to calculate the Skyrme fields for special instantons where the holonomy is abelian along each time-line. Expressions (8) and (9) are examples. For the general $SO(3)$ symmetric 2-instanton one can also give an expression for the profile $f(r)$. The potential ρ depends only on τ and $r = |\mathbf{x}|$,

$$\rho(\tau, r) = \frac{\lambda_1}{(\tau - \tau_1)^2 + r^2} + \frac{\lambda_2}{(\tau - \tau_2)^2 + r^2} + \frac{\lambda_3}{(\tau - \tau_3)^2 + r^2} , \tag{18}$$

and it may be written as the ratio of two polynomials in τ and r. The numerator is quartic in τ (and in r), and since ρ is positive for real τ its roots are two complex conjugate pairs $a \pm ib$ and $c \pm id$, with b and d positive. Then the Skyrme profile is

$$f(r) = \pi(b + d) . \tag{19}$$

This is not a simple expression, because b and d depend in a complicated (but algebraic) way on r. An explicit formula could be found, since the quartic equation is solvable by radicals. Things simplify when there is time-reversal symmetry and the quartic reduces to a quadratic in r^2.

Another special 2-instanton generates a good approximation to the minimal energy Skyrme field of degree 2. The conics associated with this instanton are a pair of concentric circles in a spatial plane (perpendicular to the time-lines) with the ratio of the radii equal to 2. The triangles tangent to the inner circle, with vertices on the outer, are all equilateral. The Skyrme field generated from this 2-instanton is axisymmetric about the spatial line perpendicular to the plane of the circles and passing through their centres. It has not been possible to compute the Skyrme field at a general point because the holonomy is non-abelian, but on the axis of symmetry and in the plane of the circles the holonomy is abelian and can be computed straightforwardly. Suppose the field is in its standard position and orientation, with the x_3-axis as symmetry axis, and suppose the outer circle has radius R. Let $x_1 = r \cos\phi$, $x_2 = r \sin\phi$ and $x_3 = z$. Then, on the axis

$$U(0,0,z) = \exp i\pi \left[\left(1 + \frac{R^2}{z^2}\right)^{-\frac{1}{2}} - 1 \right] \sigma^3 , \tag{20}$$

and in the plane

$$U(r\cos\phi, r\sin\phi, 0) = \exp i\, f(r)\, (\sigma^1 \cos 2\phi - \sigma^2 \sin 2\phi) , \tag{21}$$

where

$$f(r) = \pi \left[\frac{r + \frac{1}{2}R}{(r^2 + rR + R^2)^{\frac{1}{2}}} - \frac{r - \frac{1}{2}R}{(r^2 - rR + R^2)^{\frac{1}{2}}} \right] . \tag{22}$$

There is qualitative agreement with the numerical solution of the Skyrme equations, if one chooses the scale R appropriately. It would be interesting to quantitatively compare the instanton-generated Skyrme field with the numerical solution, and to compute its energy.

$\tilde{M}_2$, the set of Skyrme fields of degree 2 generated from 2-instantons, appears to provide a sensible subset of Skyrme fields with which to model low-energy two nucleon interactions. However, one needs to understand the topology of $\tilde{M}_2$ better. The Skyrme energy functional can be regarded as a Morse function on $\tilde{M}_2$ and one should verify that all the critical points correspond closely to true critical points of the Skyrme model.

$\tilde{M}_2$ is a set of static fields, but if the coordinates (moduli) of $\tilde{M}_2$ vary with time, then the fields become dynamical. The Skyrme Lagrangian, restricted to $\tilde{M}_2$, defines a Lagrangian on $\tilde{M}_2$, but the computations which are necessary to find and solve the equations of motion are heavy. Since the Skyrme model is itself only an approximation, it would be natural to seek a Lagrangian on $\tilde{M}_2$, defined directly in terms of instanton moduli, and of the same qualitative form as the Skyrme Lagrangian. This requires that one find a metric g and potential energy V directly in terms of instanton moduli. Given such data, the natural Hamiltonian to use for the quantized dynamics is $H = -\nabla^2 + V$, where ∇^2 is the Laplacian on $\tilde{M}_2$ constructed using the metric g. The wave functions are (complex) scalar functions on $\tilde{M}_2$. The novel feature of such a model in a nuclear physics context is that the curvature and the non-trivial topology of $\tilde{M}_2$ are important. Curvature alone can lead to non-trivial scattering and quantum bound states.

ACKNOWLEDGEMENTS

Section 2 is the result of joint work with Professor Sir Michael Atiyah. A fuller account of the Skyrme fields generated from 2-instantons will be forthcoming. The author is grateful for the hospitality of the Theoretical Physics Institute, University of Helsinki, where this paper was written.

REFERENCES

Adkins, G.S., Nappi, C.R. and Witten, E. (1983). 'Static Properties of Nucleons in the Skyrme Model'. *Nucl. Phys.* **B228**, 552.

Atiyah, M.F.(1979). 'Geometry of Yang-Mills Fields.' *Lezioni Fermiane, Scuola Normale Superiore, Pisa.*

Atiyah, M.F. and Manton, N.S. (1989). 'Skyrmions from Instantons.' *Phys. Lett.* **222B**, 438.

Bagger, J. Goldstein, W. and Soldate, M. (1985). 'Static Solutions in the Vacuum Sector of the Skyrme Model.' *Phys. Rev.* **D31**, 2600.

Braaten, E. and Carson L. (1988). 'Deuteron as a Toroidal Skyrmion.' *Phys. Rev.* **D38**, 3525.

Castillejo, L., Jones, P.S.J., Jackson, A.D., Verbaarschot, J.J.M. and Jackson, A. (1988). 'Dense Skyrmion Systems.' To be published.

Castillejo, L. and Kugler, M. (1987). 'The Interaction of Skyrmions.' Unpublished.

Esteban, M.J. (1986). 'A Direct Variational Approach to Skyrme's Model for Meson Fields.' *Comm. Math. Phys.* **105**, 571.

Fadeev, L.D. (1976). 'Some Comments on the Many-Dimensional Solitons.' *Lett. Math. Phys.* **1**, 289.

Hartshorne, R. (1978). 'Stable Vector Bundles and Instantons'. *Comm. Math. Phys.* **59**, 1.

Jackiw, R., Nohl, C. and Rebbi, C. (1977). 'Conformal Properties of Pseudoparticle Configurations.' *Phys. Rev.* **D15**, 1642.

Jackson, A.D. and Rho, M. (1983). 'Baryons as Chiral Solitons.' *Phys. Rev. Lett.* **51**, 751.

Jackson, A.D., Manton, N.S. and Wirzba, A. (1989). 'New Skyrmion Solutions on a 3-Sphere.' *Nucl. Phys.* **A495**, 499.

Kopeliovich, V.B. and Shtern, B.E. (1987). 'Exotic Skyrmions.' *JETP Lett.* **45**, 203.

Kugler, M. and Shtrikman, S. (1988). 'A New Skyrmion Crystal.' *Phys. Lett.* **208B**, 491.

Manton, N. S. (1987). 'Geometry of Skyrmions.' *Comm. Math. Phys.* **111**, 469.

Manton, N.S. (1988). 'Unstable Manifolds and Soliton Dynamics.' *Phys. Rev. Lett.* **60**, 1916.

Schramm, A.J., Dothan, Y. and Biedenharn, L.C. (1988). 'A Calculation of the Deuteron as a Biskyrmion.' *Phys. Lett.* **205B**, 151.

Skyrme, T.R.H. (1962). 'A Unified Theory of Mesons and Baryons.' *Nucl. Phys.* **31**, 556.

Verbaarschot, J.J.M. (1987). 'Axial Symmetry of Bound Baryon Number-Two Solution of the Skyrme Model.' *Phys. Lett.* **195B**, 235.

Wirzba, A. and Bang, H. (1989). 'The Mode Spectrum and the Stability Analysis of Skyrmions on a 3-Sphere.' To be published.

REPRESENTATIONS OF BRAID GROUPS AND OPERATORS COUPLED TO MONOPOLES

RALPH L. COHEN AND JOHN D.S. JONES

Let $\mathcal{M}_k$ be the moduli space of based $SU(2)$-monopoles in $\mathbb{R}^3$ of charge k, see for example [4]. Associated to each monopole c there is a natural real differential operator δ_c, the coupled Dirac operator. The space of solutions of the Dirac equation $\delta_c f = 0$ is k-dimensional and as c varies these k-dimensional spaces form a real vector bundle over the space of monopoles. One of the main purposes of this paper is to explain how this bundle is related to representations of the braid group.

Braid groups appear in the theory of monopoles in the following way. Let Rat_k be the space of rational functions on $\mathbb{C}$ which map infinity to zero and have k poles, counted with multiplicity. In [**10**], Donaldson showed that there is a diffeomorphism

$$\mathcal{M}_k \cong \mathrm{Rat}_k\,.$$

The topological properties of the space Rat_k are extensively studied in [**8**] and in particular it is shown that, for large enough N, there is a homotopy equivalence

$$\Sigma^N \mathrm{Rat}_k \simeq \Sigma^N B\beta_{2k}.$$

Here Σ^N means N-fold suspension, β_{2k} is the braid group on $2k$ strings, and $B\beta_{2k}$ is its classifying space, an Eilenberg-MacLane space of type $K(\beta_{2k}, 1)$. Combining these two results shows that the space of monopoles $\mathcal{M}_k$ and the space $B\beta_{2k}$ have the same homology and cohomology; indeed if E is any (generalised) cohomology theory, then $E^*(\mathcal{M}_k)$ and $E^*(B\beta_{2k})$ are isomorphic. Our aim is to investigate this isomorphism between the K-theory of $\mathcal{M}_k$ and the K-theory of $B\beta_{2k}$.

Vector bundles over a classifying space $B\pi$ arise most naturally from representations of the group π. Indeed a well-known theorem, due to Atiyah [**2**], shows that if the group π is finite then the K-theory of $B\pi$ can be computed from the representation ring of π by completion. The braid group β_{2k} is not finite and Atiyah's theorem does not hold; but nonetheless many interesting bundles arise from representations. On the other hand the moduli space $\mathcal{M}_k$ consists of analytic objects, connections. One very natural way of constructing vector bundles over $\mathcal{M}_k$ is to construct differential operators using the connections and to form the index bundles for the corresponding families of operators parametrised by $\mathcal{M}_k$. The isomorphism between the K-theory of $\mathcal{M}_k$ and the K-theory of $B\beta_{2k}$ suggests that there may be a natural correspondence between representations of β_{2k} and operators coupled

---

The first author was partially supported by grants from the NSF and PYI. Both authors were supported by NSF grant DMS-8505550

to monopoles. Here we will study the coupled Dirac operator and the permutation representation of the braid group.

It is natural to look at other representations of the braid group, in particular those used in Vaughan Jones's construction of polynomial invariants of knots and links, and to see if these representations have some interpretation in terms of the space of monopoles. We will make some comments on this project in §6. The Jones polynomials have been given a gauge theoretic interpretation by Witten [**18**] and it is natural to wonder if there is any direct connection between these various points of view.

In this article we will summarise some of the ideas involved in establishing the relationship between the permutation representation of the braid group and the index bundle for the family of Dirac operators; full details will be given in [**9**].

The authors are indebted to Michael Atiyah, Simon Donaldson, Nigel Hitchin, and Cliff Taubes, for helpful conversations and correspondence concerning this project. The first author would like to thank the mathematics departments at Oxford University and University of Paris VII for their hospitality while some of this work was being carried out.

§1 Monopoles

The purpose of this section is to summarise the basic facts concerning monopoles and Yang-Mills-Higgs theory and to show that there are many interesting topological features in the theory. Fix the structure group to be $SU(2)$ with Lie algebra $su(2)$. We use the standard invariant inner product on $su(2)$ and so $su(2)$ becomes a three dimensional Euclidean space. We study $SU(2)$-monopoles on $\mathbf{R}^3$ equipped with its standard (Euclidean) metric and orientation.

We consider pairs (A, φ) where:

(1) A is a connection on the trivial bundle $SU(2)$ on $\mathbf{R}^3$,

(2) φ is an $su(2)$ valued function on $\mathbf{R}^3$.

So A is a 1-form on $\mathbf{R}^3$ with values in $su(2)$;

$$A = \sum_{\mu=1}^{3} A_\mu(x) dx_\mu$$

where $A_\mu : \mathbf{R}^3 \to su(2)$ and the x_μ are the coordinates on $\mathbf{R}^3$. The connection A is called the **gauge potential** and φ the **Higgs field**. The pair (A, φ) is required to satisfy several conditions. First we require the Yang-Mills-Higgs action of (A, φ) to be finite, that is

$$\text{(I)} \qquad \mathcal{U}(A, \varphi) = \int_{\mathbf{R}^3} \left(\|F_A\|^2 + \|D_A\varphi\|^2\right) dvol < \infty.$$

Here $F_A = dA + A \wedge A$ is the curvature of the connection A, D_A is the covariant derivative operator defined by A, and $dvol$ is the usual volume form on $\mathbf{R}^3$. We also require a condition on the behaviour of φ at infinity. There are several different conditions which may be imposed. Here we use the weakest condition which seems to be sufficient [**17**]

$$\text{(II)} \qquad 1 - |\varphi| \in L^6(\mathbf{R}^3).$$

Notice that we do not assume any asymptotic conditions on the gauge potential A. In addition we impose a base point condition

$$\text{(III)} \qquad \lim_{t\to\infty} \varphi(t,0,0) = (1,0,0).$$

Let $\mathcal{A}$ be the space of pairs $(A,\varphi) \in \mathcal{A}$ which satsify these three conditions.

There is a natural map

$$I : \Omega^2 S^2 \longrightarrow \mathcal{A},$$

where $\Omega^2 S^2$ is the space of all smooth maps $S^2 \longrightarrow S^2$, which preserve the base point $(1,0,0)$ in S^2. This map I is defined explicitly as follows. We identify the unit sphere S^2 with the sphere in $su(2)$. Now given a map

$$\alpha : S^2 \longrightarrow S^2 \subseteq su(2)$$

define the pair $I(\alpha) = (A,\varphi)$ by the formula

$$A = \beta(|x|)\left[\alpha\left(\frac{x}{|x|}\right), d\alpha\left(\frac{x}{|x|}\right)\right]$$
$$\varphi = -\beta(|x|)\alpha\left(\frac{x}{|x|}\right).$$

In this formula $\beta : \mathbf{R} \longrightarrow [0,1]$ is a smooth cut-off function which is identically 0 if $t \leq 1/2$ and identically 1 if $t \geq 3/4$, and $[\,,\,]$ is the Lie bracket on $su(2)$.

PROPOSITION. *The map $I : \Omega^2 S^2 \longrightarrow \mathcal{A}$ is a homotopy equivalence.*

This is proved in [**17**]. It immediately shows that there is a definite topological aspect to the study of monopoles.

There is a group of **gauge transformations**

$$\mathcal{G} = \mathrm{Map}_0(\mathbf{R}^3, SU(2))$$

which acts on the space $\mathcal{A}$. Here Map_0 means those maps $g : \mathbf{R}^3 \longrightarrow SU(2)$ which satisfy the base point condition

$$\lim_{t\to\infty} g(t,0,0) = 1.$$

Because of this base point condition $\mathcal{G}$ is contractible. Gauge transformations act on $\mathcal{A}$ by

$$g^*(A,\varphi) = (g^{-1}Ag + g^{-1}dg, g^{-1}\varphi g).$$

This formula makes sense since g is a matrix valued function and A is a matrix of 1-forms. The action of $\mathcal{G}$ on $\mathcal{A}$ is free and has local slices, see [**17**], and so we form the quotient spaces

$$\mathcal{B} = \mathcal{A}/\mathcal{G}.$$

The quotient map is a principal fibre bundle with fibre $\mathcal{G}$ and since $\mathcal{G}$ is contractible we deduce that the projection $\mathcal{A} \longrightarrow \mathcal{B}$ is a homotopy equivalence. The natural maps fit into a diagram

$$\begin{array}{ccc} \mathcal{A} & \longrightarrow & \Omega^2 S^2 \\ \downarrow & & \\ \mathcal{B} & & \end{array}$$

and since each of these maps is a homotopy equivalence we deduce the following proposition.

PROPOSITION. *There is a natural homotopy equivalence*

$$\mathcal{B} \longrightarrow \Omega^2 S^2.$$

In view of these results $\mathcal{A}$ and $\mathcal{B}$ break up into components labelled by the integers $\mathbb{Z}$ and we use the subscript k to denote the k-th component. This integer k is the **charge** of the pair (A,φ). The above propositions show that

$$\mathcal{A}_k \simeq \mathcal{B}_k \simeq \Omega^2_k S^2$$

where $\Omega^2_k S^2$ means the space of all base point preserving maps $S^2 \longrightarrow S^2$ of degree k.

The Yang-Mills-Higgs functional is invariant under the gauge group $\mathcal{G}$ and so defines a function $\mathcal{U}$ on $\mathcal{B}$. The space $\mathcal{M}_k \subseteq \mathcal{B}_k$ of based $SU(2)$**-monopoles of charge** k is defined to be the space of absolute minima of $\mathcal{U}$. If $k \geq 0$ then $\mathcal{M}_k$ can be identified with the space of pairs (A,φ) which satisfy the **Bogomolnyi equation**

$$D_A\varphi = *F_A,$$

where $*$ is the Hodge star operator on $\mathbf{R}^3$, modulo gauge equivalence. If $k < 0$ then $\mathcal{M}_k$ can be identified with the space of pairs (A,φ) which satisfy the equation $D_A\varphi = - * F_A$.

We concentrate on the case where $k \geq 0$ and so study the space of solutions of the equation $D_A\varphi = *F_A$. This space $\mathcal{M}_k$ is a smooth manifold of dimension $4k$

and its geometrical properties are extensively studied in [4]. For example, when $k = 1$ a monopole is uniquely determined by its centre, which is a point in $\mathbf{R}^3$, and a "phase" parameter in S^1; thus

$$\mathcal{M}_1 = S^1 \times \mathbf{R}^3.$$

In fact monopoles can be regarded as "time invariant instantons" in the following sense. Given the pair (A, φ) we can form an $su(2)$ connection

$$\alpha = A + \varphi\, dt$$

on $\mathbf{R}^4 = \mathbf{R}^3 \times \mathbf{R}$. This connection is independent of t and it is easy to check that (A, φ) satisfies the Bogomolnyi equation if and only if α is self-dual,

$$*F_\alpha = F_\alpha$$

where $*$ is the star operator on $\mathbf{R}^4$.

The following theorem, due to Cliff Taubes [17], shows that these spaces $\mathcal{M}_k$ have some very interesting topological features.

THEOREM. *The inclusion $i : \mathcal{M}_k \to \mathcal{B}_k$ is an "asymptotic homotopy equivalence"; that is there is a function $q(k)$ with $q(k) \to \infty$ as $k \to \infty$ such that the map i induces an isomorphism of homotopy groups π_q provided $q \leq q(k)$.*

Note that the homotopy type of $\mathcal{B}_k \simeq \Omega^2_k S^2$ is independent of k so the spaces $\mathcal{M}_k$ provide finite dimensional approximations to a fixed homotopy type which become better and better approximations as $k \to \infty$.

§2 Braids

In the previous section we saw that the space of monopoles is a finite dimensional homotopical approximation to the space $\Omega^2 S^2$ of all base-point preserving maps $S^2 \to S^2$. In fact there is a more classical finite dimensional homological approximation to this space and this is where braid groups come into the picture.

Define $C_k = C_k(\mathbf{R}^2)$ to be the space of unordered k-tuples of distinct points in the plane $\mathbf{R}^2$. Then recall that the braid group β_k is the fundamental group of C_k

$$\beta_k = \pi_1(C_k).$$

Indeed it is well-known that C_k is the classifying space of the braid group β_k, that is

$$\pi_i(C_k) = 0, \qquad \text{if } i \geq 2.$$

It is usual to draw braids as follows:

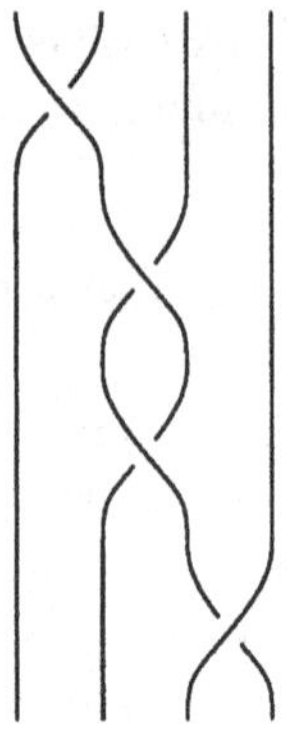

Let $\tilde{C}_k = \tilde{C}_k(\mathbf{R}^2)$ be the space of ordered k-tuples of distinct points in $\mathbf{R}^2$. The symmetric group Σ_k acts freely on $\tilde{C}_k$ by permuting points and the quotient is C_k. The covering $\tilde{C}_k \to C_k$ corresponds to a homomorphism

$$\rho : \beta_k \to \Sigma_k.$$

In terms of the diagram representing a braid this is given by mapping a braid to the permutation of its end points.

There is a natural map

$$j : C_k \to \Omega_k^2 S^2$$

defined as follows. First replace the space C_k by the homotopy equivalent space of configurations of k disjoint disks in the plane. Now given a configuration $\mathbf{D} = \{D_1, \ldots, D_k\}$ of k disjoint disks define a map $f_{\mathbf{D}} : S^2 \to S^2$ as follows. Regard the domain of $f_{\mathbf{D}}$ as $\mathbf{R}^2 \cup \infty$ and then $f_{\mathbf{D}}$ maps the complement of $D_1 \cup \cdots \cup D_k$ to the base point of S^2 and on the disk D_i it is the usual identification of $D_i/\partial D_i$ with S^2. Then $\mathbf{D} \mapsto f_{\mathbf{D}}$ gives the required map $j : C_k \to \Omega_k^2 S^2$. The main theorem relating C_k and $\Omega_k^2 S^2$ is the following result due to May, Milgram, and Segal [**14**], [**15**], [**16**].

THEOREM. *The map $j : C_k \to \Omega_k^2 S^2$ is an asymptotic homology equivalence.*

Now we see that there must indeed be a relation between the monopole space $\mathcal{M}_k$ and the space $B\beta_k = C_k$. There are maps

$$\mathcal{M}_k \xrightarrow{i} \mathcal{B}_k \simeq \Omega_k^2 S^2 \xleftarrow{j} C_k$$

and the map i is an asymptotic homotopy equivalence whereas the map j is an asymptotic homology equivalence. So the space $\mathcal{M}_k$ is a finite dimensional homotopical approximation to $\Omega_k^2 S^2$ and the space C_k is a finite dimensional homological approximation to the same space. However the precise relation between braids and monopoles is rather more subtle than the above remarks might lead one to expect and it is explained in detail in §3.

One may think of the map $j : C_k \to \Omega_k^2 S^2$ as superimposing, or gluing in, a standard map, the identity of S^2 or, to put it another way, the identification of $D^2/\partial D^2$ with the sphere S^2, at each of k distinct points in the plane. There is a similar superposition process in the theory of instantons on $\mathbf{R}^4$; in this particular case this is the 't Hooft construction of instantons on $\mathbf{R}^4$ and in the case of a general 4-manifold this is the patching process due to Taubes. If we follow the analogy with instantons, compare [**5**], we might think that there is a gluing process for monopoles which superimposes a standard $k = 1$ monopole at each of k distinct points in the plane and that this process leads to a map $\lambda : C_k(\mathbf{R}^3) \to \mathcal{M}_k$ with the property that the composite

$$C_k(\mathbf{R}^2) \to C_k(\mathbf{R}^3) \to \mathcal{M}_k \to \mathcal{B}_k \simeq \Omega_k^2 S^2$$

is homotopic to the map j used above. Suppose indeed that such a map λ exists, then computing the induced homomorphisms on fundamental groups quickly gives a contradiction;

$$\pi_1(C_k(\mathbf{R}^2)) = \beta_k, \qquad \pi_1(\Omega_k^2 S^2) = \mathbf{Z}$$

and the induced homomorphism j_* is surjective. But $\pi_1(C_k(\mathbf{R}^3))$ is the symmetric group Σ_k and the homomorphism

$$j_* : \pi_1(C_k(\mathbf{R}^2)) \to \pi_1(\Omega_k^2 S^2)$$

cannot factor through Σ_k.

This argument shows that the gluing proceedure for monopoles is considerably more delicate than that for instantons. Indeed it is possible to superimpose "well-separated monopoles" but great care must be taken with this construction and this is one of the points where the theory of monopoles is very different from the corresponding theory of instantons.

§3 Rational functions

To understand the full relation between braids and monopoles we need to use Donaldson's theorem relating $\mathcal{M}_k$ and Rat_k, the space of rational functions on $\mathbf{C}$ which map infinity to 0 and have k poles.

THEOREM. *There is a diffeomorphism* $\mathcal{M}_k \cong \mathrm{Rat}_k$.

The proof of this theorem is given in [**10**], [**11**], [**12**]. Now let Rat_k^0 be the subspace of Rat_k consisting of those rational functions with k distinct simple poles; this is the subspace of generic rational functions. If $f \in \mathrm{Rat}_k^0$ then $f(z)$ can be written in the form

$$f(z) = \sum_{i=1}^{n} \frac{a_i}{z - b_i}$$

where the b_i are distinct complex numbers and the a_i are non-zero complex numbers, therefore

$$\mathrm{Rat}_k^0 = \tilde{C}_k \times_{\Sigma_k} (\mathbf{C}^*)^k.$$

Here $\tilde{C}_k$ is the space of ordered k-tuples of distinct points in $\mathbf{C}$ and $\mathbf{C}^*$ is the space of non-zero complex numbers. The symmetric group Σ_k acts on both $\tilde{C}_k$ and $(\mathbf{C}^*)^k$ by permutations. From this we deduce that Rat_k^0 is the classifying space $B\beta_{2,k}$ of the semi-direct product

$$\beta_{2,k} = \beta_k \ltimes (\mathbf{Z})^k$$

where β_k acts on $(\mathbf{Z})^k$ by permuting factors. This group $\beta_{2,k}$ can be thought of as the group of **framed braids** and it is naturally a subgroup of the group β_{2k}. The inclusion $\beta_{2,k} \longrightarrow \beta_{2k}$ is given by the cabling process which can be described as follows. Start with k pairs of pieces of string and twist the i-th pair n_i times where $n_i \in \mathbf{Z}$. Now braid the k pairs according to the braid $b \in \beta_k$. This gives a braid on $2k$ strings and the map which sends $(\beta; n_1, ..., n_k)$ to this braid gives the inclusion $\beta_{2,k} \longrightarrow \beta_{2k}$.

We now have the following diagram.

$$\begin{array}{ccc} \mathrm{Rat}_k^0 = B\beta_{2,k} & \xrightarrow{\varphi} & B\beta_{2k} \\ \psi\big\downarrow & & \\ \mathrm{Rat}_k = \mathcal{M}_k & & \end{array}$$

and the following theorem follows directly from one of the main results of [**8**].

THEOREM. *Let E be any cohomology theory, then*

$$\varphi^* : E^*(B\beta_{2k}) \to E^*(B\beta_{2,k})$$
$$\psi^* : E^*(\mathcal{M}_k) \to E^*(B\beta_{2,k})$$

are both split injective and the splittings induce a natural isomorphism

$$E^*(\mathcal{M}_k) \cong E^*(B\beta_{2k})$$

We will use this theorem in the case where E is K-theory but before doing so we discuss some of its implications. It is a reasonably straightforward piece of group theory to check the following lemma concerning the braid group, for example see [**7**].

LEMMA. (1) *The abelianisation of the braid group β_k is the integers $\mathbf{Z}$.*
(2) *If $k \geq 5$ then the commutator subgroup $[\beta_k, \beta_k] \subseteq \beta_k$ is perfect.*

Given this lemma we can form the Quillen plus construction to kill $[\beta_{2k}, \beta_{2k}]$ to get the space $B\beta_{2k}^+$, and this space then has fundamental group $\mathbf{Z}$ and the same homology, with any twisted coefficients, as $B\beta_{2k}$. We know that $\pi_1 \mathcal{M}_k = \mathbf{Z}$ and, by the above theorem, $H_*(\mathcal{M}_k; \mathbf{Z}) \cong H_*(B\beta_{2k}; \mathbf{Z})$ and so the following question seems very natural

QUESTION. *Is there a homotopy equivalence*

$$B\beta_{2k}^{+} \simeq \mathcal{M}_k.$$

The essential difficulty is whether there is a map $B\beta_{2k} \longrightarrow \mathcal{M}_k$ which abelianises the fundamental group and induces an isomorphism in homology with any twisted coefficient system. Both spaces are as nice as possible, they are $4k$-dimensional manifolds and both have natural complex structures. A theorem due to Kan and Thurston [**13**] asserts that any space is homotopy equivalent to a space obtained by applying Quillen's plus construction to a space of the form $B\pi$. It would be rather remarkable if it were possible to obtain the monopole space by applying the plus constrution to $B\beta_{2k}$. If there is such an equivalence this would give the neatest possible way of expressing the relation between braid groups and monopoles, however for many purposes the above theorem is completely satisfactory since it provides a definite method of producing the isomorphism $E^*(\mathcal{M}_k) \longrightarrow E^*(B\beta_{2k})$ as we shall see in a specific example.

§4 THE COUPLED DIRAC OPERATOR

Now we turn to the construction of vector bundles on the space $\mathcal{M}_k$ using differential operators and the purpose of this section is to describe the basic operator, the coupled Dirac operator, and its index bundle. Let S_3 be the space of spinors on $\mathbf{R}^3$ and write E for the usual 2-dimensional complex representation of $SU(2)$. Then the Dirac operator coupled to a pair $c = (A, \varphi) \in \mathcal{A}$ is the operator

$$\partial_c : C^\infty(\mathbf{R}^3; S_3 \otimes E) \longrightarrow C^\infty(\mathbf{R}^3; S_3 \otimes E)$$

defined by the formula

$$\partial_c(f) = \sum_{i=1}^{3} (e_i \otimes 1) \cdot (D_{A,i}(f)) + (1 \otimes \varphi) f.$$

Here e_i is the i-th generator of the Clifford algebra C_3 of $\mathbf{R}^3$ and $D_{A,i}$ is covariant differentiation in the i-th direction in $\mathbf{R}^3$ defined using the spinor connection on S_3 and the given connection A on E. The e_i act on S_3 via the usual spin representation of C_3 and $su(2)$ acts on E via the standard representation of the Lie algebra $su(2)$.

In fact this operator is the time invariant Dirac operator on $\mathbf{R}^4$ coupled to a time invariant instanton in the following sense. Given the pair (A, φ) form the connection $\alpha = A + \varphi\, dt$ on $\mathbf{R}^4$, as in §1. Now we can form the Dirac operator on $\mathbf{R}^4$ coupled to the connection α

$$\partial_\alpha : C^\infty(\mathbf{R}^4; S_4^+ \otimes E) \longrightarrow C^\infty(\mathbf{R}^4; S_4^- \otimes E).$$

where $S_4^{\pm}$ are the positive and negative spinors on $\mathbf{R}^4$. We can restrict this operator to the subspace of functions on $\mathbf{R}^4 = \mathbf{R}^3 \times \mathbf{R}$ which are independent of the fourth coordinate t and since α is independent of t we get an operator

$$\partial_\alpha : C^\infty(\mathbf{R}^3; S_4^+ \otimes E) \to C^\infty(\mathbf{R}^3; S_4^- \otimes E).$$

Now using e_4 to identify S_4^+ and S_4^- we can form the operator

$$e_4 \cdot \partial_\alpha : C^\infty(\mathbf{R}^3; S_4^+ \otimes E) \to C^\infty(\mathbf{R}^3; S_4^+ \otimes E).$$

Using the usual identification of the C_3 module S_3 with S_4^+ where $e_i \in C_3$ acts as $e_4 e_i$ this operator $e_4 \cdot \partial_\alpha$ can be identified with the operator ∂_c defined above.

Now note that Spin(3) is isomorphic to $SU(2)$ and under this isomorphism the spin representation S_3 becomes the usual 2-dimensional representation E of $SU(2)$. Therefore $S_3 \otimes E$ is the 4-dimensional representation $E \otimes E$ of $SU(2) \times SU(2)$. This representation has a real structure; indeed using the fact that Spin(4) is isomorphic to $SU(2) \times SU(2)$ the underlying real representation is the double covering Spin(4) $\to SO(4)$. The outcome is that there is there is a 4-dimensional real representation R of $SU(2) \times SU(2)$ whose complexification is $E \otimes E$. Therefore the operator ∂_c has a real structure and there is an operator

$$\delta_c : C^\infty(\mathbf{R}^3; R) \to C^\infty(\mathbf{R}^3; R)$$

whose complexification is ∂_c. This operator δ_c is the one we use and we refer to it as the **real Dirac operator coupled to** (A, φ).

In [**17**] Taubes shows that this operator extends to a Fredholm operator on the appropriate Sobolev spaces. Furthermore if c_1 and c_2 are gauge equivalent the operators δ_{c_1} and δ_{c_2} are isomorphic and so we get a family of operators δ_c parametrised by the points $c \in \mathcal{B}$. Taubes also shows that this is a continuous family and therefore has an index bundle

$$\mathrm{ind}(\delta) \in KO^0(\mathcal{B}).$$

Recall briefly how this index bundle is defined. Let

$$K_c = \ker \delta_c, \qquad C_c = \operatorname{coker} \delta_c$$

so both K_c and C_c are finite dimensional real vector spaces. Suppose in fact that the spaces K_c form a vector bundle K over $\mathcal{A}$ and the spaces C_c also form a vector bundle C over $\mathcal{A}$. Then

$$\mathrm{ind}(\delta) = K - C.$$

In general the spaces K_c do not form a vector bundle, nor do the spaces C_c, since the kernel and cokernel of δ_c may jump in dimension, but we can reduce to this case by a deformation, compare [**2**].

Now we must identify this index bundle

$$\operatorname{ind}(\delta) \in KO^0(\mathcal{B})$$

using the equivalence of $\mathcal{B}$ with $\Omega^2 S^2$. There is the usual isomorphism

$$S^2 = Sp(1)/U(1)$$

and we can compose this equivalence with the stabilisation

$$Sp(1)/U(1) \longrightarrow Sp/U = \lim_{n\to\infty} Sp(n)/U(n)$$

to get a natural map $S^2 \longrightarrow Sp/U$. Now apply Ω^2 to this map and use the equivalence

$$\Omega^2(Sp/U) \simeq \mathbf{Z} \times BO$$

occuring in real Bott periodicity, and we get a map

$$\Omega^2 S^2 \longrightarrow \mathbf{Z} \times BO$$

This map gives us an element $\gamma \in KO^0(\Omega^2 S^2)$ with corresponding components $\gamma_k \in KO^0(\Omega^2_k S^2)$.

THEOREM. *Using the natural isomorphism of* $KO^0(\mathcal{B}_k)$ *with* $KO^0(\Omega^2_k S^2)$ *the index bundle* $\operatorname{ind}(\delta)$ *can be identified with* γ_k*:*

$$KO^0(\mathcal{B}_k) \cong KO^0(\Omega^2_k S^2)$$
$$\operatorname{ind}(\delta) \leftrightarrow \gamma_k$$

This theorem is proved by adapting Atiyah's proof of Bott preriodicity [**1**] using elliptic operators and it is given in [**9**]. In some sense the proof is straightforward but in detail it is quite tricky since it must necessarily involve the intricacies of 8-fold Bott periodicity in real K-theory. The complexification of the bundle γ is trivial so there is no way to avoid the extra complications of real K-theory and get non-trivial results.

The appearance of $Sp(1)/U(1)$ in the above description of γ is very natural. In the theory of monopoles for general Lie groups the asymptotic conditions imposed on the pair (A, φ) are that φ approaches a fixed orbit of the adjoint action of the group on its Lie algebra. In our case we see that using the identification of $Sp(1)$ with $SU(2)$ the unit 2-sphere in the Lie algebra $su(2)$ can be identified with the orbit of $Sp(1)$ acting on a fixed unit vector, that is $Sp(1)/U(1)$.

§5 The Dirac operator and representations of the braid group

Now we look at the operator

$$\delta_c : C^\infty(\mathbf{R}^3; R) \to C^\infty(\mathbf{R}^3; R)$$

where $c \in \mathcal{M}_k$. The first simplification is that since c now satisfies the Bogomolnyi equation we can use the vanishing theorem of [**17**].

THEOREM.

$$C_c = \mathrm{coker}(\delta_c) = 0$$

Therefore the spaces $K_c = \ker(\delta_c)$ form a k-dimensional real vector bundle over the monopole space $\mathcal{M}_k$ and this is the index bundle $\mathrm{ind}(\delta)$ of the family of operators δ_c where $c \in \mathcal{M}_k$. Now we use the theorem of [**8**] described in §3 to identify the corresponding bundle over $B\beta_{2k}$, more precisely the corresponding element in $KO^0(B\beta_k)$. To do this we must first identify $\psi^*(\mathrm{ind}(\delta))$ where

$$\psi : B\beta_{2,k} = \tilde{C}_k \times_{\Sigma_k} (\mathbf{C}^*)^k \to \mathcal{M}_k$$

is the map described in §3.

There is a natural representation

$$\pi_k : \beta_{2,k} \to O(k)$$

defined as follows. Let a_i be the generator of the i-th copy of $\mathbf{Z}$ in $\beta_{2,k} = \beta_k \ltimes (\mathbf{Z})^k$, then in the representation π_k, a_i changes the sign of the i-th basis vector in $\mathbf{R}^k$ and a braid $b \in \beta_k \subseteq \beta_{2,k}$ permutes the basis of $\mathbf{R}^k$. Another way to describe this representation is as the composition

$$\beta_{2,k} = \beta_k \ltimes (\mathbf{Z})^k \to \Sigma_k \ltimes (\mathbf{Z}/2)^k \subseteq O(k)$$

where the first homomorphism is the obvious quotient and $\Sigma_k \ltimes (\mathbf{Z}/2)^k$ is identified with the subgroup of $O(k)$ generated by the permutation matrices and the diagonal matrices with ±1's along the diagonal. This representation defines a bundle over $B\beta_{2,k}$ which will still be denoted by π_k. The corresponding bundle can be described quite explicitly as follows. Let H be the real Hopf line bundle over $\mathbf{C}^* = \mathbf{R} \times S^1$; then π_k is the bundle

$$\tilde{C}_k \times_{\Sigma_k} (H)^k \to \tilde{C}_k \times_{\Sigma_k} (\mathbf{C}^*)^k.$$

THEOREM.

$$\psi^*(\mathrm{ind}(\delta)) = \pi_k \in KO^0(B\beta_{2,k})$$

The proof of this theorem is given in [**9**] so here we try to describe the intuition behind it. First we deal with the case $k = 1$, then $\mathcal{M}_1 = \mathbf{R}^3 \times S^1$ and $\mathrm{ind}(\delta)$ is the non-trivial 1-dimensional line bundle over $\mathcal{M}_1$, that is the (real) Hopf line bundle over the circle S^1. This of course immediately verifies the theorem in the case $k = 1$ since $\beta_{2,1} = \mathbf{Z}$, $\mathcal{M}_1 = B\beta_{2,1}$ and the Hopf line bundle is just π_1. These facts have the following intuitive interpretation. A 1-monopole is determined by its centre in $\mathbf{R}^3$ and a phase, which determines its component in S^1. The space of solutions of the Dirac equation $\delta_c(f) = 0$ is one dimensional and any non-trivial solution of this equation has the property that if we rotate the phase of the monopole through 2π then this solution changes sign.

Now we look at the general case of $\mathcal{M}_k$ and restrict attention to the subspace of "well separated monopoles"; this is the subspace obtained by gluing in 1-monopoles at k distinct points in $\mathbf{R}^3$. Then it is natural to expect the solutions of the Dirac equation coupled to such a monopole to be k-dimensional with one basic solution associated to each of the k-distinct points. If the points are permuted then so is the basis and if the phase of the i-th 1-monopole is rotated through 2π then the sign of the i-th basis vector changes. Therefore the representation π_k should appear quite naturally. However as we have already pointed out the gluing process for monopoles is quite subtle so that while this heuristic argument is quite convincing, considerable care is needed to give a precise proof.

The next step is to look for the corresponding element of $KO^0(B\beta_{2k})$. First we show that there is an element $x \in KO^0(B\beta_{2k})$ with the property that

$$\varphi^*(x) = \pi_k$$

where $\varphi : B\beta_{2,k} \longrightarrow B\beta_{2k}$ is induced by the inclusion $j : \beta_{2,k} \longrightarrow \beta_{2k}$. The obvious starting point is the permutation representation $\rho_{2k} : \beta_{2k} \longrightarrow O(2k)$ and so the first step is to compute the representation $j^*(\rho_{2k})$. Let $q : \beta_{2,k} \longrightarrow \beta_k$ be the quotient homomorphism $\beta_k \ltimes (\mathbf{Z})^k \longrightarrow \beta_k$.

LEMMA.

$$j^*(\rho_{2k}) = \pi_{2k} \oplus q^*(\rho_k)$$

The proof of this lemma is quite straightforward. If we write the basis of $\mathbf{R}^{2k}$ as $e_1, \ldots, e_{2k}$ then in the representation $j^*(\rho_{2k})$ the generator a_i of the i-th factor $\mathbf{Z}$ acts by interchanging e_{2i-1} and e_{2i} and leaving the other basis vectors fixed while a braid $b \in \beta_k$ permutes the k pairs $(e_1, e_2), \ldots, (e_{2k-1}, e_{2k})$. Therefore in the basis

$$f_1 = e_1 - e_2, \ldots, f_k = e_{2k-1} - e_{2k}, f_{k+1} = e_1 + e_2, \ldots, f_{2k} = e_{2k-1} + e_{2k}$$

the representation $j^*(\rho_{2k})$ is precisely $\pi_{2k} \oplus q^*(\rho_k)$.

To find the element $x \in KO^0(B\beta_{2k})$ such that $\varphi^*(x) = \mathrm{ind}(\delta) \in KO^0(\mathcal{M}_k)$ we need to find a way to subtract off ρ_k from ρ_{2k}. In $KO^0(B\beta_{2k})$ there is a natural way to do this; at the level of representations things are more delicate and will be analysed carefully in [**9**] . The inclusion $i : \beta_k \longrightarrow \beta_{2k}$ induces a surjection

$$i^* : KO^0(B\beta_{2k}) \longrightarrow KO^0(B\beta_k)$$

and this map has a natural splitting. More precisely we use the following proposition from [**6**].

PROPOSITION. *There is a natural map*

$$\alpha : KO^*(B\beta_k) \to KO^*(B\beta_{2k})$$

such that

(1) $i^* \cdot \alpha = 1$, *where* $i : \beta_k \longrightarrow \beta_{2k}$ *is the inclusion*
(2) $j^* \cdot \alpha = q^*$ *where* $j : \beta_{2,k} \longrightarrow \beta_{2k}$ *is the inclusion.*

From this proposition we see that the element x we have been looking for is $\rho_{2k} - \alpha(\rho_k)$. Now by analysing the splittings which occur in the theorem of [**8**] we end up with the following result, proved in [**9**], which gives the precise relation between the index bundle for the family of Dirac operators coupled to monopoles and the permutation representation of the braid group.

THEOREM. *Using the natural isomorphism of* $KO^0(\mathcal{M}_k)$ *with* $KO^0(B\beta_{2k})$ *the index bundle* ind(δ) *can be identified with* $\rho_{2k} - \alpha(\rho_k)$:

$$\begin{aligned} KO^0(\mathcal{M}_k) &\cong KO^0(B\beta_{2k}) \\ \mathrm{ind}(\delta) &\leftrightarrow \rho_{2k} - \alpha(\rho_k) \end{aligned}$$

§6 Further comments

It is natural to wonder which of the other representations of the braid groups appear in index bundles of operators parametrised by the space of monopoles. The above proceedure can be generalised as follows. We can couple the Dirac operator on $\mathbf{R}^3$ to the pair (A, φ) using any representation of $SU(2)$. The Dirac operator has a quaternionic structure and so if we couple it to a representation with a quaternionic structure the corresponding index bundle will have a real structure. This gives a homomorphism

$$RSp(SU(2)) \to KO^0(B\beta_{2k})$$

where RSp is the quaternionic representation ring. Similarily we get a homomorphism

$$RO(SU(2)) \to KSp^0(B\beta_{2k})$$

where RO is the real representation ring and KSp is quaternionic K-theory. These two homomorphisms will be studied in detail in [**9**]. These constructions give the most general natural elements of K-theory which can be constructed by these methods.

The representations which are used in Vaughan Jones's work all occur in continuous one parameter families which for special values of the parameter factor through the permutation group Σ_n. The corresponding bundles are homotopic, therefore isomorphic, and so define the same element of K-theory. So K-theory is not sufficiently sensitive to detect the difference between these inequivalent representations and every K-theory class which arises from these representations factors through the permutation group. However the index bundles come as sub-bundles of a trivial bundle of Hilbert spaces and so they inherit natural connections. It seems reasonable to guess that the holonomy of these natural connections is closely related to representations of the braid groups and if true this would give a more precise way of associating representations of the braid groups to these index bundles. One final point worth mentioning is that the space of monopoles on hyperbolic space is also diffeomorphic to the space of rational functions Rat_k, see [**3**], and it may be possible to exploit this fact. We plan to return to these ideas in future work.

References

1. M.F. Atiyah, *Bott periodicity and the index of elliptic operators*, Quart. J. Math., Oxford (2) **19** (1968), 113–140.
2. M.F. Atiyah, "K-theory," W.A. Benjamin, New York, 1967.
3. M.F. Atiyah, *Magnetic monopoles in hyperbolic space*, in "Proccedings of the Bombay colloquium on vector bundles, 1984," Oxford University Press, Oxford, 1987, pp. 1–34.
4. M.F. Atiyah and N.J. Hitchin, "The Geometry and Dynamics of Magnetic Monopoles," Princeton Univ. Press, Princeton, 1988.
5. M.F. Atiyah and J.D.S. Jones, *Topological aspects of Yang-Mills theory*, Comm. Math. Phys. **61** (1978), 97–118.
6. E.H. Brown Jr. and F.P. Peterson, *On the stable decomposition of* $\Omega^2 S^{r+2}$, Trans. of the A.M.S. **243** (1978), 287–298.
7. Joan Birman, "Braids, links and, mapping class groups," Annals of mathematics studies, 82, Princeton Univ. Press, Princeton, 1974.
8. F.R. Cohen, R.L. Cohen, B.M. Mann, and R.J. Milgram, *The topology of rational functions and divisors of surfaces*, Acta Math. (to appear).
9. R.L. Cohen and J.D.S. Jones, *Monopoles braid groups and the Dirac operator*, to appear.
10. S.K. Donaldson, *Nahm's equations and the classification of monopoles*, Comm. Math. Phys. **96** (1984), 387–407.
11. N.J. Hitchin, *On the construction of monopoles*, Comm. Math. Phys. **89** (1983), 145–190.
12. J. Hurtubise, *Monopoles and rational maps: a note on a theorem of Donaldson*, Comm. Math. Phys. **100** (1985), 191–196.
13. D.M. Kan and W.P. Thurston, *Every connected space has the homology of a* $K(\pi,1)$, Topology **15** (1976), 253–259.
14. J.P. May, "The Geometry of Iterated Loop Spaces," Lecture Notes in Mathematics 271, Springer-Verlag, 1972.
15. R.J. Milgram, *Iterated loop spaces*, Ann. Math **84** (1966), 386–403.
16. G.B. Segal, *Configuration spaces and iterated loop spaces*, Invent. Math. **21** (1973), 213–221.
17. C.H. Taubes, *Monopoles and maps from* S^2 *to* S^2*; the topology of the configuration space*, Comm. Math. Phys **95** (1984), 345–391.
18. E. Witten, *Quantum field theory and the Jones polynomials*, Comm. Math. Phys. **121** (1989), 351–399.

Mathematics Department, Stanford University, Stanford California 94305
Mathematics Institute, University of Warwick, Coventry CV4 7AL England

Extremal Immersions and the Extended Frame Bundle

D H Hartley, R W Tucker

Department of Physics, University of Lancaster, UK
1989

Abstract

We present a computationally powerful formulation of variational problems that depend on the extrinsic and intrinsic geometry of immersions into a manifold. The approach is based on a lift of the action integral to a larger space and proceeds by systematically constraining the variations to preserve the foliation of a Pfaffian system on an extended frame bundle. Explicit Euler-Lagrange equations are computed for a very general class of Lagrangians and the method illustrated with examples relevant to recent developments in theoretical physics. The method provides a means of determining spatial boundary conditions for immersions with boundary and enables a construction to be made of constants of the motion in terms of Euler-Lagrange solutions and admissible symmetry vectors.

INTRODUCTION

Current trends in theoretical physics have focussed attention on the properties of spacetime immersions that extremalise various aspects of their geometrical structure. Thus string theories are based on models that extremalise the induced area of two dimensional time-like world sheets. Their generalisations to p-dimensional immersions provide a dynamical prescription for $(p-1)$-dimensional membranes. Extremalising the integral of the natural induced measure has provided a very rich phenomenological interpretation in the context of particle physics and has led to a number of speculations connecting gravitation to the other forces of Nature. In these developments the properties of the ambient embedding space for the various immersions play a minor role at the classical level. At the quantum level consistency conditions constrain their dimensionality when the ambient space is flat.

A number of recent papers [1] have begun to investigate the properties of spacetime immersions that extremalise integrals of certain of their extrinsic properties. Such an approach incorporates the ambient space into the fundamentals of the theory in a non-trivial way. At the classical level such models promise a rich and varied phenomenological interpretation and offer important challenges for the quantisation program of non-linear systems.

In order to appreciate the properties of individual models constructed from their extrinsic geometrical properties it is important to be able to study a class of models. In this manner we may put individual properties into their proper perspective. However the classical methods of the variational calculus applied to such higher-order Lagrangian systems become increasingly unwieldy for the models that we have in mind. The traditional approach is to formulate the calculus of variations on a bundle of $k-$order jets. In principle this formulation is available to us but we have found a more economical approach based on the use of exterior differential systems on an extended frame bundle. This idea has been inspired by the work of Griffiths [2] who has considered $1-$dimensional variational problems in this context. For variational problems that are concerned with the extremal properties of immersions this approach seems both natural and powerful.

We consider below a class of immersions determined from properties of their shape tensor. (The immersions with extremal volume are included as a special case.) Immersions in this class lie at the basis of a number of recent membrane models in theoretical physics and have been studied by Willmore and others in the mathematical literature [3],[4]. We believe that the approach to be described will provide a unifying route to the Euler-Lagrange equations of these and more general geometrical actions.

1. The Darboux Frame and Second Fundamental Forms

Let C be a $p-$dimensional submanifold of an $m-$dimensional (pseudo-)Riemannian manifold (M, g) with Levi-Civita connection ∇. For any point $p \in C$ write $T_pM =$

$T_pC \oplus (T_pC)^\perp$ and denote by $\bar{\nabla}$ the Levi-Civita connection of the induced metric $\bar{g}$ on C. Then

$$\nabla_Y Z = \bar{\nabla}_Y Z + h(Y,Z) \qquad \forall Y, Z \in TC \tag{1.1}$$

defines the shape tensor $h \in T^2_1(TM)$. For the ranges $a, b, c = 1, \ldots, m$; $\alpha, \beta, \gamma = 1, \ldots, p$; $i, j, k = p+1, \ldots, m$. Let $\{e^a\} = \{e^i, e^\alpha\}$ be a local g–orthonormal coframe for M such that $\{e^\alpha\}_p \in T^*_pC$ and $\{e^i\}_p \in (T^*_pC)^\perp$. We introduce the matrix $\eta^{ab} = g(e^a, e^b)$ with inverse η_{ab}. If C is the image of the embedding map

$$f : D \to M \tag{1.2}$$

then $\{e^a\}$ constitutes a local Darboux coframe adapted to C if

$$f^* e^i = 0 \qquad \forall i \tag{1.3}$$

Denoting the dual orthonormal frame as $\{X_a\} = \{\mathcal{N}_i, X_\alpha\}$ we may expand the shape tensor h in terms of the normal basis to define the set of $m - p$ second fundamental forms H^i on M:

$$h(X,Y) = H^i(X,Y)\mathcal{N}_i \qquad \forall X, Y \in TC \tag{1.4}$$

Since $\nabla_{X_\alpha} X_\beta = -\omega_\beta{}^c(X_\alpha)X_c$ in terms of the connection 1–forms $\omega_a{}^b$ in a Darboux frame:

$$g(h(X_\alpha, X_\beta), \mathcal{N}_i) = -\omega_\beta{}^c(X_\alpha) g(X_c, \mathcal{N}_i) = -\omega_{\beta i}(X_\alpha)$$

or

$$H^i(X_\alpha, X_\beta) = \omega^i{}_\alpha(X_\beta) \tag{1.5}$$

Since

$$\omega^i{}_\alpha = \omega^i{}_\alpha(X_\beta)e^\beta + \omega^i{}_\alpha(\mathcal{N}_j)e^j \tag{1.6}$$

and each e^i restricts to zero on C we see that:

$$\omega^i{}_\alpha \simeq H^i(X_\alpha, X_\beta)e^\beta \tag{1.7}$$

where here and below $\simeq$ denotes restriction by pullback. The restricted structure equations imply that each H^i is a symmetric second degree tensor.

2. Differential Ideals of Exterior Forms [5]

A set of differential forms $\{\alpha^A\}$ $(1 \leq p_A \leq n)$ with $\alpha^A \in \Lambda^{p_A}(T^*B)$ on an $n-$dimensional smooth manifold B generates a *differential ideal* $\mathcal{I}(B)$ in the exterior algebra $\Lambda(T^*B)$ consisting of all forms of the type $\pi_A \wedge \alpha^A$ where $\pi_A \in \Lambda(T^*B)$. The differential ideal is *closed* if for all A, $d\alpha^A = \rho^A{}_B \wedge \alpha^B$ for some $\rho^A{}_B \in \Lambda(T^*B)$. Two sets of differential forms $\{\alpha^A\}$ and $\{\beta^B\}$, not necessarily with the same number of elements, are said to be *algebraically equivalent* if they generate the same ideal: in which case any subspace of $T(B)$ that annuls all forms in one set will also annuls all forms in the equivalent set. In this case we write $\{\alpha^A\} \equiv \{\beta^B\}$. The *associated space* of an ideal $\mathcal{I}(B)$ generated by $\{\alpha^A\}$ is the smallest subspace $Q^* \subset T^*B$ such that the exterior algebra generated by the 1−forms in Q^* contains a subset that generates $\mathcal{I}(B)$. The *associated Pfaff system* of $\{\alpha^A\}$ is this set $\{\theta^1, \theta^2, \dots \theta^r\}$ of 1−forms that span Q^* and the rank of the ideal $\mathcal{I}(B)$ is defined as the dimension of Q^*. In the following we are mainly concerned with ideals generated by a set of 1−forms, known as a Pfaffian system. Then any Pfaffian system of rank r is algebraically equivalent to a set of r linearly independent 1−forms. An integral manifold of an ideal $\mathcal{I}(B)$ is a mapping

$$\mathcal{F} : D \rightarrow B \tag{2.1}$$

of maximal rank such that all forms in the ideal vanish when restricted to D, i.e.

$$\mathcal{F}^*\alpha = 0 \qquad \forall \alpha \in \mathcal{I}(B) \tag{2.2}$$

The *closure* of a Pfaffian system $\{\theta^A\}$ is the ideal containing $\{\theta^A, d\theta^A\}$. Since d commutes with pull backs, integral manifolds of the Pfaffian system $\{\theta^A\}$ are also integral manifolds of its closure and conversely. Clearly the closure of a Pfaffian system is closed. The system is said to be *completely integrable* if it is algebraically equivalent to a set of r exact 1−forms where r is the rank of the system. This occurs iff the system is (locally) closed. In this case one has a (local) foliation on B whose leaves $\xi^A = \mathit{constant}$ are the integral manifolds of $\{\theta^A\} \equiv \{d\xi^A\}$ More generally a system S of forms $\{\alpha^A\}$, not containing 0−forms, has as its *characteristic system* $Ch(S)$ the associated Pfaff system of its closure $\bar{S} = \{\alpha^A, d\alpha^A\}$ [6]. If r is the rank of this associated Pfaff system then $Ch(S)$ always possesses local integral manifolds of dimension $dim\, B - r$. These are the *characteristic manifolds* for S, and can be used to determine integral manifolds of S itself. In the following we have been able to bypass the explicit construction of $Ch(S)$ by examining directly the structure of $\bar{S}$ in order to determine the appropriate integral manifold for our particular problem.

3. Variations of Exterior Systems

A regular exterior differential system on a smooth manifold B consists of a differential ideal $\mathcal{I}(B) \subset \Lambda(T^*B)$ and a preferred set $\{\Omega^\alpha\}$ of linearly independent 1–forms on B. An integral manifold of such a system consists of a smooth mapping

$$\tilde{f} : D \to B \qquad 3.1$$

such that $\tilde{f}$ is an integral manifold of $\mathcal{I}(B)$ together with the condition that $\{\tilde{f}^*\Omega^\alpha\}$ is non-zero. The problem discussed below entails finding an appropriate space B for an ideal $\mathcal{I}(B)$ in formulating our variational principle.

In the first instance we deal with an integral $\int_C \beta$ of a p–form β on a manifold M. In our applications we have in mind that M is a spacetime. Thus we wish to find the extremum of the integral

$$\Lambda(C, f) = \int_C \beta \qquad 3.2$$

where $f : D \to M$ is a p–dimensional immersion, $C = f(D)$ and β some prescribed form constructed from properties of the immersion and the ambient space M. In terms of the Hodge p–form, $\hat{*}1$, defined with respect to the induced metric on C we define the Lagrangian 0–form $\mathcal{L}$ by

$$\beta = \mathcal{L}\hat{*}1 \qquad 3.3$$

A variation of f is a map

$$F : D \times [-\epsilon, \epsilon] \to M$$

such that $f(p) = F(p, 0)$ for all $p \in D$. This defines a family of immersions $f_t : D \times \{t\} \to M$ with $f_t(D) = C_t, \quad t \in [-\epsilon, \epsilon]$, and we seek a critical immersion $f = f_0$ such that

$$\frac{d\Lambda_t}{dt}|_{t=0} = \frac{d}{dt}\int_{C_t} \beta|_{t=0} = 0 \qquad 3.4$$

subject to conditions on ∂C_0. Denoting $\mathbf{V} = F_* \frac{\partial}{\partial t}$ to be a variational vector field on M we have in terms of the Lie derivative

$$\delta\Lambda[\mathbf{V}] \equiv \frac{d\Lambda_t}{dt}|_{t=0} = \int_{C_0} \mathcal{L}_{\mathbf{V}}\beta \qquad 3.5$$

The problem is to compute $\mathcal{L}_{\mathbf{V}}\beta$ when β depends on properties of the immersion. A relatively simple case occurs when $\beta = \hat{*}1$ and $p = m - 1$. In this case suppose $\mathbf{V} = \mathcal{N}$, the normal vector field to the varied hypersurfaces and choose $\beta = i_{\mathcal{N}} * 1$ in terms of the Hodge m−form on M, since this pulls back to the volume form on D. Clearly $\mathcal{L}_{\mathcal{N}}\beta = i_{\mathcal{N}}d\beta = i_{\mathcal{N}}d * \tilde{\mathcal{N}}$, where $\tilde{\mathcal{N}}$is the metric dual of $\mathcal{N}$, so using

$$d * \tilde{\mathcal{N}} = -(Tr\, H) * 1 \qquad 3.6$$

we have

$$\delta\Lambda[\mathbf{V}] = -\int_{C_0} (TrH)\hat{*}1 \qquad 3.7$$

Hence closed immersions of zero mean curvature ("minimal immersions") will extremalise our integral. To accommodate the induced variations in β in the more general case we reformulate our problem by lifting f to a map $\tilde{f} : D \to B$ for some suitable space B such that $\tilde{C} = \tilde{f}(D)$ becomes a p−dimensional integral manifold of a regular Pfaffian differential system $\{\theta^A, d\theta^A\}$,$\{\Omega^\alpha\}$ on B. We now regard $\tilde{\beta}$ as a p−form on B such that $f^*\beta = \tilde{f}^*\tilde{\beta}$ and look for integral manifolds $\tilde{f}$ that extremalise the integral

$$\Lambda(\tilde{C}, \tilde{f}) = \int_{\tilde{C}} \tilde{\beta} \qquad 3.8$$

for admissible variations $\tilde{F} : D \times [-\epsilon, \epsilon] \to B$, generated by the vector field $\tilde{\mathbf{V}} = \tilde{F}_* \frac{\partial}{\partial t}$ on B, and appropriate conditions on $\partial\tilde{C}$. By an admissible variation we shall mean that the various maps $\tilde{f}_t : D \times \{t\} \to B$ are also integral manifolds of our system $\{\theta^A, d\theta^A\}$. This implies that admissible variations are generated by fields $\tilde{\mathbf{V}}$ that satisfy:

$$\tilde{f}^* \mathcal{L}_{\tilde{\mathbf{V}}}\theta^A = 0 \qquad \forall A \qquad 3.9$$

i.e.

$$\mathcal{L}_{\tilde{\mathbf{V}}}\theta^A \simeq 0 \qquad 3.10$$

Since

$$\mathcal{L}_{\tilde{\mathbf{V}}}(\lambda_A \wedge \theta^A) = \mathcal{L}_{\tilde{\mathbf{V}}}\lambda_A \wedge \theta^A + \lambda_A \wedge \mathcal{L}_{\tilde{\mathbf{V}}}\theta^A \qquad 3.11$$

for any forms λ_A it follows that admissible variations satisfy

$$\mathcal{L}_{\tilde{\mathbf{V}}}(\lambda_A \wedge \theta^A) \simeq 0 \qquad 3.12$$

Thus given some $p-$form β on M we have elevated the variational problem to the determination of an integral manifold $\tilde{f}$ of a differential exterior system such that for all variations satisfying 3.12 we have

$$\int_{\tilde{C}} \mathcal{L}_{\tilde{\mathbf{V}}}\tilde{\beta} = 0 \qquad 3.13$$

modulo boundary data.

Let us first consider the case where $\tilde{\mathbf{V}}$ is chosen to vanish on $\partial\tilde{C}$. (One might equivalently postulate $\partial\tilde{C} = 0$ although in the classical spacetime context this is a somewhat unphysical condition.) Then since $\mathcal{L}_{\tilde{\mathbf{V}}} = i_{\tilde{\mathbf{V}}}d + di_{\tilde{\mathbf{V}}}$ equation 3.13 is equivalent to

$$\int_{\tilde{C}} i_{\tilde{\mathbf{V}}}d\tilde{\beta} = 0 \qquad 3.14$$

Note this statement is invariant under $\tilde{\beta} \to \tilde{\beta} + \lambda_A \wedge \theta^A$ since $d\theta^A$ pulls back to zero on D, and under $\tilde{\beta} \to \tilde{\beta} + d\rho$. Furthermore 3.14 is satisfied if

$$i_{\tilde{\mathbf{V}}}d\tilde{\beta} = d\xi \qquad 3.15$$

provided $\tilde{\mathbf{V}}$ is admissible and ξ vanishes on $\partial\tilde{C}$.

Now suppose we seek a set of $(p-1)-$forms λ_A so that for all $\tilde{\mathbf{V}}$, $(\tilde{\mathbf{V}}|_{\partial\tilde{C}} = 0)$

$$\xi = (-1)^{p-1}\lambda_A \wedge i_{\tilde{\mathbf{V}}}\theta^A \qquad 3.16$$

Then 3.15 becomes with the aid of 3.12

$$i_{\tilde{\mathbf{V}}}d(\tilde{\beta} + \lambda_A \wedge \theta^A) \simeq 0 \qquad 3.17$$

Such an equation respects the invariances above and will be adopted as the set of Euler-Lagrange equations for the integral manifolds that extremalise 3.8 for all variations that vanish on $\partial\tilde{C}$.

When $\tilde{C}$ has a boundary on which we relax the condition $\tilde{\mathbf{V}}|_{\partial\tilde{C}} = 0$ the above conclusions need refining. In general we do not expect to be able to impose either arbitrary or unique boundary conditions that determine $\tilde{C}$ from the Euler-Lagrange equations. On the other hand we do not expect that 3.17 will need modification in the presence of boundaries. Thus we adopt 3.17 as before but re-examine

$$\delta\Lambda[\tilde{\mathbf{V}}] \equiv \int_{\tilde{C}} \mathcal{L}_{\tilde{\mathbf{V}}}\tilde{\beta} \qquad 3.18$$

$$= \int_{\tilde{C}} i_{\tilde{\mathbf{V}}} d\tilde{\beta} + \int_{\partial\tilde{C}} i_{\tilde{\mathbf{V}}}\tilde{\beta} \qquad 3.19$$

where we have used Stoke's theorem in the second integral. At this point we can no longer assert that Λ be stationary under arbitrary variations. However from 3.17 we may write

$$\delta\Lambda[\tilde{\mathbf{V}}] = \int_{\partial\tilde{C}} i_{\tilde{\mathbf{V}}}\tilde{\beta} - \int_{\tilde{C}} i_{\tilde{\mathbf{V}}} d(\lambda_A \wedge \theta^A) \qquad 3.20$$

or using 3.10

$$\delta\Lambda[\tilde{\mathbf{V}}] = \int_{\partial\tilde{C}} i_{\tilde{\mathbf{V}}}(\tilde{\beta} + \lambda_A \wedge \theta^A) \qquad 3.21$$

for admissible variations. At this point it is useful to introduce the notion of an admissible symmetry vector. $\tilde{\mathbf{V}}$ is an admissible symmetry vector, denoted $\tilde{\Gamma}$, if in addition to 3.12

$$\mathcal{L}_{\tilde{\Gamma}}\tilde{\beta} \simeq 0 \qquad 3.22$$

If such a vector field exists

$$\delta\Lambda[\tilde{\Gamma}] = 0 = \int_{\partial\tilde{C}} i_{\tilde{\Gamma}}(\tilde{\beta} + \lambda_A \wedge \theta^A)$$

or

$$\int_{\partial\tilde{C}} j_{\tilde{\Gamma}} = 0 \qquad 3.23$$

where

$$j_{\tilde{\Gamma}} = i_{\tilde{\Gamma}}(\tilde{\beta} + \lambda_A \wedge \theta^A) \qquad 3.24$$

When M has a Lorentzian structure we may classify subspaces of T_pD according to the signature of the restriction of f^*g to T_pD. (Recall $f : D \to M$.) We shall suppose that for physical applications $f(D)$ is an orientable time-like (or null) $p-$chain representing the history in M of some spacelike $(p-1)-$chain. (For timelike signature f^*g contains one timelike eigenvalue.) Then suppose we write

$$\partial\tilde{C} = \tilde{f}(D^+ + D^- + D_\tau) \qquad 3.25$$

where $D^{\pm}$ are spacelike $(p-1)$–chains (corresponding to parameter time slices that "cap" D) and D_τ is a timelike $(p-1)$–chain (corresponding to the history of a spacelike $(p-2)$–chain). Conditions on $D^{\pm}$ constitute initial and final data. Thus we adopt *spatial boundary conditions* that ensure:

$$\int_{\tilde{f}(D_\tau)} i_{\tilde{\mathbf{V}}}(\tilde{\beta} + \lambda_A \wedge \theta^A) = 0 \qquad 3.26$$

for all admissible $\tilde{\mathbf{V}}$, so that

$$\delta\Lambda[\tilde{\mathbf{V}}] = \int_{\tilde{C}} \mathcal{L}_{\tilde{\mathbf{V}}}\tilde{\beta} = \int_{\tilde{f}(D^+ + D^-)} i_{\tilde{\mathbf{V}}}(\tilde{\beta} + \lambda_A \wedge \theta^A) \qquad 3.27$$

for admissible $\tilde{\mathbf{V}}$.

It follows from the global boundary condition 3.26 that if $\tilde{\Gamma}$ is an admissible symmetry vector

$$Q(\tilde{\Gamma}) = \int_{\mathbf{D}_\sigma} \tilde{f}^* j_{\tilde{\Gamma}} \qquad 3.28$$

is a constant of the motion for *any* spacelike $(p-1)$–chain $\mathbf{D}_\sigma \subset D$ with boundary in D_τ.

4. The Extended Frame Bundle and its Darboux Leaves

Having set up a general variational formalism we wish to apply it to the problem of extremalising 3.2 where $f : D \to M$ is an immersion and β depends on properties of its extrinsic geometry.

Consider the orthonormal frame bundle (OM, π) of M, elements of which are $\{p, \{X_a\}\}$ for $p \in M$. Extend this to the bundle $B = (OM \times \mathbf{R}^k, \pi)$ for some k. A local coordinate system for $U_B \subset B$ is given by $\{x^a, \alpha^{ab}, H^i{}_{\alpha\beta}\}$ where $\{x^a\}$ are chart functions for $U_M \subset M$, $\alpha^{ab} = -\alpha^{ba}$ parameterise the m-dimensional (pseudo)-orthogonal group and $H^i{}_{\alpha\beta} = H^i{}_{\beta\alpha}$ are coordinates for $\mathbf{R}^k$. With the index ranges declared in section 1, we have $k = \frac{1}{2}p(p+1)(m-p)$ and $dim\, B = m + \frac{1}{2}m(m-1) + k$. Let us denote a local coframe by a set of 1–forms

$$\{e^a, \omega_{ab}, dH^i{}_{\alpha\beta}\} \in T^*B \qquad 4.1$$

Indices attached to forms on B will be raised and lowered using the matrix η_{ab} introduced in section 1. A variation of C gives rise to a local foliation of M by leaves C_t. Our first aim is to construct a local foliation $\mathcal{D}$ of B by subbundles $(\mathcal{D}(C_t), \pi)$ over C_t such that local sections $\sigma_M : U_M \to B$ give rise to a choice of a local Darboux frame field adapted to each leaf C_t. Having established a set of "Darboux leaves", $\mathcal{D}(C_t)$, we shall lift our variational problem into B in order to determine a critical leaf $\mathcal{D}(C_0)$ by the Euler-Lagrange equations on B. From such a leaf we may obtain the solution C_0 on M of our original problem by projection. Any section σ_M that restricts to $\sigma_{C_0} : C_0 \to \mathcal{D}(C_0)$ fixes a Darboux frame adapted to C_0.

Thus we seek an exterior differential system on B whose integral manifolds include the leaves $\mathcal{D}(C_t)$. Consider then the system $S = \{\theta^i, \theta^i{}_\alpha\}$ where in terms of the coframe 4.1

$$\theta^i = e^i \qquad 4.2$$

$$\theta^i{}_\alpha = \omega^i{}_\alpha - H^i{}_{\alpha\beta} e^\beta \qquad 4.3$$

together with the preferred forms $\Omega^\alpha = e^\alpha$.

We claim that S possesses as a local integral manifold the leaf $\mathcal{D}(C)$ where C is a $p-$dimensional submanifold of M and that locally

$$\mathcal{D}(C) \simeq C \times SO(C) \times (SO(C))^\perp \qquad 4.4$$

where $SO(C)$ is the structure group for the orthonormal frame bundle of C and $(SO(C))^\perp$ the structure group for the bundle of orthonormal frames normal to C. To justify this assertion we must obtain from S its closure $\bar{S}$ and examine its integral manifolds. From the first structure equations:

$$d\theta^i = -\omega^i{}_a \wedge e^a = -\omega^i{}_\alpha \wedge e^\alpha - \omega^i{}_j \wedge e^j \qquad 4.5$$

$$= -\omega^i{}_j \wedge \theta^j + e^\alpha \wedge \theta^i{}_\alpha \qquad 4.6$$

using the symmetry of $H^i{}_{\alpha\beta}$. Furthermore

$$d\theta^i{}_\alpha = d\omega^i{}_\alpha - dH^i{}_{\alpha\beta} \wedge e^\beta - H^i{}_{\alpha\beta} \wedge de^\beta \qquad 4.7$$

$$= R^i{}_\alpha - \omega^i{}_b \wedge \omega^b{}_\alpha - dH^i{}_{\alpha\beta} \wedge e^\beta + H^i{}_{\alpha\beta} \wedge \omega^\beta{}_c \wedge e^c \qquad 4.8$$

using both structure equations. Splitting

$$R^i{}_\alpha = \frac{1}{2} R^i{}_{\alpha bc} e^b \wedge e^c$$

into different index ranges and using 4.2,4.3 gives:

$$d\theta^i{}_\alpha = \frac{1}{2} R^i{}_{\alpha\beta\gamma} e^\beta \wedge e^\gamma + R^i{}_{\alpha\beta j} e^\beta \wedge \theta^j + \frac{1}{2} R^i{}_{\alpha jk} \theta^j \wedge \theta^k$$

$$-(\theta^i{}_\beta + H^i{}_{\beta\gamma} e^\gamma) \wedge \omega^\beta{}_\alpha - \omega^i{}_j \wedge (\theta^j{}_\alpha + H^j{}_{\alpha\beta} e^\beta)$$

$$-dH^i{}_{\alpha\beta} \wedge e^\beta + H^i{}_{\alpha\beta} \omega^\beta{}_\gamma \wedge e^\gamma - H^i{}_{\alpha\beta} (\theta_j{}^\beta + H_j{}^\beta{}_\gamma \wedge e^\gamma) \wedge \theta^j \qquad 4.9$$

Under pull back to integral manifolds we may work modulo products of the forms 4.2,4.3 hence:

$$d\theta^i{}_\alpha \simeq (\frac{1}{2} R^i{}_{\alpha\beta\gamma} e^\beta + H^i{}_{\beta\gamma} \omega^\beta{}_\alpha - \omega^i{}_j H^j{}_{\alpha\gamma} - dH^i{}_{\alpha\gamma} + H^i{}_{\alpha\beta} \omega^\beta{}_\gamma) \wedge e^\gamma$$

$$+(R^i{}_{\alpha\beta j} e^\beta - H^i{}_{\alpha\beta} H_j{}^\beta{}_\gamma e^\gamma) \wedge \theta^j + (\delta^i_j \omega^\beta{}_\alpha - \delta^\beta_\alpha \omega^i{}_j) \wedge \theta^j{}_\beta \qquad 4.10$$

In the first term the expression in brackets is $\alpha\gamma$ symmetric except for the curvature term. However using the symmetry properties

$$R^i{}_{\gamma\alpha\beta} = -R^i{}_{\gamma\beta\alpha} \qquad 4.11$$

$$R^i{}_{\alpha\beta\gamma} + R^i{}_{\beta\gamma\alpha} + R^i{}_{\gamma\alpha\beta} = 0 \qquad 4.12$$

$$\frac{1}{3}(R^i{}_{\alpha\beta\gamma} + R^i{}_{\gamma\beta\alpha}) e^\beta \wedge e^\gamma = \frac{1}{2} R^i{}_{\alpha\beta\gamma} e^\beta \wedge e^\gamma \qquad 4.13$$

we may write

$$d\theta^i{}_\alpha \simeq -\xi^i{}_{\alpha\gamma} \wedge e^\gamma + (R^i{}_{\alpha\beta j} e^\beta - H^i{}_{\alpha\beta} H_j{}^\beta{}_\gamma e^\gamma) \wedge \theta^j + (\delta^i_j \omega^\beta{}_\alpha - \delta^\beta_\alpha \omega^i{}_j) \wedge \theta^j{}_\beta \quad 4.14$$

in terms of the 1−forms:

$$\xi^i{}_{\alpha\gamma} = \xi^i{}_{\gamma\alpha} = -\frac{1}{3}(R^i{}_{\alpha\beta\gamma} + R^i{}_{\gamma\beta\alpha}) e^\beta + dH^i{}_{\alpha\gamma} - \omega^\beta{}_\alpha H^i{}_{\beta\gamma} - \omega^\beta{}_\gamma H^i{}_{\alpha\beta} + \omega^i{}_j H^j{}_{\alpha\gamma} \qquad 4.15$$

The closure of S is the system $\bar{S} = \{\theta^i, \theta^i{}_\alpha, d\theta^i, d\theta^i{}_\alpha\}$. From 4.6 and 4.14 we see that integral manifolds of the system

$$Pf(S) = \{\theta^i, \theta^i{}_\alpha, \xi^i{}_{\alpha\beta}\} \qquad 4.16$$

are integral manifolds of $\bar{S}$. The rank of $Pf(S)$ is the number of linearly independent 1−forms in it. Hence:

$$rank\, Pf(S) = (m-p)+p(m-p)+\frac{1}{2}(m-p)p(p+1) = (m-p)(1+p+\frac{1}{2}p(p+1)) \quad 4.17$$

It follows that the integral manifolds of $Pf(S)$ give rise to a local foliation $\mathcal{D}$ in B whose leaves have codimension equal to the rank of $Pf(S)$. Since $dim\, B = \frac{1}{2}m(m+1) + \frac{1}{2}(m-p)p(p+1)$ we have:

$$dim\,(leaf\, \mathcal{D}) = dim\, B - rank\, Pf(S) = p + \frac{1}{2}p(p-1) + \frac{1}{2}(m-p)(m-p-1) \quad 4.18$$

Thus the dimension of $leaf\, \mathcal{D}$ coincides with that of a bundle locally isomorphic to $C \times SO(C) \times (SO(C))^{\perp}$. We may reorganise the local coframe for B as $\{e^{\alpha}, \omega^{\alpha\beta}, \omega^{ij}, \theta^{j}, \theta^{j}{}_{\alpha}, \xi^{i}{}_{\alpha\beta}\}$ and we observe that a suitable coframe for $leaf\, \mathcal{D}$ would be $\{e^{\alpha}, \omega^{\alpha\beta}, \omega^{ij}\}$. Thus the maximal integral manifolds of $Pf(S)$ are just the Darboux bundles $(\mathcal{D}(C), \pi)$ which are those subbundles of B whose cotangent spaces are locally spanned by such coframes. Since integral manifolds of $Pf(S)$ are integral manifolds of $\bar{S}$ and hence S we identify the bundles $(\mathcal{D}(C), \pi)$ as leaves of the foliation associated with system 4,2,4.3. We have already seen in section 1 that the equations $\theta^{i} \simeq 0$, $\theta^{i}{}_{\alpha} \simeq 0$ (pulled back to $C = \pi(\mathcal{D}(C)) \subset M$ enable us to identify the pull backs of $\{e^{i}\}$ and $\{H^{i}{}_{\alpha\beta}\}$ as the normal coframe and second fundamental forms of C respectively.

It is of interest to examine the pull back equation $\xi^{i}{}_{\alpha\beta} \simeq 0$. This equation may be written in terms of a pull back of the exterior covariant derivative $\bar{D}$ on sections of $\Lambda(T^{*}C)$ as

$$\xi^{i}{}_{\alpha\gamma} = \frac{1}{3}(R^{i}{}_{\alpha\beta\gamma} + R^{i}{}_{\gamma\beta\alpha})e^{\beta} + \bar{D}H^{i}{}_{\alpha\gamma} + \omega^{i}{}_{j}H^{j}{}_{\alpha\gamma} \quad 4.19$$

Thus from 4.11,4.12,4.13

$$\xi^{i}{}_{\alpha\beta} \wedge e^{\gamma} = -R^{i}{}_{\alpha\beta\gamma} + (i_{\beta}\bar{D}H^{i}{}_{\alpha\gamma} - i_{\gamma}\bar{H}^{i}{}_{\alpha\beta})$$

$$+(i_{\beta}\omega^{i}{}_{j}H^{i}{}_{\alpha\gamma} - i_{\gamma}\omega^{i}{}_{j}H^{j}{}_{\alpha\beta})e^{\beta} \wedge e^{\gamma} = 0 \quad 4.20$$

In terms of the curvature operator R_{XY} on M:

$$R^{i}{}_{\alpha\beta\gamma} = e^{i}(R_{X_{\beta}X_{\gamma}}X_{\alpha}) \quad 4.21$$

Furthermore:

$$i_\beta \bar{D} H^i{}_{\alpha\beta} = (\bar{\nabla}_{X_\beta} H^i)(X_\alpha, X_\gamma) \qquad 4.22$$

$$\nabla^\perp_{X_\alpha} X_i = e^j(\nabla_{X_\alpha} X_i) X_j \qquad 4.23$$

in terms of the connection $\bar{\nabla}$ induced on TC and the normal connection $\nabla^\perp$ on $(TC)^\perp$. Thus

$$i_\beta \omega^i{}_j = e^i(\nabla_{X_\beta} X_j) \equiv e^i(\nabla^\perp_{X_\beta} X_j) \qquad 4.24$$

and 4.20 becomes:

$$e^i(R_{X_\beta X_\gamma} X_\alpha) = (\bar{\nabla}_{X_\beta} H^i)(X_\alpha, X_\gamma) - (\bar{\nabla}_{X_\gamma} H^i)(X_\alpha, X_\beta)$$

$$+ e^i(\nabla^\perp_{X_\beta} X_j) H^j{}_{\alpha\gamma} - e^i(\nabla^\perp_{X_\gamma} X_j) H^j{}_{\alpha\beta} \qquad 4.25$$

which is just the Codazzi equation for $C \subset M$.

5. The Variational Problem on the Extended Frame Bundle

In section 4 we established a set $\mathcal{D}$ of integral manifolds for the differential system $(S, \{\Omega^\alpha\})$ on B. Leaves $\mathcal{D}(C_t)$ of its foliation contain the orthonormal frame bundle for a p–dimensional submanifold $C_t \subset M$ together with the bundle of normal frames. For any C_t let

$$\sigma' : C_t \to \mathcal{D}(C_t) \qquad 5.1$$

be any section of $(\mathcal{D}(C_t), \pi)$. Extend the map arbitrarily to give a local section

$$\sigma_M : U_M \to B \qquad 5.2$$

of B such that σ_M restricts to σ' on C_t. The section σ_M defines an orthonormal coframe $\{\sigma_M^* e^a\}$ and connection forms $\{\sigma_M^* \omega^a{}_b\}$ on M together with a set of functions $\{\sigma_M^* H^i{}_{\alpha\beta}\}$ which we have seen in 4 are the components of the shape tensor of $C_t \subset M$. The map σ_M also defines a lift $\tilde{C}_t$ of C_t to B:

$$\tilde{C}_t = \sigma_M(C_t) = \sigma'(C_t) \qquad 5.3$$

and since $C_t = f_t(D)$, a lift $\tilde{f}_t : D \to B$ of the immersion $f_t = F|_t : D \times \{t\} \to M$;

$$\tilde{f}_t = \sigma_M \circ f_t \qquad 5.4$$

so that

$$\tilde{C}_t = \tilde{f}_t(D) \qquad 5.5$$

We may now lift our original variational problem onto B. For any $p-$form $\tilde{\beta}$ on B such that

$$\sigma_M^* \tilde{\beta} = \beta \qquad 5.6$$

we have

$$\Lambda(C, f) = \int_C \beta = \int_C \sigma_M^* \tilde{\beta} = \int_{\tilde{C}} \tilde{\beta} \equiv \Lambda(\tilde{C}, \tilde{f}) \qquad 5.7$$

We now seek to extremalise $\Lambda(\tilde{C}, \tilde{f})$ under variations that preserve the foliation $\mathcal{D}$ and are subject to conditions on $\partial\tilde{C}$. Thus for variations that vanish on $\partial\tilde{C}$ we seek an immersion

$$\tilde{f}_0 : D \to B \qquad 5.8$$

such that

$$\frac{d}{dt} \int_{\tilde{C}_t} \tilde{\beta}|_{t=0} = 0 \qquad 5.9$$

where $\tilde{C}_t = \tilde{f}_t(D)$ and the variations preserve the exterior differential system (S, Ω^α):

$$\tilde{f}_t^* \theta^i = \tilde{f}_t^* \theta^i{}_\alpha = 0 \qquad \forall t \in [-\epsilon, \epsilon] \qquad 5.10$$

Then the solution $(f_0, C_0 = f_0(D))$ to the original problem is obtained by simple projection $\pi : \mathcal{D}(C_0) \to M$.

We illustrate this procedure with

$$\tilde{\beta} = \mathcal{L}(g(Tr\, h, Tr\, h))\Omega \qquad 5.11$$

where on B

$$g(Tr\, h, Tr\, h) = \eta_{ij} \eta^{\alpha\alpha'} \eta^{\beta\beta'} H^i{}_{\alpha\alpha'} H^j{}_{\beta\beta'}$$

corresponds to the square of the length of the trace of the shape tensor and $\Omega = \Pi_\alpha (e^\alpha \wedge)$ so $\sigma_M^* \Omega = \hat{*}1$. According to 3.17 we must compute;

$$i_{\tilde{V}} d(\tilde{\beta} + \lambda_i \wedge \theta^i + \lambda^{i\alpha} \wedge \theta_{i\alpha}) \simeq 0 \qquad 5.12$$

where

$$\theta^i = e^i \qquad 5.13$$

$$\theta^i{}_\alpha = \omega^i{}_\alpha - H^i{}_{\alpha\beta} e^\beta \qquad 5.14$$

and the $(p-1)$–forms λ_i and $\lambda^{i\alpha}$ are to be determined. We have

$$d(\tilde{\beta} + \lambda_i \wedge \theta^i + \lambda^{i\alpha} \wedge \theta_{i\alpha}) = d\tilde{\beta} + d\lambda_i \wedge \theta^i + \eta\lambda_i \wedge d\,\theta^i$$

$$+ d\lambda^{i\alpha} \wedge \theta_{i\alpha} + \eta\lambda^{i\alpha} \wedge d\theta_{i\alpha} \qquad 5.16$$

where $\eta\psi = (-1)^q\psi$ for any q–form ψ. Using equations 4.6 and 4.14 we have

$$d(\tilde{\beta} + \lambda_i \wedge \theta^i + \lambda^{i\alpha} \wedge \theta_{i\alpha}) \simeq d\tilde{\beta} + \{d\lambda_j - \omega^i{}_j \wedge \lambda_i + \eta\lambda^{i\alpha} \wedge (R_{i\alpha\beta j} - H_{i\alpha\gamma} H_j{}^\gamma{}_\beta) e^\beta\} \wedge \theta^j$$

$$+ \{e^\beta \wedge \lambda_j + d\lambda_j{}^\beta + (\delta^i_j \omega^\beta{}_\alpha - \delta^\beta_\alpha \omega^i{}_j) \wedge \lambda_i{}^\alpha\} \wedge \theta^j{}_\beta + \{e^\beta \wedge \lambda_i{}^\alpha\} \wedge \xi^i{}_{\alpha\beta} \qquad 5.17$$

"Normal" variations of B are maps between leaves of the foliation $\mathcal{D}$ generated by the vector fields $\{\mathbf{X}_i, \mathbf{X}_i{}^\alpha, \mathbf{X}_i{}^{\alpha\beta}\}$ on B, dual to the system $\{\theta^i, \theta^i{}_\alpha, \xi^i{}_{\alpha\beta}\}$. Contracting 5.17 successively with these vectors yields:

$$\tilde{f}^*\{\eta i_{\mathbf{X}_j} d\tilde{\beta} + d\lambda_j - \omega^i{}_j \wedge \lambda_i + \eta\lambda^{i\alpha} \wedge (R_{i\alpha\beta j} - H_{i\alpha\gamma} H_j{}^\gamma{}_\beta) e^\beta\} = 0 \qquad 5.18$$

$$\tilde{f}^*\{\eta i_{\mathbf{X}^j{}_\beta} d\tilde{\beta} + e_\beta \wedge \lambda^j + (\delta^j_i \omega_\beta{}^\alpha - \delta^\alpha_\beta \omega_i{}^j) \wedge \lambda^i{}_\alpha\} = 0 \qquad 5.19$$

$$\tilde{f}^*\{\eta i_{\mathbf{X}^i{}_{\alpha\beta}} d\tilde{\beta} + e_\beta \wedge \lambda^i{}_\alpha\} = 0 \qquad 5.20$$

These are the Euler-Lagrangian equations for any Lagrangian β on C. From the structure equations it follows that

$$d\Omega = -H_i{}^\alpha{}_\alpha \eta\Omega \wedge \theta^i - i_\alpha\Omega \wedge \theta_i \wedge \theta^{i\alpha} \qquad 5.21$$

so for our choice 5.11 of $\tilde{\beta}$

$$i_{\mathbf{X}_j} d\tilde{\beta} \simeq -\mathcal{L} H_j{}^\alpha{}_\alpha \Omega \qquad 5.22$$

$$i_{\mathbf{X}^j{}_\beta} d\tilde{\beta} \simeq 0 \qquad 5.23$$

$$i_{\mathbf{X}^i{}_{\alpha\beta}} d\tilde{\beta} \simeq 2\mathcal{L}' H^{i\gamma}{}_\gamma \eta_{\alpha\beta} \Omega \qquad 5.24$$

where $\mathcal{L}' \equiv \frac{d\mathcal{L}(x)}{dx}$. The equations 5.18-5.20 are all pulled back to D. In the following we shall write equations on C that are readily pulled back to D by applying maps induced from f. Thus the Euler-Lagrange equations may be written

$$(-1)^{(p-1)} \mathcal{L} H_j{}^\alpha{}_\alpha \hat{*}1 + D^\perp \lambda_j + \eta\lambda^{i\alpha} \wedge (R_{i\alpha\beta j} - H_{i\alpha\gamma} H_j{}^\gamma{}_\beta) e^\beta = 0 \qquad 5.25$$

$$e_\beta \wedge \lambda^j + \bar{D}\lambda^j{}_\beta + \omega^j{}_i \wedge \lambda^i{}_\beta = 0 \qquad 5.26$$

$$(-1)^{(p-1)} 2\mathcal{L}' H^{i\gamma}{}_\gamma \eta_{\alpha\beta} \hat{*} 1 + e_\beta \wedge \lambda^i{}_\alpha = 0 \qquad 5.27$$

where for any form $\mu_{\alpha i}$

$$\bar{D}\mu_{\alpha i} = d\mu_{\alpha i} - \omega^\beta{}_\alpha \wedge \mu_{\beta i} \qquad 5.28$$

is the exterior covariant on sections of $\Lambda(T^*C)$ and

$$D^\perp \mu_{\alpha i} = d\mu_{\alpha i} - \omega^j{}_i \wedge \mu_{\alpha j} \qquad 5.29$$

is the exterior covariant derivative on sections of $\Lambda((T^*C)^\perp)$ induced from the normal bundle. From 5.27,5.34

$$\lambda^i{}_\alpha = (-1)^{(p-1)} 2\mathcal{L}' H^{i\gamma}{}_\gamma \hat{*} e_\alpha \qquad 5.30$$

By the first structure equation, $\bar{D}\hat{*}e_\alpha = 0$, so

$$\bar{D}\lambda^i{}_\alpha = (-1)^{(p-1)} 2d(\mathcal{L}' H^{i\gamma}{}_\gamma) \hat{*} e_\alpha \qquad 5.31$$

Then 5.26 becomes

$$e^\beta \wedge \lambda^j + (-1)^{(p-1)} 2D^\perp(\mathcal{L}' H^{j\gamma}{}_\gamma) \hat{*} e^\beta = 0 \qquad 5.32$$

with solution

$$\lambda^j = (-1)^p \hat{*} D^\perp(\mathcal{L}' H^{j\gamma}{}_\gamma) \qquad 5.33$$

We have the relations

$$e^\beta \wedge \hat{*} e^\alpha = \eta^{\alpha\beta} \hat{*} 1 \qquad 5.34$$

$$D^\perp \hat{*} D^\perp \Phi = \Delta^\perp \Phi \hat{*} 1 \qquad 5.35$$

where $\Delta^\perp$ is the Laplacian induced from the connection in the normal bundle on any 0−form Φ:

$$\Delta^\perp = \nabla^\perp_{X^\beta} \nabla^\perp_{X_\beta} - \nabla^\perp_{\mathcal{A}} \qquad 5.36$$

and

$$\mathcal{A} = \bar{\nabla}_{X_\alpha} X^\alpha \qquad 5.37$$

Inserting λ_i and $\lambda^i{}_\alpha$ into 5.25 and using 5.34 and 5.35 we obtain finally

$$\Delta^\perp(\mathcal{L}' H_i{}^\alpha{}_\alpha) - \frac{1}{2}\mathcal{L} H_i{}^\alpha{}_\alpha - \mathcal{L}' H_j{}^\alpha{}_\alpha (R^{j\beta}{}_{\beta i} - H^{j\beta\gamma} H_{i\beta\gamma}) = 0 \qquad 5.38$$

This is the Euler-Lagrange equation for the immersion C into M which extremalises

$$\int_C \mathcal{L}(g(Tr\,h, Tr\,h))\hat{*}1$$

A few specific examples may serve to illustrate the scope of this result. Consider first

$$\mathcal{L} = 1 \qquad 5.39$$

The action above is the $p-$dimensional volume of C and since $\mathcal{L}' = 0$ the Euler-Lagrange equation is simply:

$$H^{i\alpha}{}_{\alpha} = 0 \qquad 5.40$$

This is the familiar equation for an immersion with extremal volume.

As a more complicated example take

$$\mathcal{L} = 1 + \kappa H^{i\alpha}{}_{\alpha} H_i{}^{\beta}{}_{\beta} \qquad 5.41$$

for some constant κ. This is the generalisation to $p-$dimensional immersions into a curved spacetime of Polyakov's "rigid string "Lagrangian density [7]. We have

$$\mathcal{L}' = \kappa \qquad 5.42$$

so the Euler-Lagrangian equation is:

$$\kappa\Delta^{\perp} H_i{}^{\alpha}{}_{\alpha} - \frac{1}{2}(1 + \kappa H^{j\beta}{}_{\beta} H_j{}^{\gamma}{}_{\gamma}) H_i{}^{\alpha}{}_{\alpha} - \kappa H_j{}^{\alpha}{}_{\alpha}(R^{j\beta}{}_{\beta i} - H^{j\beta\gamma} H_{i\beta\gamma}) = 0 \qquad 5.43$$

Finally take the real root

$$\mathcal{L} = (\pm g(Tr\,h, Tr\,h))^{c/2} \equiv |Tr h|^c \qquad 5.44$$

for some real value $c \geq 1$. Then 5.38 becomes

$$c\Delta^{\perp}\{|Tr\,h|^{c-2} H_i{}^{\alpha}{}_{\alpha}\} - |Tr\,h|^c H_i{}^{\alpha}{}_{\alpha} - c|Tr\,h|^{c-2} H_j{}^{\alpha}{}_{\alpha}(R^{j\beta}{}_{\beta i} - H^{j\beta\gamma} H_{i\beta\gamma}) = 0 \quad 5.46$$

This equation agrees with that obtained by Chen and Willmore [3] for the special case $p = m - 1$ with M flat, and with that obtained by Weiner [4] for the case $c = p = 2$ in a space of constant sectional curvature.

6. Boundary Conditions and Constants of the Motion

In this section we seek admissible variation vector fields $\tilde{\mathbf{V}} \in TB$ for the differential system S defined in section 4. This will enable us to specify appropriate boundary conditions and derive conserved quantities for the variational problem discussed in 5. Adopt the coframe $\{e^\alpha, \omega^{\alpha\beta}, \omega^{ij}, \theta^i, \theta^i{}_\alpha, \xi^i{}_{\alpha\beta}\}$ for B and let $\{\mathbf{X}_\alpha, \mathbf{W}_{\alpha\beta}, \mathbf{W}_{ij}, \mathbf{X}_i, \mathbf{W}_i{}^\alpha, \mathbf{W}_i{}^{\alpha\beta}\}$ be the dual frame with $\mathbf{W}_{\alpha\beta} = -\mathbf{W}_{\beta\alpha}, \mathbf{W}_{ij} = -\mathbf{W}_{ji}, \mathbf{W}_i{}^{\alpha\beta} = \mathbf{W}_i{}^{\beta\alpha}$. Any vector $\mathbf{V} \in TB$ can be written

$$\tilde{\mathbf{V}} = V^\alpha \mathbf{X}_\alpha + V^{\alpha\beta}\mathbf{W}_{\alpha\beta} + V^{ij}\mathbf{W}ij + V^i \mathbf{X}_i + V^{i\alpha}\mathbf{W}_{i\alpha} + V^{i\alpha\beta}\mathbf{W}_{i\alpha\beta} \qquad 6.1$$

with appropriate index symmetries understood in the summation. As described in section 3 $\tilde{\mathbf{V}}$ generates an admissible variation iff

$$\mathcal{L}_{\tilde{\mathbf{V}}}\theta^i \simeq \mathcal{L}_{\tilde{\mathbf{V}}}\theta^i{}_\alpha \simeq 0 \qquad 6.2$$

Writing $\mathcal{L}_{\tilde{\mathbf{V}}} = di_{\tilde{\mathbf{V}}} + i_{\tilde{\mathbf{V}}}d$, we can use the $d\theta^i$ and $d\theta^i{}_\alpha$ given in 4.6 and 4.14 to find

$$\mathcal{L}_{\tilde{\mathbf{V}}}\theta^i \simeq dV^i + \omega^i{}_j V^j - e^\alpha V^i{}_\alpha \simeq D^\perp V^i - e^\alpha V^i{}_\alpha \qquad 6.3$$

and

$$\begin{aligned}\mathcal{L}_{\tilde{\mathbf{V}}}\theta^i{}_\alpha &\simeq dV^i{}_\alpha - \omega^\beta{}_\alpha V^i{}_\beta + \omega^i{}_j V^j{}_\alpha \\ &\quad -(R^i{}_{\alpha\beta j} - H^i{}_{\alpha\gamma}H_j{}^\gamma{}_\beta)e^\beta V^j - e^\beta V^i{}_{\alpha\beta} \\ &\simeq \hat{D}V^i{}_\alpha - (R^i{}_{\alpha\beta j} - H^i{}_{\alpha\gamma}H_j{}^\gamma{}_\beta)e^\beta V^j - e^\beta V^i{}_{\alpha\beta}\end{aligned} \qquad 6.4$$

where the mixed exterior derivative $\hat{D}$ is

$$\hat{D}V^i{}_\alpha = dV^i{}_\alpha - \omega^\beta{}_\alpha V^i{}_\beta + \omega^i{}_j V^j{}_\alpha \qquad 6.5$$

All the equations 6.2 to 6.5 hold under restriction by pullback to D or equivalently to M. We shall suppress again pullback maps and write $\{e^\alpha, e^i\}$ as a Darboux frame for the immersion of D in M; with dual frame $\{X_\alpha, \mathcal{N}_i\}$. Putting 6.3 and 6.4 into 6.2 gives

$$V^i{}_\alpha = (D^\perp V^i)(X_\alpha) \qquad 6.6$$

$$V^i{}_{\alpha\beta} = \hat{D}V^i{}_\alpha - (R^i{}_{\alpha\beta j} - H^i{}_{\alpha\gamma}H_j{}^\gamma{}_\beta)V^j \qquad 6.7$$

These are the constraints which must be satisfied by $\tilde{\mathbf{V}}$ of the form 6.1 if it is to generate admissible variations. We note that the $V^\alpha, V^{\alpha\beta}$ and V^{ij} components are unconstrained. They merely induce leaf preserving diffeomorphisms on $\mathcal{D}(C)$. Furthermore we see that 6.6 and 6.7 completely determine $V^i{}_\alpha$ and $V^i{}_{\alpha\beta}$ in terms of the V^i. Thus apart from leaf preserving diffeomorphisms an admissible $\tilde{\mathbf{V}}$ can be given in terms of a single vector $V^{\mathcal{N}} \in (TC)^\perp$ given by

$$V^{\mathcal{N}} = V^i X_i \tag{6.8}$$

In terms of the connections $\bar{\nabla}$, $\nabla^\perp$, 6.6 and 6.7 can be written

$$V^i{}_\alpha = e^i(\nabla^\perp_{X_\alpha} V^{\mathcal{N}}) \tag{6.9}$$

$$V^i{}_{\alpha\beta} = e^i(\nabla^\perp_{X_\beta}\nabla^\perp_{X_\alpha} V^{\mathcal{N}} - \nabla^\perp_{\bar{\nabla}_{X_\beta}X_\alpha} V^{\mathcal{N}}) - (R^i{}_{\alpha\beta j} - H^i{}_{\alpha\gamma}H_j{}^\gamma{}_\beta)V^j \tag{6.10}$$

It is interesting to note that the left-hand side of 6.10 is symmetric in $\alpha\beta$ whilst the right-hand side appears at first to be asymmetric. However, the antisymmetric part of the right-hand side vanishes by virtue if the Ricci equation:

$$R^i{}_{j\alpha\beta} = R^{\perp i}{}_{j\alpha\beta} - (H^i{}_{\alpha\gamma}H_j{}^\gamma{}_\beta - H^i{}_{\beta\gamma}H_j{}^\gamma{}_\alpha) \tag{6.11}$$

Having established the admissible variation vectors $\tilde{\mathbf{V}} \in TB$, we can now seek spatial boundary conditions. In principle such conditions need not bear any relation to an extremalisation procedure. However the determination of the critical immersion by such a process always involves integrals over possible boundaries and it is therefore natural to elevate the variational principle in such a manner that both the Euler-Lagrange equations and compatible boundary conditions be simultaneously determined. It is worth noting that for the higher order Lagrangian systems considered in this paper, a determination of boundary conditions from a prescription that considers only admissible variations vanishing on a boundary is no longer sufficient to exhaust all possibilities. Specifying the derivatives of admissible boundary variations (arbitrarily) enlarges the space of possible boundary conditions. In this paper we have adopted the view that for immersions into a Lorentzian spacetime the variational procedure determines spatial boundary conditions in any manner that results in the (lifted) variational integral being stationary modulo integrals over (lifted) space-like chains. Thus from 3.26 the conditions should ensure that

$$\int_{\tilde{f}(D_\tau)} j_{\tilde{\mathbf{V}}} = 0 \tag{6.12}$$

where

$$\begin{aligned} j_{\tilde{\mathbf{V}}} &= i_{\tilde{\mathbf{V}}}(\tilde{\beta} + \lambda_i \wedge \theta^i + \lambda_{i\alpha} \wedge \theta^{i\alpha}) \\ &\simeq i_{\tilde{\mathbf{V}}}\tilde{\beta} + \eta\lambda_i \wedge i_{\tilde{\mathbf{V}}}\theta^i + \eta\lambda_{i\alpha} \wedge i_{\tilde{\mathbf{V}}}\theta^{i\alpha} \end{aligned} \tag{6.13}$$

For our choice $\tilde{\beta} = \mathcal{L}\Omega$

$$i_{\tilde{\mathbf{V}}}\tilde{\beta} = \mathcal{L}V^\alpha i_{\mathbf{X}_\alpha}\Omega \simeq \mathcal{L}\hat{*}\check{V}^T \tag{6.14}$$

where the vector $V^T \in TC$ is defined by

$$V^T = V^\alpha X_\alpha \tag{6.15}$$

and $\check{V}^T$ is the dual form related to it by the metric $\bar{g}$. Substituting λ_i, $\lambda_{i\alpha}$ from 5.33, 5.30 and using 1.6 $j_{\tilde{\mathbf{V}}}$ becomes

$$j_{\tilde{\mathbf{V}}} \simeq \hat{*}\{\mathcal{L}\check{V}^T + 2(\mathcal{L}'H^{i\beta}{}_\beta D^\perp V_i - V^i D^\perp(\mathcal{L}'H_i{}^\beta{}_\beta))\} \tag{6.16}$$

Thus

$$\int_{\tilde{f}(D_\tau)} j_{\tilde{\mathbf{V}}} = \int_{f(D_\tau)} \hat{*}\{\mathcal{L}\check{V}^T + 2(\mathcal{L}'H^{i\beta}{}_\beta D^\perp V_i - V^i D^\perp(\mathcal{L}'H_i{}^\beta{}_\beta))\} \tag{6.17}$$

If the spatial boundary $f(D_\tau)$ of $f(D)$ has a unit normal field $n \in TC$ then the induced volume element on $f(D_\tau)$ is

$$\#1 = \hat{*}\check{n} \tag{6.18}$$

The integral 6.17 can be rewritten in terms of n, $\nabla^\perp$ and the vector $Tr\,h = H^{i\alpha}{}_\alpha \mathcal{N}_i$, so that the boundary condition 6.12 becomes

$$\int_{f(D_\tau)} \{\mathcal{L}\check{V}^T(n) + 2\{\mathcal{L}'Tr\,\check{h}(\nabla^\perp_n V^{\mathcal{N}}) - \check{V}^{\mathcal{N}}(\nabla^\perp_n(\mathcal{L}'Tr\,h))\}\}\#1 = 0 \tag{6.19}$$

To satisfy these boundary conditions for arbitrary V^T, $V^{\mathcal{N}}$ and $dV^{\mathcal{N}}$ we can take either:

$$\#1 = 0 \qquad \text{on} \qquad f(D_\tau) \tag{6.20}$$

or

$$\mathcal{L} = 0 \qquad and \qquad \nabla_{\vec{n}}^{\perp}(\mathcal{L}' Tr\, h) = 0 \qquad and \qquad \mathcal{L}' Tr\, h = 0 \qquad on \qquad f(D_\tau) \tag{6.21}$$

For example, for an open membrane in Lorentzian spacetime with the Lagrangian 5.39 the boundary condition on D_τ is

$$\#1 = 0 \tag{6.22}$$

Condition 6.22 implies that the induced metric on the spatial boundary is degenerate. For strings, this is the familiar condition that the ends of the string move at the speed of light. For $(p-1)$–dimensional membranes $(p > 2)$, it implies that the spatial boundary sweeps out a null surface, possessing a single null direction and $(p-2)$ space-like directions. This is a stronger condition than requiring the boundary to move at the speed of light (although the latter still applies).

For the Lagrangian with 5.41 the conditions 6.21 are incompatible for an immersion with boundary and 6.22 may be imposed in the Lorentzian context. On the other hand for admissible variations that vanish on $f(D_\tau)$, but not their derivatives we see from 6.19 that an alternative boundary condition is

$$\mathcal{L}'\, Tr\, h = 0 \tag{6.23}$$

Thus for the Lagrangian 5.41 with 5.42 we may impose

$$Tr\, h = 0 \qquad on \qquad f(D_\tau) \tag{6.24}$$

For the Lagrangian 5.44 we have the alternative boundary conditions on $f(D_\tau)$:

$$Tr\, h = \nabla_{\vec{n}}^{\perp}(Tr\, h) = 0 \tag{6.25}$$

To find the constants of the motion given by 3.28 we need to determine the admissible symmetry vectors $\tilde{\Gamma}$. Recall that $\tilde{\mathbf{V}} = \tilde{\Gamma}$ is an admissible symmetry vector if it satisfies 3.12 and 3.22. Thus for solutions of 3.17 the current $j_{\tilde{\Gamma}} = i_{\tilde{\Gamma}}(\tilde{\beta} + \lambda_A \wedge \theta^A)$ satisfies

$$d\, j_{\tilde{\Gamma}} \simeq 0 \tag{6.26}$$

Since d commutes with pull backs this may be written on C as

$$d\hat{*}\{\mathcal{L}\check{\Gamma}^T + 2(\mathcal{L}'H^{i\beta}{}_\beta D^\perp\Gamma_i - \Gamma^i D^\perp(\mathcal{L}'H_i{}^\beta{}_\beta))\} = 0 \qquad 6.27$$

Noting that

$$d\hat{*}(\mathcal{L}'H^{i\beta}{}_\beta D^\perp\Gamma_i) = D^\perp(\mathcal{L}'H^{i\beta}{}_\beta\hat{*}D^\perp\Gamma_i) = D^\perp(\mathcal{L}'H^{i\beta}{}_\beta)\wedge\hat{*}D^\perp\Gamma_i + \mathcal{L}'H^{i\beta}{}_\beta\Delta^\perp\Gamma_i\hat{*}1 \qquad 6.28$$

and that

$$d\hat{*}(\mathcal{L}\check{\Gamma}^T) = \bar{\nabla}\cdot(\mathcal{L}\Gamma^T)\hat{*}1 \qquad 6.29$$

the symmetry vector condition 6.26 may be further rewritten

$$\bar{\nabla}\cdot(\mathcal{L}\Gamma^T) + 2(\mathcal{L}'H^{i\beta}{}_\beta\Delta^\perp\Gamma_i - \Gamma^i\Delta^\perp(\mathcal{L}'H_i{}^\beta{}_\beta)) = 0 \qquad 6.30$$

If this condition is satisfied, then by an argument similar to the one in section 3 the quantity

$$Q(\tilde{\Gamma}) = \int_{\mathbf{D}_\sigma} f^*\{\mathcal{L}\check{\Gamma}^T(n) + 2\{\mathcal{L}'Tr\,\check{h}(\nabla_n^\perp\Gamma^{\mathcal{N}}) - \check{\Gamma}^{\mathcal{N}}(\nabla_n^\perp(\mathcal{L}'Tr\,h))\}\}\#1 \qquad 6.31$$

is a constant of the motion for any space like $(p-1)$–chain $\mathbf{D}_\sigma$ (with boundary in D_τ) with unit (timelike) normal vector field $n \in f(TD)$ and:

$$\#1 = \hat{*}\check{n} \qquad 6.32$$

Conclusions

As can be seen from the preceeding calculations, the differential system of constraints on the extended frame bundle that we have used provides a convenient framework for tackling a wide class of higher-order variational problems depending on the extrinsic geometry of immersions. In this approach the Euler-Lagrange equations are expressed in terms of geometrical quantities relevant to the problem. As illustrated in section 6, the "natural" boundary conditions which arise for higher-order problems depend on a somewhat arbitrary choice of details in the variational principle. We have adopted boundary conditions at the boundaries of space-like chains only, thus allowing for the specification of initial and final conditions. This has also enabled us to construct constants of the motion in the presence of symmetry vectors.

The general approach is both powerful and flexible, and clearly lends itself to a number of generalisations.

Acknowledgement

D H Hartley acknowledges support from the SERC.

References

T L Curtright, G I Ghandour and C K Zachos, Phys. Rev. **D34** (1986) 3811.
T L Curtright and P van Nieuwenhuizen, Nucl. Phys. **B294** (1987) 125.
U Lindström, M Roček and P van Nieuwenhuizen, Phys. Lett. **199B** (1987) 219.
M Mukherjee and R W Tucker, Class. Quantum Grav. **5** (1988) 849.
M Önder and R W Tucker, Phys. Lett. **202B** (1988) 501.
M Önder and R W Tucker, J. Phys. A: Math. Gen. **21** (1988) 3423.
U Lindström, Int. J. Mod. Phys. **A10** (1988) 2410.
R W Tucker, "The Motion of Membranes in spacetime", Lancaster preprint, 1989.
D H Hartley, M Önder and R W Tucker, "On the Einstein-Maxwell equations for a "stiff" membrane", to appear in Class. Quantum Grav.

P A Griffiths, "Exterior Differential Systems and the Calculus of Variations", Birkhäuser, 1983.

B-Y Chen, J. London Math. Soc. **6** (1973) 321.
T J Willmore, "Total curvature in Riemannian geometry", Ellis Horwood, 1982.

J L Weiner, Indiana Univ. Math. J. **27** (1978) 19.

[5] Y Choquet-Bruhat, C DeWitt-Morette and M Dillard-Bleick, "Analysis, Manifolds and Physics", North-Holland, 1982.

[6] E Cartan, "Les systèmes différentielles extérieurs et leurs applications géométriques", Herman, 1945.

[7] A Polyakov, Nucl. Phys. **B268** (1986) 406.

Minimal surfaces in quaternionic symmetric spaces

F.E. BURSTALL

University of Bath

We describe some birational correspondences between the twistor spaces of quaternionic Kähler compact symmetric spaces obtained by Lie theoretic methods. By means of these correspondences, one may construct minimal surfaces in such symmetric spaces. These results may be viewed as an explanation and a generalisation of some results of Bryant **[1]** concerning minimal surfaces in S^4.

This represents work in progress in collaboration with J.H. Rawnsley and S.M. Salamon.

BACKGROUND

This work has its genesis in our attempt to understand the following result of Bryant **[1]**:

Theorem. *Any compact Riemann surface may be minimally immersed in* S^4.

To prove this, Bryant considers the Penrose fibration $\pi : \mathbf{C}P^3 \to S^4 = \mathbf{H}P^1$. The perpendicular complement to the fibres (with respect to the Fubini-Study metric) furnishes $\mathbf{C}P^3$ with a holomorphic distribution $\mathcal{H} \subset T^{1,0}\mathbf{C}P^3$ and it is well-known that a holomorphic curve in $\mathbf{C}P^3$ tangent to $\mathcal{H}$ (a *horizontal* holomorphic curve) projects onto a minimal surface in S^4. Bryant gave explicit formulae for the horizontality condition on an affine chart which enabled him to integrate it and provide a "Weierstraß formula" for horizontal curves. Indeed, if f, g are meromorphic functions on a Riemann surface M then the curve $\Phi(f,g) : M \to \mathbf{C}P^3$ given on an affine chart by

$$\Phi(f,g) = (f - \tfrac{1}{2}g(dg/df), g, \tfrac{1}{2}(df/dg))$$

is an integral curve of $\mathcal{H}$. For suitable f, g, $\Phi(f,g)$ is an immersion (indeed, an embedding) and the theorem follows.

In **[6]**, Lawson gave an interesting interpretation of Bryant's method by introducing the flag manifold $\mathbf{D}^3 = \mathrm{U}(3)/\mathrm{U}(1) \times \mathrm{U}(1) \times \mathrm{U}(1)$ which may be viewed as the twistor space of $\mathbf{C}P^2$ (the twistor fibring being the non-$\pm$holomorphic homogeneous fibration of $\mathbf{D}^3$ over $\mathbf{C}P^2$). Again we have a holomorphic horizontal distribution $\mathcal{K}$

perpendicular to the fibres of the twistor fibration but, this time, horizontal curves are easy to construct. Indeed, viewing $\mathbf{D}^3$ as $P(T^{1,0}\mathbf{C}P^2)$, $\mathcal{K}$ is just the natural contact distribution and a holomorphic curve in $\mathbf{C}P^2$ has a canonical horizontal lift into $\mathbf{D}^3$ given by its tangent lines.

The remarkable fact, implicit in Bryant's work and brought to the fore by Lawson, is

Theorem. *There is a birational correspondence $\Phi : \mathbf{D}^3 \to \mathbf{C}P^3$ mapping $\mathcal{K}$ into $\mathcal{H}$.*

Recall that a birational correspondence of projective algebraic varieties is a holomorphic map which is defined off a set of co-dimension 2 and biholomorphic off a set of co-dimension 1.

Thus it suffices to produce horizontal curves in $\mathbf{D}^3$ which avoid the singular set of Φ and this may be done by taking the lifts of suitably generic curves in $\mathbf{C}P^2$.

Lawson gave an analytic expression for Φ but a geometrical interpretation of the map seemed quite hard to come by. An algebro-geometric interpretation has been given by Gauduchon [4] but it is our purpose here to show how this map arises naturally from Lie theoretic considerations.

QUATERNIONIC SYMMETRIC SPACES

The 4-sphere and $\mathbf{C}P^2$ may be viewed as the 4-dimensional examples of the quaternionic Kähler compact symmetric spaces. These are $4n$-dimensional symmetric spaces N with holonomy contained in $\mathrm{Sp}(1)\mathrm{Sp}(n)$. Geometrically, this means that there is a parallel subbundle E of $\mathrm{End}(TN)$ with each fibre isomorphic to the imaginary quaternions. There is one such symmetric space for each simple Lie group; the classical ones in dimension $4n$ being

$$\mathbf{H}P^n, \quad G_2(\mathbf{C}^{n+2}), \quad G_4(\mathbf{R}^{n+4}).$$

Following [7], we consider the *twistor space* Z of N which is the sphere bundle of E or, equivalently,

$$Z = \{j \in E: \quad j^2 = -1\}.$$

This twistor space is a Kähler manifold, indeed a projective variety, and once more the perpendicular complement to the fibres $\mathcal{H}$ is a holomorphic subbundle which is called the *horizontal distribution.* Our main theorem is then

Theorem. *Let N_1, N_2 are compact irreducible quaternionic Kähler symmetric spaces of the same dimension with twistor spaces Z_1, Z_2. Then there is a birational correspondence $Z_1 \to Z_2$ which preserves the horizontal distributions.*

For this we must study the homogeneous geometry of the twistor spaces: if N is the symmetric space G/K then G acts transitively on Z and, moreover, this action extends to a holomorphic action of the complexified Lie group $G^{\mathbf{C}}$. Further, the horizontal distribution is invariant under this $G^{\mathbf{C}}$ action. In fact, Z is a special kind of $G^{\mathbf{C}}$-space: it is a flag manifold, that is, of the form $G^{\mathbf{C}}/P$ where P is a parabolic subgroup.

For any flag manifold $G^{\mathbf{C}}/P$, let $\mathbf{p}$ be the Lie algebra of P. We have a decomposition of the Lie algebra of $G^{\mathbf{C}}$

$$\mathbf{g}^{\mathbf{C}} = \mathbf{p} \oplus \overline{\mathbf{n}}$$

where $\mathbf{n}$ is the nilradical of $\mathbf{p}$ so that $\overline{\mathbf{n}} \cong T^{1,0}_{eP} G^{\mathbf{C}}/P$ is a nilpotent Lie algebra. Let $\overline{N}$ be the corresponding nilpotent Lie group and consider the $\overline{N}$ orbits in $G^{\mathbf{C}}/P$. The orbit Ω of the identity coset is a Zariski dense open subset of $G^{\mathbf{C}}/P$ (it is the "big cell" in the Bruhat decomposition of the flag manifold). In fact, the map $\overline{\mathbf{n}} \to \Omega$ given by

$$\xi \mapsto \exp \xi \cdot P$$

is a biholomorphism with polynomial components (since $\overline{\mathbf{n}}$ is nilpotent) and so extends to give a birational correspondence of $G^{\mathbf{C}}/P$ with $P(\overline{\mathbf{n}} \oplus \mathbf{C})$. Thus $G^{\mathbf{C}}/P$ is a rational variety: a classical result of Goto [5]. However, more is true: let $G_1^{\mathbf{C}}/P_1$ and $G_2^{\mathbf{C}}/P_2$ be flag manifolds and suppose that the nilradicals $\mathbf{n}_1$ and $\mathbf{n}_2$ are isomorphic as complex Lie algebras. Then we have an isomorphism $\phi : \overline{\mathbf{n}}_1 \to \overline{\mathbf{n}}_2$ which we may exponentiate to get an isomorphism of Lie groups $\Phi : \overline{N}_1 \to \overline{N}_2$ and thus a biholomorphism $\Omega_1 \to \Omega_2$ which extends to a birational correspondence between the flag manifolds. Moreover, on Ω_1, this biholomorphism is $\overline{N}_1$-equivariant and so will preserve any invariant distribution so long as ϕ does when viewed as a map $T^{1,0}_{eP_1} G_1^{\mathbf{C}}/P_1 \to T^{1,0}_{eP_2} G_2^{\mathbf{C}}/P_2$.

We now specialise to the case at hand: if Z is the twistor space of a quaternionic Kähler compact irreducible symmetric space then Z is a rather special kind of flag manifold. In fact, Wolf [8] has shown that here $\overline{\mathbf{n}}$ is two-step nilpotent with 1-dimensional centre and so is precisely the complex Heisenberg algebra. Thus any two of our twistor spaces of the same dimension have isomorphic $\overline{\mathbf{n}}$ and so the main theorem follows.

APPLICATIONS TO MINIMAL SURFACES

The relevance of these constructions to minimal surface comes from the well-known fact that, just as in the 4-dimensional case, horizontal holomorphic curves in Z project onto minimal surfaces in N. Moreover, in some of the classical cases, horizontal holomorphic curves are quite easy to come by. For example, the twistor space of $G_2(\mathbf{C}^{n+2})$ is the flag manifold $Z = \mathrm{U}(n+2)/\mathrm{U}(1) \times \mathrm{U}(n) \times \mathrm{U}(1)$ which we may realise as the set of flags

$$\{\ell \subset \pi \subset \mathbf{C}^{n+2} : \dim \ell = 1, \dim \pi = n+1\}.$$

Horizontal, holomorphic curves in this setting are just a special kind of ∂'-pair in the sense of Erdem-Wood [3] and may be constructed as follows: if $f : M \to \mathbf{C}P^{n+1}$ is a holomorphic curve, we may construct the associated holomorphic curves $f_r : M \to G_{r+1}(\mathbf{C}^{n+2})$ given locally by

$$f_r = f \wedge \frac{\partial f}{\partial z} \wedge \ldots \wedge \frac{\partial^r f}{\partial z^r}.$$

Generically, f is full so that $f_1, \ldots, f_n$ are defined and then the map $\psi : M \to Z$ given by

$$\psi = (f \subset f_n)$$

is horizontal and holomorphic. Note that for $n = 1$, ψ is just the lift of $f : M \to \mathbf{C}P^2$ discussed above. Composing these curves with the birational correspondences of the previous section, we then have horizontal holomorphic curves in all the other twistor spaces of the same dimension so long as we can ensure that the curves avoid the singular sets of the correspondences. Thus, for example, one has the possibility of constructing minimal surfaces in the 8-dimensional exceptional quaternionic symmetric space $G_2/SO(4)$ from holomorphic curves in $\mathbf{C}P^3$.

However, to carry out such a programme, a rather more detailed analysis of these singular sets is required so as to ensure that they are avoided for suitably generic f. Work is still in progress on this issue.

EXTENSIONS

Many parts of the above development apply to arbitrary generalised flag manifolds. Burstall-Rawnsley [2] have shown that any flag manifold fibres in a canonical way over a Riemannian symmetric space of compact type and, moreover, any such symmetric space with inner involution is the target of such a fibration. In this setting, the perpendicular complement to the fibres is not in general holomorphic but there is a sub-distribution thereof, the *superhorizontal distribution*, which is holomorphic and $G^{\mathbf{C}}$-invariant. Again, holomorphic integral curves of this distribution project onto minimal surfaces in the symmetric space.

The above discussion applies so that isomorphisms of nilradicals exponentiate to give birational correspondences of flag manifolds which preserve the super-horizontal distributions. However, apart from the quaternionic symmetric case, we have not yet found any examples of differing flag manifolds with isomorphic nilradicals.

REFERENCES

1. R. L. Bryant, *Conformal and minimal immersions of compact surfaces into the 4-sphere*, J. Diff. Geom. **17** (1982), 455–473.

2. F. E. Burstall and J. H. Rawnsley, *Twistor theory for Riemannian symmetric spaces with applications to harmonic maps of Riemann surfaces*, Bath-Warwick preprint, 1989.

3. S. Erdem and J. C. Wood, *On the construction of harmonic maps into a Grassmannian*, J. Lond. Math. Soc. **28** (1983), 161–174.

4. P. Gauduchon, *La correspondance de Bryant*, Astérisque **154–155** (1987), 181–208.

5. M. Goto, *On algebraic homogeneous spaces*, Amer. J. Math. **76** (1954), 811–818.

6. H. B. Lawson, *Surfaces minimales et la construction de Calabi-Penrose*, Astérisque **121–122** (1985), 197–211.

7. S. M. Salamon, *Quaternionic Kähler manifolds*, Invent. Math. **67** (1982), 143–171.

8. J. A. Wolf, *Complex homogeneous contact manifolds and quaternionic symmetric spaces*, J. Math. Mech. **14** (1965), 1033–1047.

Three-dimensional Einstein-Weyl Geometry

K.P.TOD

The Mathematical Institute, Oxford

Abstract. *I review what is known about 3-dimensional Einstein-Weyl spaces.*

1. Introduction.

The Einstein-Weyl equations are a naturally-arising, conformally-invariant, generalisation of the Einstein equations. In the special case of 3-dimensions, the Einstein equations on a space force it to have constant curvature. The metric is then characterised locally by a single number, the Ricci scalar. By contrast, the Einstein-Weyl equations allow some functional freedom locally. Further, the equations fall into the small class of non-linear partial differential equations which can be solved by the twistor correspondence.

In this article I shall review the definition of an Einstein-Weyl structure (§2) and its twistor correspondence (§3). I also give some examples of Einstein-Weyl spaces (§4), and describe some general and some particular properties (§5,§6).

I am very grateful to the organisers of the LMS Durham Symposium for inviting me to take part in a most stimulating and worthwhile conference.

2. Einstein-Weyl spaces.

A *Weyl space* is a smooth (real or complex) manifold $\mathcal{W}$ equipped with

(1) a conformal metric
(2) a symmetric connection or torsion-free covariant derivative (the *Weyl connection*)

which are *compatible* in the sense that the connection preserves the conformal metric.

This compatability ensures in particular that orthogonal vectors stay orthogonal when parallel propogated in the Weyl connection and that the null geodesics, which can be defined just given a conformal metric, are also geodesics of the Weyl connection.

In local co-ordinates (or using the "abstract index convention" [**8**]) we write a chosen representative g for the conformal metric as g_{ab} and we write the Weyl covariant derivative as D_a. The compatability condition becomes

$$D_a g_{bc} = \omega_a g_{bc} \tag{2.1}$$

for some 1-form $\omega = \omega_a dx^a$. Under change of representative metric we have

$$\begin{aligned} g \to \hat{g} &= \Omega^2 g \\ \omega \to \hat{\omega} &= \omega + 2\frac{d\Omega}{\Omega} \end{aligned} \tag{2.2}$$

where Ω is a smooth, strictly positive, function on $\mathcal{W}$.

Note that the 1-form ω encodes the difference between the Weyl connection and the Levi-Civita connection of the chosen representative metric. Thus we can think of a Weyl space as a pair (g, ω) modulo (2.2).

The Weyl connection has a curvature tensor and hence a contracted curvature tensor or Ricci tensor. The skew part of the Ricci tensor is a 2-form which is automatically a multiple of $d\omega$. To impose the *Einstein condition* on $\mathcal{W}$ we require that the symmetric part of the Ricci tensor be proportional to the conformal metric, [4]. In local co-ordinates (or abstract indices) the Einstein condition is

$$W_{(ab)} = \Lambda g_{ab} \quad \text{some } \Lambda \tag{2.3}$$

where W_{ab} is the Ricci tensor of the Weyl connection. When $\mathcal{W}$ is 3-dimensional, as we shall assume from now on, we also have the identity

$$W_{[ab]} = -3F_{ab} \tag{2.4}$$

where F_{ab} is the 2-form $d\omega$. (As is customary in the relativity literature we denote symmetrisation and anti-symmetrisation by round and square brackets respectively.)

A Weyl space satisfying the Einstein condition we shall call an *Einstein-Weyl space*.

Since the Weyl connection can be written in terms of the Levi-Civita connection of the chosen representative metric g and the 1-form ω, we can rewrite (2.3) as an equation on g_{ab} and ω_a. We find that the condition is

$$R_{ab} - \frac{1}{2}\nabla_{(a}\omega_{b)} - \frac{1}{4}\omega_a\omega_b \propto g_{ab} \tag{2.5}$$

where R_{ab} is the Ricci tensor of the Levi-Civita connection of g and ∇_a is the corresponding covariant derivative.

Note that if the chosen representative metric g is actually an Einstein metric on $\mathcal{W}$ then (2.5) is satisfied with the 1-from ω vanishing. Since (2.5) is invariant under (2.2), any $\mathcal{W}$ conformal to an Einstein space is an Einstein-Weyl space. These examples have F_{ab} in (2.4) vanishing and, locally at least, are characterised by this property. Thus *the Einstein-Weyl equation (2.3) is a natural conformally-invariant equation which generalises the Einstein equation.*

That the Einstein-Weyl equation has a direct geometrical content is shown by the following

PROPOSITION 2.1 [2]. *If $\mathcal{W}$ is 3-complex dimensional and satisfies the Einstein-Weyl equation then $\mathcal{W}$ admits a 2-complex dimensional family of totally geodesic null hypersurfaces.*

(That is, the hypersurfaces are totally geodesic with respect to the Weyl connection and have normal which is of zero length or null with respect to the conformal metric. Cartan calls the hypersurfaces "isotropes".) If $\mathcal{W}$ is 3 real dimensional we need to suppose that it is real analytic and then complexify it to use this Proposition. We shall return to this point in §5. This result was re-interpreted by Hitchin [4] in a way that we may conveniently call a corollary.

COROLLARY 2.2 [4]. *Such a $\mathcal{W}$ admits a twistor construction.*

To say what this means we shall consider twistor constructions in the next section.

To conclude this section we note that there is another characterisation of the Einstein-Weyl condition equivalent to Proposition (2.1) and obtainable by a consideration of the geodesic deviation equation in $\mathcal{W}$:

PROPOSITION(2.3) [9]. *The Einstein-Weyl condition is equivalent to the existence of a complex structure on the space of Jacobi fields along each geodesic.*

We shall see the connection between these two characterisations below.

3. Twistor correspondences.

The idea of *twistor constructions* or *twistor correspondences*, as I shall use the term here, is that one has two manifolds with a non-local correspondence; one space has some differential geometric objects satisfying local equations; the other does not but is a complex manifold, and under the correspondence the local equations on one space are solved automatically by virtue of the the complex analyticity of the other space.

The first example, which I shall call the *Penrose correspondence* [6], begins with a 4-complex dimensional manifold $\mathcal{M}$ with a conformal structure. The conformal curvature C of the conformal structure on $\mathcal{M}$ is required to be anti-self-dual as a 2-form:

$$*C = -C.$$

This condition is the integrability condition for $\mathcal{M}$ to admit a 3-complex dimensional family of 2-dimensional surfaces on which the conformal metric degenerates completely ("totally null" surfaces). Call these *α-planes*, then the space of α-planes is a 3-dimensional complex manifold $\mathcal{G}$, the *twistor space* of $\mathcal{M}$. A point p of $\mathcal{M}$ is represented in $\mathcal{G}$ by the α-planes through it, which define a holomorphic curve $\mathcal{L}_p$ in $\mathcal{G}$. The curve $\mathcal{L}_p$ is a copy of $\mathbf{CP}^1$ and automatically has normal bundle $N \cong \mathcal{O}(1) \oplus \mathcal{O}(1)$. This gives the non-local correspondence between $\mathcal{G}$ and $\mathcal{M}$.

To define the conformal metric on $\mathcal{M}$ it is sufficient to be able to recognise which vectors are null. A tangent vector v at a point p in $\mathcal{M}$ corrresponds to a section L_v of the normal bundle N of $\mathcal{L}_p$. Define this vector to be null if L_v has a zero. From the character of N this is a quadratic condition. Remarkably, the conformal metric obtained in this way automatically has anti-self-dual conformal curvature tensor.

This is the *twistor correspondence*; it becomes a *twistor construction* if one can find a way of building complex manifolds like $\mathcal{G}$, since these will in turn generate "space-times"$\mathcal{M}$ which automatically satisfy local equations.

The second example, which I shall call the *Hitchin correspondence* [4], relates a 3- complex- dimensional Einstein-Weyl space $\mathcal{W}$ to the 2-complex -dimensional manifold $\mathcal{T}$ whose points are the totally geodesic null hypersurfaces in $\mathcal{W}$, the existence of which is guaranteed by Proposition 2.1. ($\mathcal{T}$ is sometimes called the mini-twistor space of $\mathcal{W}$, by analogy with the previous case.) A point p of $\mathcal{W}$ is represented by the set of surfaces through it, which forms a holomorphic curve $\mathcal{L}_p$ in $\mathcal{T}$. Again, $\mathcal{L}_p$ is a copy of $\mathbf{CP}^1$, but now the normal bundle is $N \cong \mathcal{O}(2)$. This defines the non-local correspondence between $\mathcal{W}$ and $\mathcal{T}$.

To define the conformal metric on $\mathcal{W}$, we note as before that a vector v at a point p of $\mathcal{W}$ defines a section L_v of the normal bundle N of $\mathcal{L}_p$. We define v to be null if L_v has *coincident zeros*. Again, this is a quadratic condition by virtue of the character of N.

To define the Weyl connection on $\mathcal{W}$ is more complicated. The idea is as follows: take a curve $\mathcal{L}_p$ in $\mathcal{T}$ corresponding to a point p in $\mathcal{W}$; now take two points σ_1, σ_2 on $\mathcal{L}_p$ and consider all the curves $\mathcal{L}_q$ which meet $\mathcal{L}_p$ in σ_1 and σ_2. The corresponding point q in $\mathcal{W}$ lies on a geodesic $\gamma = \gamma(\sigma_1, \sigma_2)$. Geometrically, we can think of γ as being the intersection of the two null hypersurfaces in $\mathcal{W}$ corresponding to σ_1 and σ_2 in $\mathcal{T}$. (This intersection is automatically a geodesic, since the hypersurfaces are totally geodesic.)

If the Einstein-Weyl space $\mathcal{W}$ is the complexification of a real space with positive-definite conformal structure then one has an alternative description of $\mathcal{T}$. A real geodesic γ will lie in the intersection of a complex conjugate pair of totally geodesic null hypersurfaces, so symbolically $\sigma_2 = \overline{\sigma}_1$. We can define an orientation on γ and associate σ_1 with one orientation and σ_2 with the opposite orientation. Then $\mathcal{T}$ appears as the space of *oriented geodesics* of $\mathcal{W}$. (This picture of $\mathcal{T}$ arises readily from Proposition 2.3.)

The two twistor correspondences can be fitted into a larger picture (Figure 3.1), which we summarise as follows.

PROPOSITION 3.1.

(1) *Suppose $\mathcal{M}$ has a 1-parameter family of conformal isometries (by which we mean self-diffeomorphisms preserving the conformal structure). The infinitesimal generator X is called a conformal Killing field. The space of trajectories of X automatically has a conformal metric, and also has a natural Weyl connection which satisfies the Einstein condition ,[**5**]. The vector field X induces a holomorphic vector field on the twistor space $\mathcal{G}$, and the quotient is $\mathcal{T}$.*

(2) *If $\mathcal{W}$ is an Einstein-weyl space then one can construct conformally anti-self-dual spaces $\mathcal{M}$, as fibre bundles over $\mathcal{W}$, from solutions (V, α) of the monopole equation*

$$dV = *d\alpha.$$

*Here V is a function and α is a 1-form on $\mathcal{M}$ (more generally we can work with a connection on a circle bundle over $\mathcal{W}$ [**5**]).*

(3) *Given a mini-twistor space $\mathcal{T}$, the projective tangent bundle $\mathbf{P}(T\mathcal{T})$ is the twistor space for an anti-self-dual manifold $\mathcal{M}$, and the boundary ∂M at infinity is the Einstein-Weyl 3-manifold associated to $\mathcal{T}$,[**4**].*

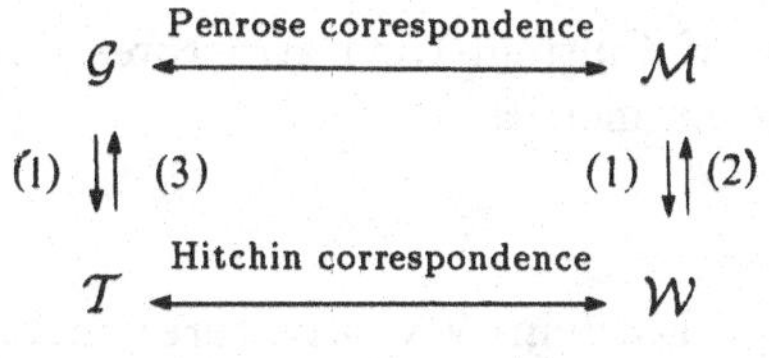

Figure 3.1 *Fitting together the Penrose and Hitchin correspondences. The numerals refer to Proposition 3.1.*

4. Examples.

4.1 *The quadric.*

The simplest example of the Hitchin correspondence has for $\mathcal{T}$ the complex quadric $\mathbf{Q}$ in $\mathbf{CP}^3$. The curves $\mathcal{L}_p$ are the conics in $\mathbf{Q}$ and a geodesic of the Weyl structure is defined by the set of conics meeting a fixed line in $\mathbf{CP}^3$. The Einstein-Weyl space $\mathcal{W}$ is the complexification of S^3 or the hyperbolic space $\mathbf{H}^3$. Suitable choices of reality structure give one or the other. Choosing S^3 and applying the construction of (3.1)(3) fills in S^3 with the 4-dimensional hyperbolic space $\mathbf{H}^4$.

This example essentially just gives an Einstein space. The following Proposition suggests that Einstein-weyl spaces may be rare :

PROPOSITION 4.2, [**4**].

If $\mathcal{T}$ is an open set in a compact surface then $\mathcal{T}$ is $\mathbf{Q}$ or a cone in $\mathbf{CP}^3$, and $\mathcal{W}$ is an Einstein space. (The cone gives flat space $\mathbf{R}^3$.)

However, one can analyse an initial value problem, locally, to find:

PROPOSITION 4.3, [**2**], 4. *An Einstein-weyl space $\mathcal{W}$ is determined by an initial value problem with data 4 functions of two variables, or by a characteristic initial value problem with data 2 functions of 2 variables.*

4.4 *Einstein-Weyl spaces from S^4.*

To find some more examples explicitly we may follow the suggestion of Proposition 3.1(1). The Einstein metric on S^4 is conformally flat and has many conformal Killing fields. In this way we obtain two classes of Einstein-Weyl spaces [**9**] :

Class 1

A 3-parameter family of Einstein-Weyl structures on S^3, generalising the 1-parameter family of Einstein metrics.

Class 2

A 2-parameter family of Einstein-Weyl structures on $\mathbf{R}^3$.

The corresponding conformal Killing fields are shown diagrammatically in Figure 4.1. In both classes we can take the equatorial S^3 as a representative for $\mathcal{W}$, although in Class 2 we must remove a point.

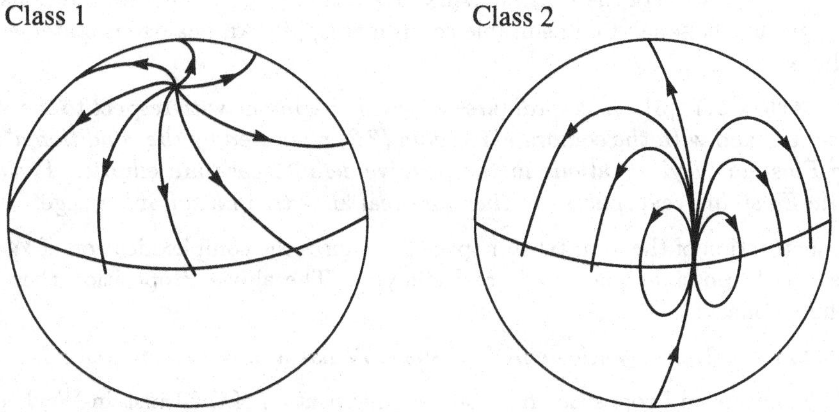

Figure 4.1

4.5 $S^1 \times S^2$

This is a manifold which admits no Einstein metric. It can be given an Einstein-Weyl structure by rescaling the flat metric in $\mathbf{R}^3$ following (2.2), [**9**]:

$$\hat{g} = d\chi^2 + d\theta^2 + sins^2\theta d\phi^2$$
$$\hat{\omega} = -2d\chi,$$

with $\chi = \log r$ and $\Omega = r^{-1}$, and imposing periodicity in χ.

4.6 *Other examples.*

By *ad hoc* methods one finds that there are Einstein-Weyl structures on three more of Thurston's eight homogeneous geometries [**10**], in which the metric is closely related to the natural one. These are $\mathbf{R} \times \mathbf{H}^2$, twisted $\mathbf{R} \times \mathbf{H}^2$ (or the universal

cover of $SL(2,\mathbf{R})$) and twisted $\mathbf{R}\times\mathbf{R}^2$ (or "Nil"). However in all three of these examples the metric is an indefinite version of the natural one [**9**].
It is also possible to produce these examples by constructing the mini-twistor space $\mathcal{T}$ directly, [**7**],[**9**].

5. Ellipticity and analyticity.

An attractive feature of the Einstein equations is that positive-definite solutions are automatically real-analytic in suitable co-ordinates, [**3**]. An analogous statement is true here:

PROPOSITION 5.1, [**9**]. *In co-ordinates which are harmonic with respect to the Weyl connection, and with the conformal freedom (2.2) restricted by the condition* $d^*\omega = 0$ *the Einstein-Weyl equations in the positive-definite case are elliptic. Positive-definite Einstein-Weyl spaces are therefore real analytic in a suitable gauge.*

The construction of the mini-twistor space $\mathcal{T}$ requires the complexification of $\mathcal{W}$ and hence is only possible when $\mathcal{W}$ is real analytic. The above Proposition therefore has the corollary:

COROLLARY 5.2. *Every Einstein-Weyl space* $\mathcal{W}$ *has a mini-twistor space* $\mathcal{T}$.

It also follows from Proposition 5.1 that the linearisation of the Einstein-Weyl equations about, say, the Einstein metric on S^3 gives an elliptic system for the perturbation. In this way one can show the following:

PROPOSITION 5.3 , [**9**].
The Einstein-Weyl structures found in Class 1 of Section 4.4 include all the solutions obtained by linearising about the Einstein metric.

Note however that there exists another family, disconnected from the Einstein metric and bifurcating from Class 1.

6. Geodesics.

In an Einstein space the behavior of geodesics is closely regulated by the Ricci scalar, which is automatically constant, [**1**]. From a study of the geodesic deviation equation one finds that a positive Ricci scalar forces all the geodesics to have conjuagate points, and ultimately forces the space to be compact, while a negative Ricci scalar forces geodesics to diverge from each other exponentially rapidly. In an Einstein-Weyl space different things can happen, as we shall see from two examples. The first is $S^1\times S^2$ from Section 4.5. We draw this as the region between two concentric spheres in $\mathbf{R}^3$ (Figure 6.1). Geodesics are straight lines, and must reappear on the inner sphere on the same radius as they meet the outer sphere. Therefore any non-radial geodesic will close in asymptotically on the radial one parrallel to it, [**9**].

The second example is the "Berger sphere" [**5**], which lies in Class 1 of Section 4.4. This has

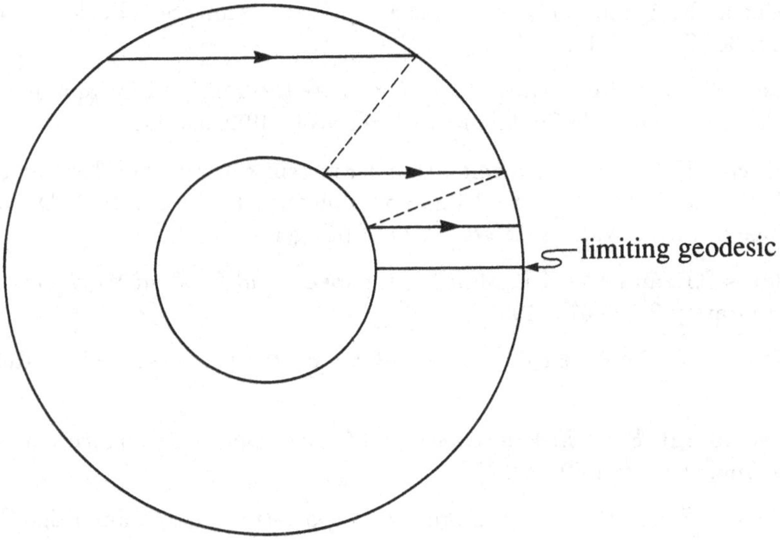

Figure 6.1

$$g = \sigma_1^2 + \sigma_2^2 + \lambda^2\sigma_3^2$$
$$\omega = \lambda(1-\lambda^2)^4\sigma_3,$$

where σ_i are the standard basis of left-invariant 1-forms on S^3.

By a direct integration of the geodesic equations one finds that the σ_3 directions form geodesics, and that geodesics in all other directions close in asymptotically on these ones [**9**].

In these two examples the space of geodesics, which is the mini-twistor space $\mathcal{T}$, will be *non-Hausdorff*.

What one finds from a study of the geodesic deviation equation in a 3-dimensional Einstein-Weyl space is that the space of Jacobi fields along each geodesic admits a complex structure (Proposition 2.3).. This is the fundamental property which leads to the existence of the twistor correspondence.

REFERENCES

[1] Besse,A.L. *Einstein Manifolds* Springer Verlag 1987

[2] Cartan,E. *Sur une classe d'espace de Weyl* Ann. Sc. Ecole Normale Superieure, 3e. Série 60 1-16, 1943

[3] Deturck, D. and Kazdan, J. *Some regularity theorems in Riemannian geometry* Ann. Sc. Ecole Normale Superieure, 4e. Série 14 249-260, 1981

[4] Hitchin,N.J. *Complex manifolds and Einstein equations* in *Twistor geometry and Nonlinear Systems*, Proc., Primorsko, Bulgaria, 1980. Ed. H.D. Doebner and T.D. Palev *Lecture Notes in Maths. 970* , Springer 1982

[5] Jones,P.E. and Tod,K.P. *Mini-twistor spaces and Einstein-Weyl spaces* Class. Quant. Gravity 2 565-577, 1985

[6] Penrose,R. *Nonlinear gravitons and curvedctwistor theory* Gen. Rel. Grav. 7 31-52, 1976

[7] Pedersen,H. *Einstein-Weyl spaces and (1,n) curves in the quadric surface* Ann. Global Anal. Geom. 4 89-120, 1986

[8] Penrose,R and Rindler,W. *Spinors and Space-time, Vol I* Cambridge U.P. 1984

[9] Pedersen,H. and Tod,K.P. *Three-dimensional Einstein-Weyl geometry* Advances in Maths. *To appear*

[10] Thurston,W.P. *Three dimensional manifolds, Kleinian groups and hyperbolic geometry* AMS Proc. Symp. Pure Maths. 39 1 87-111, 1983

HARMONIC MORPHISMS, CONFORMAL FOLIATIONS AND SEIFERT FIBRE SPACES

JOHN C. WOOD
University of Leeds

1 INTRODUCTION

A *harmonic morphism* is a mapping between Riemannian manifolds which preserves Laplace's equation in the sense that it pulls back harmonic functions on the codomain to harmonic functions on the domain (see below). The aim of this paper is to introduce harmonic morphisms to the non-specialist and show how they occur naturally in low dimensional geometry. In particular, we shall show that each of Thurston's geometries apart from Sol has a natural map which can be characterized as the unique non-constant harmonic morphism to a surface (§3), and in §4, we shall show that, for any closed Seifert fibre space M^3, there is a harmonic morphism with fibres equal to the fibres of M^3. Finally, in §5 we use these ideas to give a simple explanation of why the product of a circle and a closed oriented Seifert fibre space with oriented fibres is naturally an elliptic surface.

The author wishes to thank several participants at the conference for useful conversations about this work, especially K.P. Tod, J.H. Rubinstein and J.D.S. Jones. Most of the work in this paper represents work of the author with Paul Baird. It is hoped that this informal account will prove useful for those wishing to taste the flavour of the subject but unable to read the full account in (Baird & Wood, 1988, 1989a, 1989b).

2 DEFINITIONS AND BASIC PROPERTIES

2.1 Harmonic morphisms

Let M^m and N^n be C^∞ Riemannian manifolds of dimensions m and n respectively, and let $\varphi: M^m \to N^n$ be a C^∞ mapping. Then φ is called a *harmonic morphism* if, for any harmonic function $f: U \to \mathbb{R}$

defined on an open subset U of N with $\varphi^{-1}(U)$ non-empty, $f\circ\varphi:\varphi^{-1}(U)\to\mathbb{R}$ is also a harmonic function. Because of the existence of harmonic coordinates (Greene & Wu, 1962) any harmonic morphism is necessarily C^{∞}. Harmonic morphisms can be characterized in terms of harmonic maps as follows: Say that φ is *horizontally (weakly) conformal* if, at each point $x\in M^m$ where $d\varphi_x\neq 0$, writing $V_x=\ker d\varphi_x$ and $H_x=V_x^{\perp}\cap T_xM$, $d\varphi_x|_{H_x}$ maps H_x conformally onto $T_{\varphi(x)}N$, that is, for all $v\in H_x$, $\|d\varphi_x(v)\| = \lambda(x)\|v\|$ for some $\lambda(x)>0$. Setting $\lambda(x)=0$ at points where $d\varphi_x\neq 0$ (called *critical points*) we obtain a continuous function $\lambda:M\to\mathbb{R}$, with λ^2 smooth, called the *dilation* of φ. Note that a horizontally weakly conformal map with $\lambda\equiv 1$ is just a *Riemannian submersion.* We have

Theorem 2.1 (Fuglede, 1978; Ishihara, 1979)
A smooth map $\varphi:M^m\to N^n$ is a harmonic morphism if and only if it is (*a*) a harmonic map and (*b*) horizontally weakly conformal.

We remark that in the case $M=\mathbb{R}^3$, $N=\mathbb{R}^2=\mathbb{C}$, this says that $\varphi:\mathbb{R}^3\to\mathbb{C}$ is a harmonic morphism if and only if it satisfies

$$\frac{\partial^2\varphi}{\partial x_1{}^2}+\frac{\partial^2\varphi}{\partial x_2{}^2}+\frac{\partial^2\varphi}{\partial x_3{}^2}=0 \qquad (1)$$

$$\left(\frac{\partial\varphi}{\partial x_1}\right)^2+\left(\frac{\partial\varphi}{\partial x_2}\right)^2+\left(\frac{\partial\varphi}{\partial x_3}\right)^2=0 \qquad (2)$$

the first equation expressing harmonicity of φ and the second, horizontal conformality.

In particular, if φ is non-constant, $m\geq n$. Further, if $m=n=2$, then φ is a harmonic morphism if and only if it is a weakly conformal map, and if $m=n>2$, the only harmonic morphisms are homothetic maps. Thus we concentrate on the case $m>n$. Harmonic morphisms to *surfaces* (i.e. two-dimensional Riemannian manifolds) have many nice properties For example we have

Theorem 2.2 (Baird & Eells, 1981)
Let $\varphi: M^m \to N^2$ be a submersion to a surface. Then φ is a harmonic morphism if and only if it is horizontally conformal and its fibres are minimal submanifolds.

In the case that $m = 3$, *minimal fibres* means that the fibres are geodesics. Another property of harmonic morphisms to a surface is that composition of such a map with a conformal map of surfaces gives another harmonic morphism. In particular, the concepts of *harmonic morphism to a* C^∞ *2-manifold with a conformal structure (i.e. an equivalence class of smooth metrics)* and *harmonic morphism to a Riemann surface (i.e. a one-dimensional complex manifold)* are well-defined without specifying a particular Hermitian metric on the surface. For this reason, by *surface* we shall henceforth mean a C^∞ 2-manifold with a conformal structure.

We can now give some examples of harmonic morphisms, concentrating on ones from 3-manifolds to surfaces.

(1*a*) Let $\mathbb{H}^3$ denote hyperbolic 3-space thought of as the open unit disc $D^3 = \{(x_1, x_2, x_3) : |x|^2 < 1\}$ of $\mathbb{R}^3$ (where $|x|^2 = x_1^2 + x_2^2 + x_3^2$) with the metric $4\sum dx_i^2/(1-|x|^2)^2$. Then geodesics are circular arcs hitting the boundary ∂D^3 orthogonally. Identify the hyperbolic plane $\mathbb{H}^2$ with the equatorial disc $x_3 = 0$ of D^3. Draw all the geodesics of $\mathbb{H}^3$ which cut this disc orthogonally. Define a map $\varphi: \mathbb{H}^3 \to \mathbb{H}^2$ by projection down these geodesics.

(1*b*) As a limiting case of (3), choose a point $x_0 \in \partial D^3$. Then, for each $x \in \mathbb{H}^3$, let $\varphi(x)$ = the point of intersection of the geodesic through x with ∂D^3. This defines a map $\varphi: \mathbb{H}^3 \to \partial D^3 \setminus \{x_0\} \cong \mathbb{C}$ where we identify $\partial D^3 \setminus \{x_0\} = S^2 \setminus \{x_0\}$ with $\mathbb{C}$ by stereographic projection.

(2) Orthogonal projection $\mathbb{R}^3 \to \mathbb{R}^2$, $(x_1, x_2, x_3) \longmapsto (x_1, x_2)$

(3) The Hopf map $S^3 \to S^2$. With $S^3 = \{(z_1, z_2) \in \mathbb{C}^2 :$

$|z_1|^2+|z_2|^2 = 1\}$. This can be defined as $(z_1, z_2) \longmapsto z_1/z_2 \in \mathbb{C}\cup\{\infty\}$ followed by the inverse of stereographic projection $\sigma: S^2 \to \mathbb{C}\cup\{\infty\}$.

These are essentially the only harmonic morphisms from a simply-connected three-dimensional space form to a surface. Precisely,

Theorem 2.3. (Baird & Wood, 1988, 1989a)
Up to isometries of the domain, any non-constant harmonic morphism of $\mathbb{H}^3$, $\mathbb{R}^3$ or S^3 to a surface is one of the above four examples followed by a weakly conformal mapping of the codomain to a surface.

In particular, the only solutions $\varphi:\mathbb{R}^3 \to \mathbb{C}$ to the equations (1) and (2) above are linear ones $\varphi(x_1, x_2, x_3) = a_1x_1+a_2x_2+a_3x_3$ where the $a_i \in \mathbb{C}$ satisfy $a_1^2+a_2^2+a_3^2 = 0$, composed with a weakly conformal self-mapping of $\mathbb{C}$. More interestingly from the geometrical point of view, the Hopf map $S^3 \to S^2$ is characterized as the unique non-constant harmonic morphism from S^3 to S^2 (up to conformal self mappings of S^2 .

2.2 Conformal foliations

A smooth foliation on a Riemannian manifold is called *conformal* (resp. *Riemannian*) if its leaves are locally the fibres of a horizontally conformal (resp. Riemannian) submersion. Let $\varphi: M^3 \to N^2$ be a submersive harmonic morphism of a Riemannian 3-manifold to a surface. Then, by Theorem 2.2, the fibres give a conformal foliation of M^3 by geodesics. Conversely, given a conformal foliation by geodesics of a suitable subset U of M^3 (for example, geodesically convex), we can form its leaf space, a surface, in fact a Riemann surface if M^3 and the foliation are oriented. Then the canonical projection of U onto its leaf space is a harmonic morphism. In particular, we can translate the above results into the theorem that, *up to isometries,* $\mathbb{R}^3$ *and* S^3 *have just one conformal foliation by geodesics and* $\mathbb{H}^3$ *has two.*

The above results on the three space forms are *global*; indeed, there is a large supply of harmonic morphisms (or conformal foliations by geodesics) defined on open subsets of a space form. For let $\mathcal{F}$ be a conformal foliation by geodesics of a geodesically convex subset of a space form $\mathbb{E}^3 = \mathbb{R}^3$, S^3 or $\mathbb{H}^3$ and let N^2 be its leaf space. Then we have an injective map $i: N^2 \longrightarrow G_{\mathbb{E}^3}$ into the space of all geodesics which describes the foliation. Now $G_{\mathbb{E}^3}$ can be given the structure of a complex surface (Hitchin, 1982; Baird & Wood, 1989a). Then, because the foliation is conformal, i is holomorphic. Conversely, a holomorphic injective map defines a conformal foliation by geodesics on a suitable open subset of $\mathbb{E}^3$. Choosing this subset to be geodesically convex, the natural projection of it onto its leaf space is a harmonic morphism. For example, if $\mathbb{E}^3 = \mathbb{R}^3$, $G_{\mathbb{E}^3}$ may be identified with the unit tangent bundle of S^2. Using stereographic projection σ from a point x_0 and its differential as a chart $TS^2|_{S^2\setminus\{x_0\}} \longrightarrow \mathbb{C}\times\mathbb{C}$, a tangent vector at a point $\sigma^{-1}(x) \in S^2\setminus\{x_0\}$ is represented by a pair $(x, v) \in \mathbb{C}\times\mathbb{C}$. We can then, for example, define a holomorphic map $\mathbb{C} \longrightarrow TS^2$ by $z \longrightarrow (z, e^{i\theta} z)$ where θ is a constant. This defines a 1-parameter family of conformal foliations by geodesics of dense open subsets of $\mathbb{R}^3$. For example, if $\theta = 0$ this is the foliation of $\mathbb{R}^3\setminus\{0\}$ by radii and if $\theta = \pi$ it is a foliation which twists through the unit disc (see Baird & Wood, 1988; Baird, 1987). The corresponding harmonic morphisms are radial projection $\mathbb{R}^3\setminus\{0\} \longrightarrow S^2$ and a harmonic morphism $\mathbb{R}^3\setminus\overline{D}^2 \longrightarrow D^2$ where D^2 is the unit disc in the (x_1, x_2)-plane. Similar constructions provide harmonic morphisms and conformal foliations by geodesics of open subsets of S^3 and $\mathbb{H}^3$ (Baird & Wood, 1988, 1989a).

We note that in relativity theory, a conformal foliation by geodesics is a geodesic shear-free congruence (Penrose & Rindler, 1984; Huggett & Tod, 1985) and there are analogies between the theory of conformal foliations by geodesics of a space form and that of geodesic shear-free congruences of *null* geodesics in space-time.

3. HARMONIC MORPHISMS AND THURSTON'S GEOMETRIES

By a *geometry* we shall mean a simply-connected C^∞ 3-dimensional Riemannian homogeneous space which admits a compact quotient, see (Scott, 1983). Thurston showed that there are exactly eight geometries. Three of these are the space forms (1) $\mathbb{H}^3$, (2) $\mathbb{R}^3$, (3) S^3; we have described canonical harmonic morphisms of these in §1 ($\mathbb{H}^3$ having two), we shall now list the other geometries together with a canonical metric and canonical harmonic morphism π to a surface (which are all Riemannian submersions with geodesic fibres). The fibres of the latter give a canonical conformal foliation by geodesics of the geometry, in fact all these foliations are Riemannian.

(4) $S^2\times\mathbb{R}$ with the product of standard metrics and π = projection onto the first factor.

(5) $\mathbb{H}^2\times\mathbb{R}$ with the product of standard metrics and π = projection onto the first factor.

(6) Nil ; this can be identified with $\mathbb{R}^3$ with the metric $ds^2 = dx^2+dy^2+(dz-xdy)^2$, and then $\pi:\mathrm{Nil}\longrightarrow\mathbb{R}^2$ is defined by $(x,y,z) \longmapsto (x,y)$.

(7) The universal cover $\mathrm{SL}_2(\mathbb{R})^\sim$ of $\mathrm{SL}_2(\mathbb{R})$; this can be identified with the upper half-plane $\mathbb{R}^3_+ = \{(x,y,z)\in\mathbb{R}^3 : y>0\}$ with the metric $ds^2 = (dx^2+dy^2)/y^2 + (dx/y+dz)^2$, and then $\pi:\mathrm{SL}_2(\mathbb{R})^\sim \longrightarrow \mathbb{H}^2$ is defined by $(x,y,z)\longmapsto(x,y)$.

Note that the last two examples, as for the Hopf map $S^3\longrightarrow S^2$, have non-integrable horizontal spaces and these three geometries with their canonical metrics and foliations defined above are naturally Sasakian manifolds. We have omitted the last geometry (8) Sol because of the following:

Theorem 3.1 (Baird-Wood 1989b)
Any non-constant harmonic morphism from an open subset of a 3-dimensional geometry not of constant curvature to a surface is the restriction of one of the four maps π listed above in (4)-(7) fol-

lowed by a weakly conformal map of its codomain to a surface. In particular, there is no harmonic morphism to a surface from any open subset of Sol .

Theorem 3.2 (loc.cit.)
Any conformal foliation by geodesics of an open subset of a 3-dimensional geometry not of constant curvature is the restriction of one of the four standard examples induced by the maps π in (4)-(7) as explained above. In particular, no open subset of Sol supports a conformal foliation by geodesics.

The last result is shown in the following way: Given any conformal foliation by geodesics, let U, X, Y be an orthonormal frame with U along the foliation. Set $Z = X + iY$. Then from the Jacobi equation we can show that

$$\mathrm{Ricci}(Z,Z) = 0 \tag{3}$$

But for this equation, one of the following holds: (*i*) if the eigenvalues of the Ricci curvature are all distinct there is essentially one solution Z ; (*ii*) if two of the eigenvalues of the Ricci curvature are the same there are essentially two solutions ; (*iii*) if all three eigenvalues of the Ricci curvature are the same, in which case the 3-manifold is of constant curvature, any Z is a solution . In the case of a Riemannian product $M^2 \times \mathbb{R}$ or a Sasakian 3-manifold, it can be shown that case (*i*) applies, in particular this applies to the geometries (4)-(8). Thus, in these cases there is only one possible direction for a conformal foliation by geodesics. In the cases (4)-(7) this really does give a conformal foliation by geodesics, however in the case of Sol the foliation is not, in fact, conformal.

Note that the local uniqueness (Theorem 3.2) of a conformal foliation by geodesics of a geometry not of constant curvature is in contrast to the abundance of such foliations on an open subset of 3-dimensional space form (see §2) where Theorem 2.3 asserts global uniqueness. Furthermore, any isometry of a geometry (4)-(7) must preserve any conformal foliation by geodesics, this is not the case for the space forms. Lastly note that the analysis above, together with a unique continuation theorem for harmonic morphisms of conformal foliations by geodesics, shows that no C^∞ Riemannian 3-manifold of non-constant curvature can support more than two distinct non-constant harmonic morphisms to surfaces or conformal foliations by geodesics.

4. HARMONIC MORPHISMS AND SEIFERT FIBRE SPACES

By a *fibred solid torus* $T_{p,q}$ we mean $D^2\times I$ with the end $D^2\times\{1\}$ identified to the end $D^2\times\{0\}$ by a twist through $2\pi q/p$ for some integers p,q with $p\neq 0$. If $q=0$ this gives the solid torus $D^2\times S^1$ with its product foliation by circles. By a *fibred solid Klein bottle* we mean $D^2\times I$ with its ends identified by a reflection. By a *Seifert fibre space (without reflections)* we mean a 3-manifold with a decomposition into circles (called *fibres*) such that each circle has a neighbourhood and a fibre preserving homeomorphism to a fibred solid torus $T_{p,q}$. If $q=0$ the fibre is called *regular*, otherwise we shall call it *singular*; the integers (p,q) are called the *(unnormalized) orbit* or *Seifert invariants* of the fibre. If we also allow fibred solid Klein bottles we obtain a *Seifert fibre space with reflections*. It is result of Epstein that any foliation by circles of a compact 3-manifold is a Seifert fibre space possibly with reflections. If the manifold is orientable, this is necessarily *without* reflections. We shall here concentrate on the latter sort of Seifert fibration for reasons explained after Theorem 4.1. As an example, for any non-zero co-prime integers p,q , consider the foliation $\mathcal{F}_{p,q}$ by circles of $S^3 = \{(z_1,z_2)\in \mathbb{C}^2 : |z_1|^2+|z_2|^2=1\}$ given by

$$z_1^{p}/z_2^{q} = \text{constant} \tag{4}$$

For $p = q = 1$ this is the foliation associated to the Hopf fibration. For $p, q > 1$, the circles $z_1 = 0$ and $z_2 = 0$ are singular fibres with (unnormalized) orbit invariants (p,q) and (q,p) respectively.

Now it follows from Scott (1983) that any closed Seifert fibre space is of the form $\mathbb{E}^3/\Gamma$ where $\mathbb{E}^3$ is one of the geometries (2)-(7) with its canonical foliation, or $\mathbb{E}^3 = S^3$ with the foliation $\mathcal{F}_{p,q}$ for some p, q, and Γ is a group of isometries which acts freely and properly discontinuously on $\mathbb{E}^3$. Note that $\mathbb{H}^3$ and Sol do not give rise to any Seifert fibre spaces. For example, if we factor the geometry $S^2 \times \mathbb{R}$ by the group generated by (α, β) where α is a rotation of S^2 by an angle $2\pi q/p$ about the diameter joining points $\pm a$ and β is translation of $\mathbb{R}$ by 1 unit, we obtain a Seifert fibration on $S^2 \times S^1$ with singular fibres $\{a\} \times S^1$ and $\{-a\} \times S^1$ both with orbit invariants (p,q).

Associated to any Seifert fibre space without reflections is the natural projection onto its leaf space $\mathcal{L}$. The latter is an orbifold with cone points of angle $2\pi/p$ corresponding to singular fibres with orbit invariants (p,q). For example, for the foliation $\mathcal{F}_{p,q}$ on S^3 above, the leaf space is homeomorphic to S^2 but with two cone points of angles $2\pi/p$ and $2\pi/q$ respectively; in Scott's notation, $\mathcal{L} = S^2(p,q)$. For the foliation on $S^2 \times S^1$ described above, the leaf space is $S^2(p,p)$. Now in all cases apart from $\mathcal{F}_{p,q}$, the metric induced from the standard metric on the geometry makes the Seifert fibre space $\mathbb{E}^3/\Gamma$ a Riemannian foliation by geodesics. In the case of $\mathcal{F}_{p,q}$ we can give S^3 a metric g such that this becomes a Riemannian foliation by geodesics. Indeed, we let (S^3, g) be the ellipsoid $Q^3_{p,q} = \{(z_1, z_2) \in \mathbb{C}^2 : |z_1|^2/p^2 + |z_2|^2/q^2 = 1\}$ with the metric induced from the ambient Euclidean space (see Eells & Ratto, 1988; Baird & Wood, 1989b). The metric on the Seifert fibre space induces a metric, and thus a conformal structure on the leaf space $\mathcal{L}$ except at the cone points. In fact we can give the leaf space a conformal structure even at the cone points as follows: At a cone point, the leaf space is of the form $D^2/\mathbb{Z}_p$ where D^2 is a disc with a rotationally symmetric metric g. Then by the Uniformization Theorem (Ahlfors, 1979), there is a conformal diffeomorphism $(D^2, g) \longrightarrow (D^2, \text{standard})$. After composing this with the map $z \longmapsto z^p$ of the codomain it factors to a homeomorphism $D^2/\mathbb{Z}_p \longrightarrow (D^2, \text{standard})$ which we use to give the leaf space a conformal structure at the cone points. Our construction gives a homeomorphism h from the leaf space $\mathcal{L}$ to a smooth surface with conformal structure $\mathcal{L}_s$ which we call the *smoothed leaf space.* We can use this smoothing procedure to establish the following:

Theorem 4.1 (Baird & Wood, 1989b)
Let M^3 be a closed Seifert fibre space without reflections. Then there exists a smooth metric on M^3 and a smooth harmonic morphism φ to a smooth surface with conformal structure such that the fibres of φ coincide with the fibres of M^3.

Proof. Let φ be the composition of the natural projection onto the leaf space with the map h above. This certainly has geodesic fibres and is horizontally conformal away from critical fibres; it is thus a harmonic morphism off the polar set given by those fibres. It is also continuous everywhere; it follows (cf. Baird & Wood, 1989*a*) that it is a smooth harmonic morphism everywhere.

Remarks. Note that each point of a singular fibre of M^3 is a critical point of φ. In fact, at such a point, φ has the form of a submersion followed by $z \longmapsto z^p$. Note also that, given a Seifert fibre space M^3 with reflections, there cannot be any harmonic morphism $\varphi: M^3 \longrightarrow N^2$ to a surface, with fibres coinciding with the fibres of M^3, for, again, all points of the singular fibres would have to be critical points of φ. But now the singular fibres corresponding to fibred solid Klein bottles are whole surfaces and so not polar contradicting the fact (Fuglede, 1978) that the critical points of a harmonic morphism form a polar set.

As an example, consider the foliation $\mathcal{F}_{p,q}$ on the ellipsoid $Q^3_{p,q}$. The leaf space is the orbifold $S^2(p,q)$ with a rotationally symmetric metric. Composing with the smoothing map $h: S^2(p,q) \longrightarrow S^2$, which will be of the form (*) $(r,\theta) \longmapsto (f(r),\theta)$ in spherical polar coordinates, gives (up to conformal self-mappings of S^2 of the form (*)) the harmonic morphism $Q^3_{p,q} \rightarrow S^2$ of Eells & Ratto (1988). Note that our method does not require discussion of any ordinary differential equation, this being implicit in the use of the Uniformization Theorem.

The ellipsoid $Q^3_{p,q}$ actually has *two* Riemannian foliations, the second one given by replacing q by $-q$ in (4). These foliations are interchanged under the reflection $(z_1,z_2) \longmapsto (z_1,\bar{z}_2)$. Now for any non-zero coprime integers (r,s) we have a free properly discontinuous action of $\mathbb{Z}_r$ on $Q^3_{p,q}$ given by $(z_1,z_2) \longmapsto (z_1 e^{2\pi i/r}, z_2 e^{2\pi i s/r})$ which preserves these foliations. Then the foliations factor to foliations on the lens space with ellipsoidal metric $L(r,s) = Q^3_{p,q}/\mathbb{Z}_r$. These last foliations are not isometrically related unless $r=2$ in which case the lens space is real projective 3-space. In (Baird & Wood, 1989b) the closed oriented Riemannian 3-manifolds supporting a pair of oriented Riemannian foliations by geodesics are classified. They are the 3-torus T^3, S^3, any lens space and $S^2 \times S^1$, all with one of an infinite family of non-standard metrics. The constructions above then yield harmonic morphisms in various homotopy classes from these manifolds to 2-tori and the 2-sphere.

5 THE PRODUCT OF A SEIFERT FIBRE SPACE WITH A CIRCLE

Let M^3 be an oriented Riemannian 3-manifold with an oriented conformal foliation by geodesics. Then there is a natural almost Hermitian structure J on the product $M^3 \times S^1$ described as follows. At any point $(x,y) \in M^3 \times S^1$ let e_1, e_2, e_3 be an oriented orthonormal basis for $T_x M^3$ with e_3 along the foliation and let e_4 be the unit positive tangent to S^1 at y. Then set $J(e_1)=e_2$, $J(e_2)=-e_1$, $J(e_3)=e_4$ and $J(e_4)=-e_3$. Then a straightforward calculation of the Nijenhuis tensor shows that J is integrable. Now suppose that M^3 is a closed oriented Seifert fibre space with oriented fibres. Then, as in §4, it can be given the structure of a conformal (in fact Riemannian) foliation by geodesics and we can construct a harmonic morphism $\varphi: M^3 \to N^2$ to a Riemann surface whose fibres are the fibres of M^3. The composition $M^3 \times S^1 \to M^3 \to N^2$ of the projection onto the first factor with φ is clearly holomorphic and has tori as fibres, so we have shown in a very natural way the well-known:

Theorem 5.1
The product of a closed oriented Seifert fibre space with oriented fibres and the circle S^1 is naturally an elliptic surface.

As an example, start with the Hopf map $S^3 \to S^2$. Then we get a Hermitian structure on $S^3 \times S^1$, the well-known Hopf surface (see, for example, Besse, 1981), together with a surjective holomorphic map $S^3 \times S^1 \to S^2$. If we take, instead, the Eells-Ratto harmonic morphism $Q^3_{p,q} \to S^2$, we get a Hermitian structure on $Q^3_{p,q} \times S^1$ which gives $S^3 \times S^1$ a non-standard metric and Hermitian structure as an elliptic surface. Similarly, using the harmonic morphisms discussed at the end of §4, we obtain non-standard metrics and Hermitian structures on $S^2 \times S^1 \times S^1$, $S^1 \times S^1 \times S^1 \times S^1$ and $\mathrm{L}(p,q) \times S^1$ together with holomorphic maps $S^2 \times S^1 \times S^1 \to S^2$, $S^1 \times S^1 \times S^1 \times S^1 \to S^1 \times S^1$ and $\mathrm{L}(p,q) \times S^1 \to S^2$.

REFERENCES

Ahlfors, L.V. (1979). *Complex Analysis*, 3rd edn. London: McGraw-Hill.

Baird, P. (1987). Harmonic morphisms onto Riemann surfaces and generalized analytic functions. *Ann. Inst. Fourier, Grenoble*, 37,

135-173.

Baird, P. & Eells, J. (1981). A conservation law for harmonic maps. In *Geometry Symposium, Utrecht* 1980, ed. E. Looijenga, D. Siersma, F. Takens, pp. 1-25. Lecture Notes in Math. 894, Berlin: Springer Verlag.

Baird, P. & Wood, J.C. (1988). Bernstein theorems for harmonic morphisms from $\mathbb{R}^3$ and S^3. *Math. Ann.*, 280, 579-603.

Baird, P. & Wood, J.C. (1989a). *Harmonic morphisms and conformal foliations by geodesics of three-dimensional space forms.* Preprint, University of Leeds.

Baird, P. & Wood, J.C. (1989b). *Harmonic morphisms and conformal foliations by geodesics of arbitrary three-dimensional manifolds.* Preprint, University of Leeds.

Besse, A. (1981). *Géométrie riemannienne en dimension* 4. *Seminaire Arthur Besse* 1978/79. Paris: Cedic/Fernand Nathan.

Eells, J. & Ratto, A. (1988). *Harmonic maps between spheres and ellipsoids.* Preprint no. 88/31, Institut des Hautes Etudes Scientifiques.

Fuglede, B. (1978). Harmonic morphism between Riemannian manifolds. *Ann. Inst. Fourier, Grenoble*, 28, 107-144.

Greene, R.E. & Wu, H. (1962). Embeddings of open Riemannian manifolds by harmonic functions. *Ann. Inst. Fourier, Grenoble*. 12, 415-571.

Hitchin, N. (1982). Monopoles and geodesics. *Comm. Math. Phys.* 83, 579-602.

Huggett, S.A. & Tod, K.P. (1985). *An Introduction to Twistor Theory*, London Math. Soc. Student Texts no.4, Cambridge: University Press.

Ishihara, T. (1979). A mapping of Riemannian manifolds which preserves harmonic functions. *J. Math. Kyoto University*, 19, 215-229.

Penrose, R. & Rindler, W. (1984) *Spinors and space-time vols.* I & II. Cambridge Monogaphs on Mathematical Physics. Cambridge: University Press.

Scott, P. (1983) The geometries of 3-manifolds. *Bull. London Math. Soc.* 15, 401-487.

Printed in the United States
By Bookmasters